In the Eye of t

The Science of Face Perception

VICKI BRUCE
University of Stirling
and
ANDY YOUNG
University of York

OXFORD
UNIVERSITY PRESS

Great Clarendon Street, Oxford OX2 6DP

Oxford University Press is a department of the University of Oxford.
It furthers the University's objective of excellence in research, scholarship,
and education by publishing worldwide in

Oxford New York

Athens Auckland Bangkok Bogotá Buenos Aires Calcutta
Cape Town Chennai Dar es Salaam Delhi Florence Hong Kong Istanbul
Karachi Kuala Lumpur Madrid Melbourne Mexico City Mumbai
Nairobi Paris São Paulo Singapore Taipei Tokyo Toronto Warsaw

with associated companies in Berlin Ibadan

Oxford is a registered trade mark of Oxford University Press
in the UK and in certain other countries

Published in the United States by Oxford University Press Inc., New York

First published 1998
Reprinted new as paperback 2000

A catalogue record for this book is available from the British Library

Library of Congress Cataloguing in Publication Data

(Data available)

ISBN 0 19 852440 4 (Hbk)
0 19 852439 0 (Pbk)
0 19 852455 2 (TSP)

Printed in Great Britain
on acid-free paper by
Butler and Tanner Ltd,
Frome and London

Preface

This book was written to accompany an exhibition on 'The Science of the Face', held at the Scottish National Portrait Gallery in Spring 1998. It provides a non-technical introduction to the science of the human face and to the psychology of face perception, with illustrations and demonstrations drawn from science and from the art of portraiture. We have tried to write the book in a style which will be accessible to a wide audience; more technical material has been placed in boxed sections within each chapter.

A particular feature of the book is the juxtaposition of science and illustrations based on images taken from the Scottish National Portrait Gallery's extensive collections—with several featuring state-of-the-art computer-manipulated graphics to demonstrate the scientific message. James Holloway, the Keeper of the Scottish National Portrait Gallery, has been immensely supportive throughout, and contributed hugely with his ideas and enthusiasm for the project. It was James who urged us to make use of images from the portrait gallery, and who offered advice on important artistic matters. He also kindly helped us choose and granted permission for us to reproduce these images here as well as in the exhibition. For this reason, our illustrations are dominated by famous Scots whose images were available in the gallery! However, James also alerted us to the (apparently) disproportionate number of Scottish scientists whose work in past centuries is relevant to this theme, and whose work also gets special emphasis here. We would also like to acknowledge the further help of Deborah Hunter and her colleagues in the picture library and elsewhere within the National Galleries of Scotland, who located images of the portraits for us, sometimes at extremely short notice.

The timing of the exhibition in the Spring of 1998 allowed it to coincide with the 1998 Edinburgh International Science Festival, whose Director, Dr Simon Gage, also supported our project by including Faces as a theme in the festival, and scheduling a number of lectures to coincide with the exhibition.

One reason for our own enthusiasm for the exhibition was that it provided a wonderful opportunity to disseminate our science to a wider audience than we usually encounter. Oxford University Press have helped this process still further by agreeing to publish this extended version in book form, and this offer undoubtedly enhanced the appeal of the project in its early stages of discussion with interested parties. We extend most sincere thanks to Vanessa Whitting, who encouraged and helped us generously and cheerfully in the early stages of the project, and then, appropriately, switched to take a tougher line when the deadline for the manuscript loomed. Her colleagues at OUP went to enormous lengths to deliver this book on an extremely tight schedule, for which we are very grateful.

The research which we describe in this book has been conducted over a number of centuries by scientists from many different disciplines. We delve into biological, medical, clinical, and engineering research as well as our own specialist field of psychology. Our own research has been funded at various different times during the past twenty years by the UK Medical Research Council, the Economic and Social Research Council, the Science and Engineering Research Council, ATR Human Information Processing Laboratories, and the Nuffield Foundation. The book draws on the results of numerous grants from these agencies, and was drafted while Andy Young was working in Cambridge as an employee of the Medical Research Council.

The most important ingredients of this book are the illustrations, and we gratefully acknowledge all those who have allowed us to reproduce their images here, particularly Nicholas Wade, whose ingenious portraits feature in several places in the book, John Liggett, whose earlier book *The human face* inspired many of our themes and provided illustrations for some of them, Simon Baron-Cohen, Ed Bullmore, Andy Calder, Gemma Calvert, Derek Carson, Ben Craven, Paul Ekman, Mark Johnson, Brian Hill, Judith Langlois, Florence Lebert, Trish le Gal, Andy Meltzoff, John Morris, Christoph Nothdurft, Gill Rhodes, John Shepherd, Sarah Stevenage, and Christopher Tyler who all supplied us with illustrations from their work. Andy Ellis drew our attention to important historical source material used in some of our illustrations, and this help was much appreciated. Bergit Arends and Mike Page helped us out with information about contemporary artists working with faces. We also thank especially those who have put a great deal of effort into the creation of new illustrations specifically for our book and exhibition: Richard Kemp at the University of Westminster, Helmut Leder at the University of Fribourg, Peter Hancock, Harold Hill, and Roger Watt at the University of Stirling, Mike Burton, Aude Oliva, and Philippe Schyns at the University of Glasgow, Alf Linney at University College London, and Michael Burt, Rachel Edwards, Kieran Lee, David Perrett, and Duncan Rowland at the University of St Andrews. We hope all these many contributors will be pleased with our use of their efforts, but of course we must take full responsibility for rough edges or inaccuracies that may remain in the text. Some of these rough edges were smoothed as a result of Mike Burton's careful reading of the manuscript in draft form.

Our families deserve a special mention. The extremely tight schedule for the book meant a great deal of pressure on them, as well as ourselves. Thanks to Mavis, Alexandra, Josephine, and Mike (again).

Finally, we must record our appreciation of the people who did the 'behind-the-scenes' work; Gary Jobe and Jill Keane in Cambridge, Elaine Stewart and Bob Lavery at Stirling, all of whom were inundated with requests to prepare figures, check obscure sources, and send the large number of letters which accompany a project of this type.

This has been a project dear to our hearts, and along the way it has allowed us to learn about many interesting highways and byways of science and art, both past and present. We hope it is just a fraction as enjoyable to read as it has been fun to do.

July 1997

V.B.
A.Y.

Contents

1 The face: organ of communication

1.1 Introduction and overview

The human face provides a bewildering variety of important social signals which can be detected and interpreted, very often correctly, by another human—the 'beholder' of our title. A face tells us if its bearer is old or young, male or female, sad or happy, whether they are attracted to us or repulsed by us, interested in what we have to say or bored and anxious to depart.

Perhaps because we must watch faces so closely for all these signs and messages, we are able to perceive the often tiny variations between individual faces which can be used to identify them. Personal identity is a further, and important message conveyed by the human face, and our own individual identities are bound up with our faces in a way which makes facial injury particularly traumatic to deal with.

This book is about the science of face perception. Recent years have seen huge advances in our understanding of the physiological and psychological processes involved in face perception, and here we describe many of these discoveries about how people decipher all these different messages from faces.

In reviewing scientific findings, we will illustrate these with a variety of pictures of faces, including some of famous people. While some of these illustrations use photographs, many of the pictures we will use are portraits, reflecting the origin of this book in an exhibition held at the Scottish National Portrait Gallery, Edinburgh, in 1998.

The interplay of science and art was a theme which had been explored in the nineteenth century, often by Scottish scientists and artists. Sir Charles Bell added the subtitle 'as connected with the fine arts' to his book on *The anatomy and philosophy of expression* (Bell 1844). His aim was to put the practice of drawing and painting onto a more secure anatomical basis, especially in the portrayal of the face and facial expressions. Bell began with a chapter on ancient and modern (early nineteenth century) art, and the Italian masters. This was followed by no less than seven detailed chapters on the anatomy of the head and face and the different types of facial expression. These were a technical *tour de force*, containing many important and insightful observations and delightful illustrations, some of which are reproduced here. In the last two chapters, Bell returned to his consideration of art. He described how he thought greater knowledge of anatomy might be useful to artists, and ended with a somewhat undiplomatic foray into the faults he considered to follow from a lack of such knowledge.

Bell's ideas had a direct impact on his fellow scientists, and Charles Darwin drew heavily on his insights in his own book on *The expression of the emotions in man and animals* (Darwin 1872). Through the paintings of Sir David Wilkie, Bell's ideas also began to influence artists.

The phrenologist George Combe held similar overall views to Bell on the importance of grounding art in science, but he gave them a broader treatment—covering all forms of fine art, not just the representation of the face. Combe's 1855 *Phrenology applied to painting and sculpture* reprinted letters he had previously published in the *Phrenological Journal*, which he edited from Edinburgh. His view was that:

> Phrenology may be useful, *first*, in helping the observer to distinguish the character of his own mind, and to appreciate its powers and qualities as an instrument of observation and judgment in art. ... *secondly*, it may be useful in enabling him to analyze and understand the different kinds of interest which may be felt by the same, or by different individuals, in painting and sculpture. (Combe 1855, p.6)

Unfortunately for Combe, the claims of phrenology have subsequently been found wanting in comparison to those of anatomy (see Chapter 7). For example when it came to stating just what were the different kinds of interest inherent in painting and sculpture to which he alluded, Combe saw these as corresponding to the phrenological faculties of form, size, colouring, locality, and order. The problem is that this simply reified what we might all consider a reasonable list of some of the things which matter in a work of art. It lacked the power of Bell's analysis because it introduced nothing new; there are no novel insights to be derived from such a list. It is therefore perhaps just as well that Combe's prescriptive agenda was more modest than Bell's, and less assertive of the need to bring science directly into the training of artists:

> The professional artist and instructed amateur will readily perceive the absence of a technical knowledge of art in the following pages. In relation to art, I feel myself to stand in a position similar to that of the scientific chemist in reference to the brewer and baker. He may be unacquainted with the practical details of these trades, and nevertheless be able to explain the laws of fermentation which the brewer and baker must observe to succeed in their manipulations. This analogy, however, is not complete; for in art, genius is indispensable to the successful application of rules. But still, one who has studied the science of man by a new method, may have become acquainted with facts and principles calculated to aid the artist in realising his own inspirations: and to present such to his notice is the aim of this publication. (Combe 1855, Preface)

In writing this book, we have become conscious of treading some of the same paths as Bell and Combe, albeit with more modern scientific insights. The difference is that we do not have a prescriptive agenda at all. We present and discuss scientific findings because we believe these can inform our understanding of what goes on when anyone (artist or non-artist) looks at a face or a realistic portrait, not because we want to constrain what artists do.

This difference in stance partly reflects changes in the world of art in the last 150 years. The collapse of the dominance of realistic representation as an artistic aim, brought about by the invention of photography (which placed realistic depiction within the scope of the purely mechanical) and other factors, means that a wider range of styles and intentions are now deployed in the creation of works of art. It therefore seems presumptuous to tell artists what they should be doing.

For the purposes of this book, we thus treat portraits as pictures of the people depicted, rather in the same way we would photographs of them, and we largely ignore the artistic and aesthetic dimensions of portraiture. At the outset, however, we must acknowledge that portraits themselves serve many purposes and convey multiple messages. Only one of these involves some notion of likeness to the person depicted:

> the notion of 'correctness', when applied to portraits, subsumes two distinct values; one is usually expressed in terms of 'the most faithful portrait', that is, to the original, as in the question 'Which portrait is most like Benjamin Franklin', the other is presented in terms of 'the best portrait', as in the statement 'the best portrait of Gertrude Stein was painted by Picasso'. (Brilliant 1991, pp. 39–40.)

The second value may have little to do with likeness, but may include the other messages of the portrait, for example what the artist is trying to say about the character or status of their sitter, or what artistic skill or style is exhibited. Portraits reveal things about the artist as well as about the portrayed person, and moreover, some of the characteristics of the portrayed person may be symbolic rather than to do with likeness. As representations of faces, portraits share some of the limitations of photographs (pictures are flat, static images, but faces are moving, three-dimensional surfaces), and add new limitations imposed by conventions of portraiture (the person's face is not usually sending any of the messages that faces convey when speaking, gesturing, and expressing). Thus, one could have a poor portrait which was a good likeness, and a good portrait which was a poor likeness.

This book is not about portraits as such, or even specifically about the perception of portraits—it is about the perception of *faces*. We have taken the liberty of using portraits as an interesting way of illustrating the faces of people, some living and some dead. We will focus on how each likeness (good or bad) may illustrate the scientific points we want to make, and do not consider the social or artistic factors involved in the interpretation of portraits at all. We realize that our use of portraits as pictures of faces is an oversimplification from the point of view of art theory, but it is not one which matters for the points we wish to make.

Before we can embark on a discussion of how we perceive faces, we must first understand what faces are for and how they may have evolved, since this places important constraints on what messages can be signalled by faces and how these may be perceived. The face plays important roles in many biological functions. Eyes and ears are spaced to allow us to perceive distance; nose and mouth are arranged to minimize choking; mouth and jaws are built for chewing and swallowing, but also, in humans, for speaking and smiling.

The biological functions of the face have produced a basic face 'template' which is remarkably similar across numerous different species of animal, but with modifications that reflect the animal's behavioural needs. The particular characteristics of human faces arise, directly or indirectly, from our large brains, our position as predators and tool users, and from adaptations for vocal language. These general and species-specific factors mean that all human faces are remarkably similar in basic form. Despite this, there are subtle differences that make every face unique so that faces play an important role in the identification of individual members of our highly social species, and systematic variations of the human face pattern inform us about mood, age, sex, and race.

1.2 The face as a biological structure

There is an almost universal face blueprint across different species of animal. It may sound odd to suggest that human faces resemble those of other animals, but at the most general level that we can use to describe faces it is true. Virtually all animals have the same kind of face: two horizontally positioned eyes above a single centrally placed nose and mouth. One fundamental similarity across all animals which is reflected in their faces is their symmetry, which matches the bilateral symmetry of their bodies. Why should it be that the vast majority of animals have symmetrical bodies and faces? Gardner (1967) argues that the near-universal bilateral symmetries of the animal world are a natural consequence of the patterns of forces which operate upon them. The force of gravity imposes a vertical differentiation (between the top and bottom of the animal), and as soon as animals evolved independent means of locomotion it was inevitable that there should be a differentiation between front (towards food) and back. However, there is nothing in the environment to force a distinction between right and left, and this is why, according to Gardner, features such as eyes and legs developed equally on each side of the animal. Intriguingly, Gardner suggests that these same considerations would mean that any animals evolving on other planets would share this same basic feature of bilateral symmetry.

Fig. 1.1 All human faces are remarkably similar in their basic form. *Bill Forsyth* by Steven Campbell. Courtesy of the Scottish National Portrait Gallery.

However, it is less clear why most animals have just two (rather than one, four, or six) eyes and ears. For some animals it would be useful to have extra eyes in the backs of their heads, but this has never evolved in vertebrates (though spiders are amongst some invertebrate exceptions). For the reasons why, we need to explore what faces are and how their different parts depend upon each other and upon the structure of the brain.

Differences in the basic arrangement of the face generally reflect *evolutionary adaptations* to different environmental and behavioural demands. The theory of evolution introduced by Charles Darwin argues that, because of natural variations within the gene pool of a species, some members will possess characteristics that make them more likely to survive and reproduce than others. This is the process of natural selection, which can result in a gradual shift in the characteristics of the species as a whole. According to this theory, it could be argued that during food shortages, those leaf-eating herbivores that could reach a little higher in the foliage would be more successful than those who could not. This process repeated through many hundreds of thousands of generations resulted in the modern giraffe. In what follows we will discuss such adaptations which have affected the shapes of faces, focusing mainly on the faces of land-living mammals, though there are similarities in overall principles of face design across birds, reptiles, and many insects too.

The basic similarities between human and other faces have often been exploited by caricature artists to make humorous or satirical points about human personality and celebrity. Della Porta (1542–1597) was particularly noted for his comparisons between individual human and animal faces (see Fig. 1.2), which he used to infer similarities between individual humans and the characteristics of the animals they resembled (see Chapter 4). In Fig. 1.3, the artist Antony Wisard has caricatured the society couple Lord and Lady Castlerosse as different breeds of dog.

In this book we use a number of examples of computer-produced manipulations of images, where the differences between one face and another are exaggerated (in computer-produced caricatures, see Chapters 5 and 6) or the differences between two faces reduced to achieve a blend of one face into another, in the process of 'morphing'.

Fig. 1.2 Engraving from G. B. della Porta, De Humana Physiognomia (Naples, 1586). V&A Picture Library.

Fig. 1.3 *Entering the Embassy: Lord and Lady Castlerosse* by Antony Wisard. From *The Tatler* 1929. Sir Osbert Lancaster Collection. V&A Picture Library.

Our first example illustrates the continuity in the face structure between man and beast, using a morph between Sir Walter Scott and his famous greyhound Maida (Fig. 1.4). Interestingly, the link between caricature and morphing has long been appreciated; for example, Sir Thomas Browne claimed in 1686 'When men's faces are drawn with resemblance to some other animals, the Italians call it caricatura', noting the origins of caricature in this particular form of exaggeration.

Eyes

The eyes are an outcrop of the brain, containing the light-sensitive retina, whose specialized 'photoreceptor' cells contain chemical compounds which respond to light (see Chapter 2). Impulses are then passed through successive layers of nerve cells ('neurones') within the retina, optic nerve, and cortical and other brain areas (see Fig. 1.5).

The parts of eyes visible to the observer are the coloured iris and part of the white outer casing of the eyeball. The colour of the iris is what gives a person their characteristic 'blue', 'green', or 'brown' eyes, but these variations are quite irrelevant to the job of seeing, though they may affect our impression of facial attractiveness. The role of the iris is to expand or contract to allow more or less light to reach the retina via the pupil, and this plays an important role in adapting vision to different kinds of lighting conditions. Pupil size also changes as a

Fig. 1.4 (a) *Sir Walter Scott with his dogs*, by Sir Francis Grant. Courtesy of the Scottish National Portrait Gallery. (b) Similarity between man and dog face depicted by morphing. *Right panel*: Sir Walter Scott's face; *left panel*: Greyhound Maida's face; *central panel*: 'Scotty Dog'—a morph between the left and right faces. Computer graphics courtesy of David Perrett, Duncan Rowland and colleagues at the University of St Andrews.

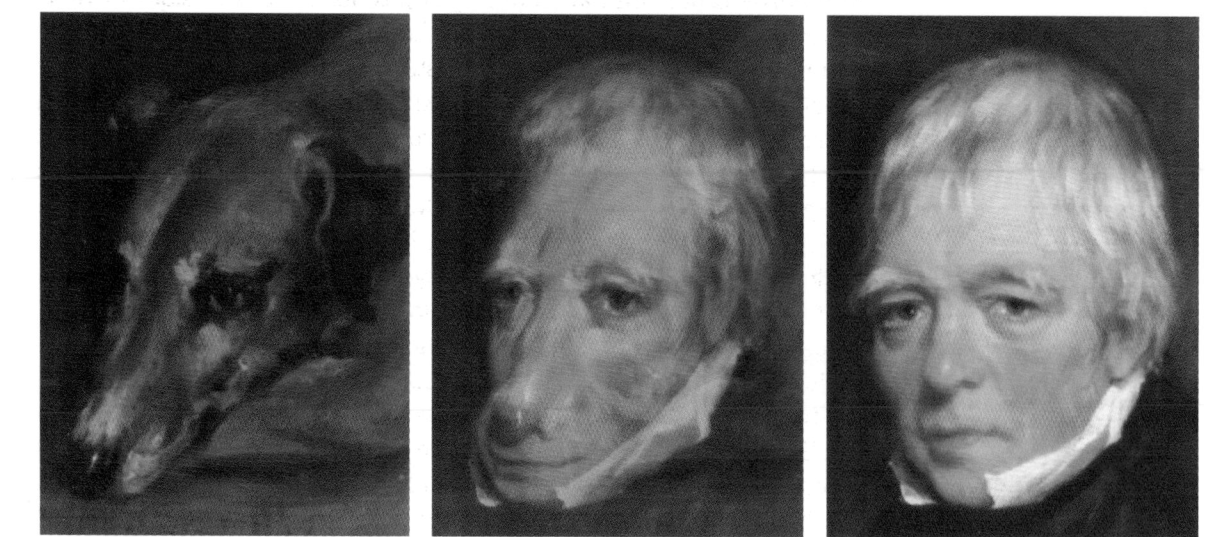

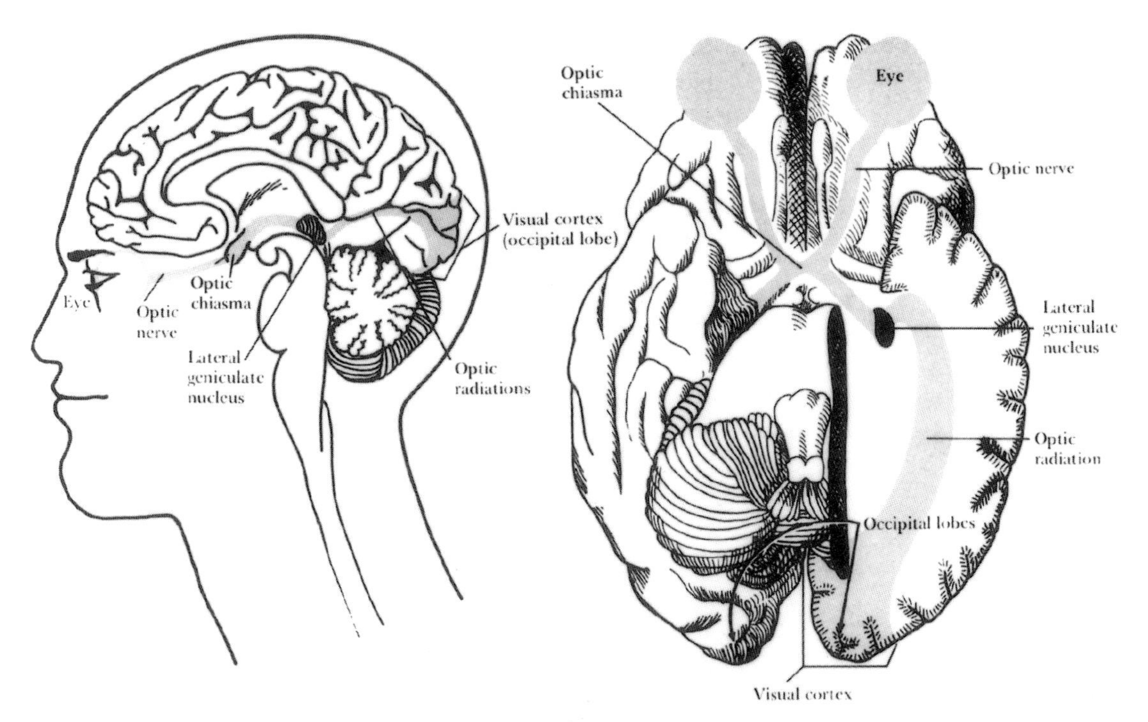

Fig. 1.5 Diagram of the human visual system, showing the primary pathway from eye, via optic nerve to the visual cortex in the occipital lobe of the brain. From P. H. Lindsay and D. A. Norman (1977).

function of arousal, and dilated pupils are seen as sexually attractive. Interestingly, sexual arousal in the cuttlefish is also accompanied by pupil dilation, an example of extraordinary convergence in two remarkably dissimilar species.

All mammals have two eyes, but the arrangement of their eyes reflects the type of animal they are. Animals which are preyed upon need the best possible early warnings that predators are approaching. Thus many herbivores, such as deer, horses, and rabbits, who may be hunted by other animals, have eyes positioned on the sides of their heads in order to keep watch over the widest possible field of vision. Animals which hunt need good distance perception. One way that vision provides information about distance is through a process called 'stereopsis', by which differences between each eye's image of the world are compared to indicate the relative distance of different objects (see Chapter 2 for a fuller account). For stereopsis to work, the visual fields seen by each eye need to overlap (Fig. 1.6). This is why carnivores such as dogs and cats tend to have their eyes at the front of their heads, so that their overlapping visual fields allow stereoscopic vision, useful for catching prey. Primates have extremely well developed stereo vision, which is particularly useful for manipulating tools.

The appearance of human eyes is rather unusual in revealing a relatively large amount of the white casing of the eye (the 'sclera') compared with other species. Hiromi Kobayashi and Shiro Kohshima (1997) examined the eyes of 88 different primate species, and showed

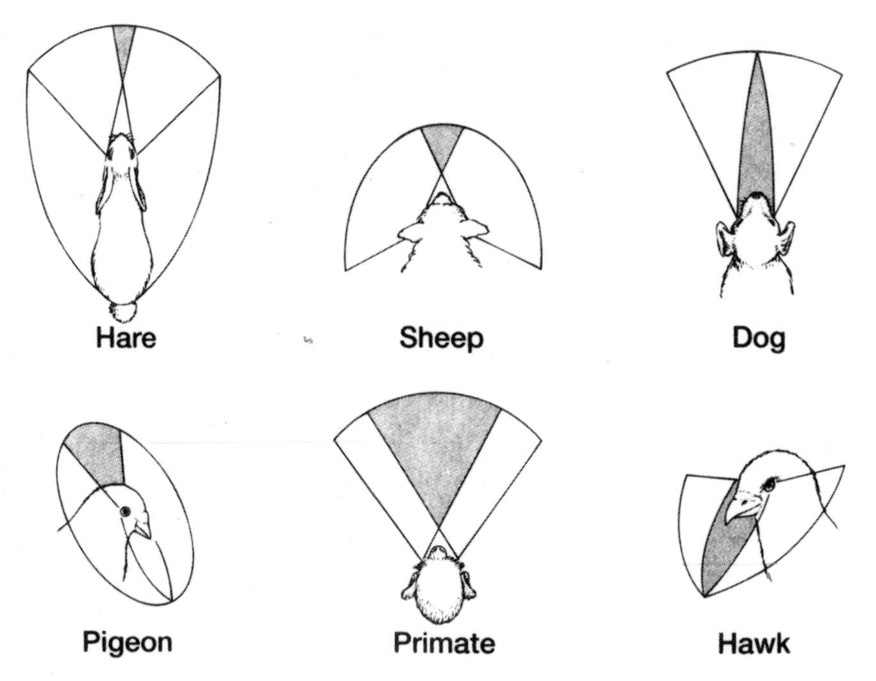

Fig. 1.6 The degree of overlap of visual fields varies considerably between different species. From K. V. Kardong (1995). *Vertebrates.* Wm. C. Brown Publishers.

that humans had the largest amount of exposed sclera, and the most horizontally elongated eyes, as well as being the only species of primates with white sclera—most others have sclera which match the skin colour around the eyes. Kobayashi and Kohshima suggest that it may be adaptive for primates to have direction of gaze camouflaged, since direct eye contact often provokes attacks. However, enhancing gaze signals, as has been achieved by the shape and colouration of the human eye, may be more useful for communication and cooperation between individuals, particularly when acting within groups.

Ears

Mammals also have two ears, and their position and shape are again evolutionary adaptations to their lifestyles. Animals which are preyed upon often have large, mobile ears, which help to collect and amplify sounds (Fig. 1.7). The physical separation of the ears helps localize sound sources through minute differences in the timing of signals arriving at each of the ears, just as differences in the spacing of visual images signal distance through stereoscopic vision.

Ears tend to be an unremarkable aspect of human facial appearance unless they are very unusual in shape. This is presumably because our interactions are typically face-to-face, where ear shape is scarcely visible, and in any case human ears are often concealed by hair. However, ears play an important role in the expressive postures of some other animals, such as dogs, cats, and horses. A dog's ears are laid back

Fig. 1.7 Rabbits have large, mobile ears, to help them detect the approach of predators. From Lockley (1964).

flat in anger and are lowered in appeasement or when the dog is miserable, and raised when alert (Fig. 1.8).

When considering the shapes of different parts of an animal's body, there is always the temptation to assume that all variations serve some adaptive purpose. That this is not necessarily the case is illustrated rather neatly when one considers variations in the shape of ears within primates (Fig. 1.9). These minor but characteristic variations from one primate species to another have never been explained in terms of differing function.

Noses

In all animals the nostrils lie above the upper jaw, thus reducing the chance of blocking airways with food. Indeed choking is almost im-

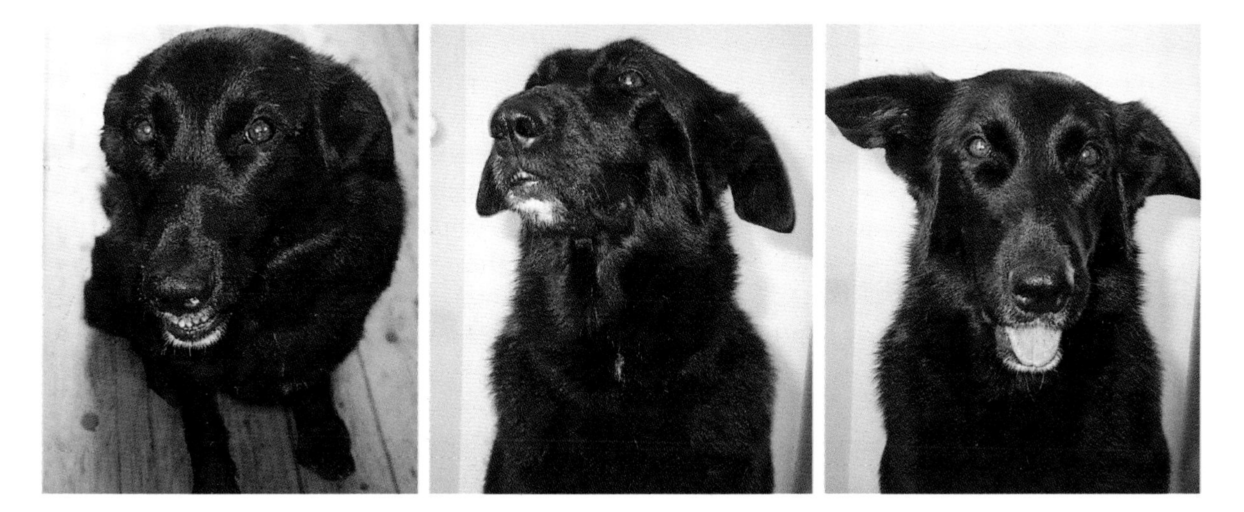

Fig. 1.8 The dog's ears can be remarkably expressive as these pictures demonstrate. Photographs courtesy of Bean.

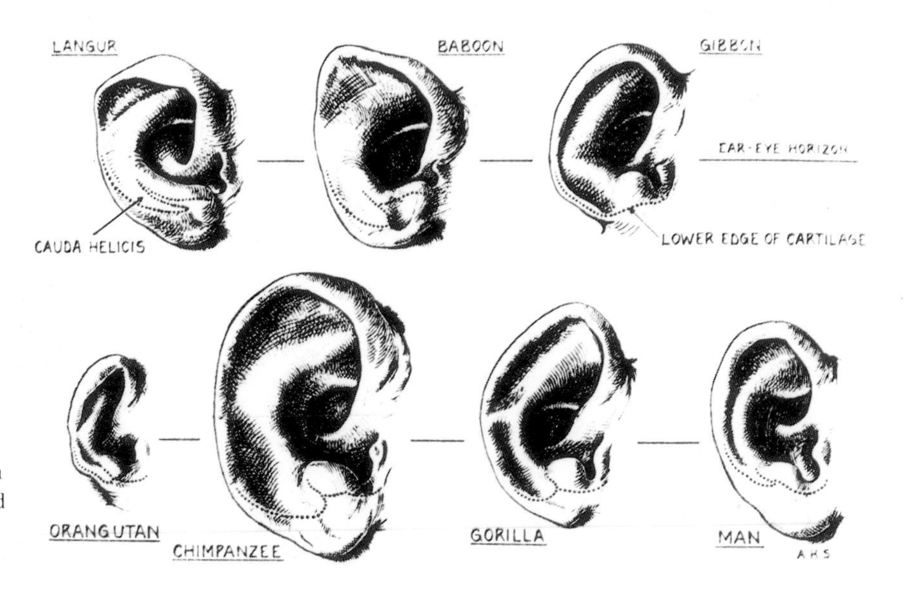

Fig. 1.9 Different shaped ears in different primate species. Adapted from Schultz (1950) by Lenneberg (1967), p. 26.

possible in species other than human, because the pharynx and oesophagus are arranged differently in humans—some of the many adaptations that facilitate vocal language. Other details of the shape of the nose and the mouth/jaw region differ a great deal depending on the role of scent for each animal, whether additional sensory whiskers are present to supplement sight and scent, and on the kind of food that the animal eats.

Most mammals other than primates have pointed faces, but the elongated fleshy 'nose' of the human is very rare. This adaptation ensures that the inflow of air is aimed upwards into the nasal chambers. The nerve endings which serve the sense of smell are located in the top of these chambers. Figure 1.10 shows the different directions in which air is inhaled in the human compared with a long-snouted mammal. In some other animals the fleshy nose serves different functions. The elongated nose of the elephant, for example, is an adaptation for foraging, but can also act as a useful hosepipe.

The horizontal positioning of the smell receptors (within the 'olfactory bulbs' of the brain, see Fig. 1.11) is itself a consequence of the size of the brain in humans compared with other mammals. The large size of the human brain has forced the olfactory bulbs to rotate downwards, and this has had a significant impact on the shape of the human face, which is arranged vertically rather than horizontally.

Jaws and mouth

The mouth serves multiple functions in humans, but its primary function in all mammals is as an entry point to the digestive tract. The

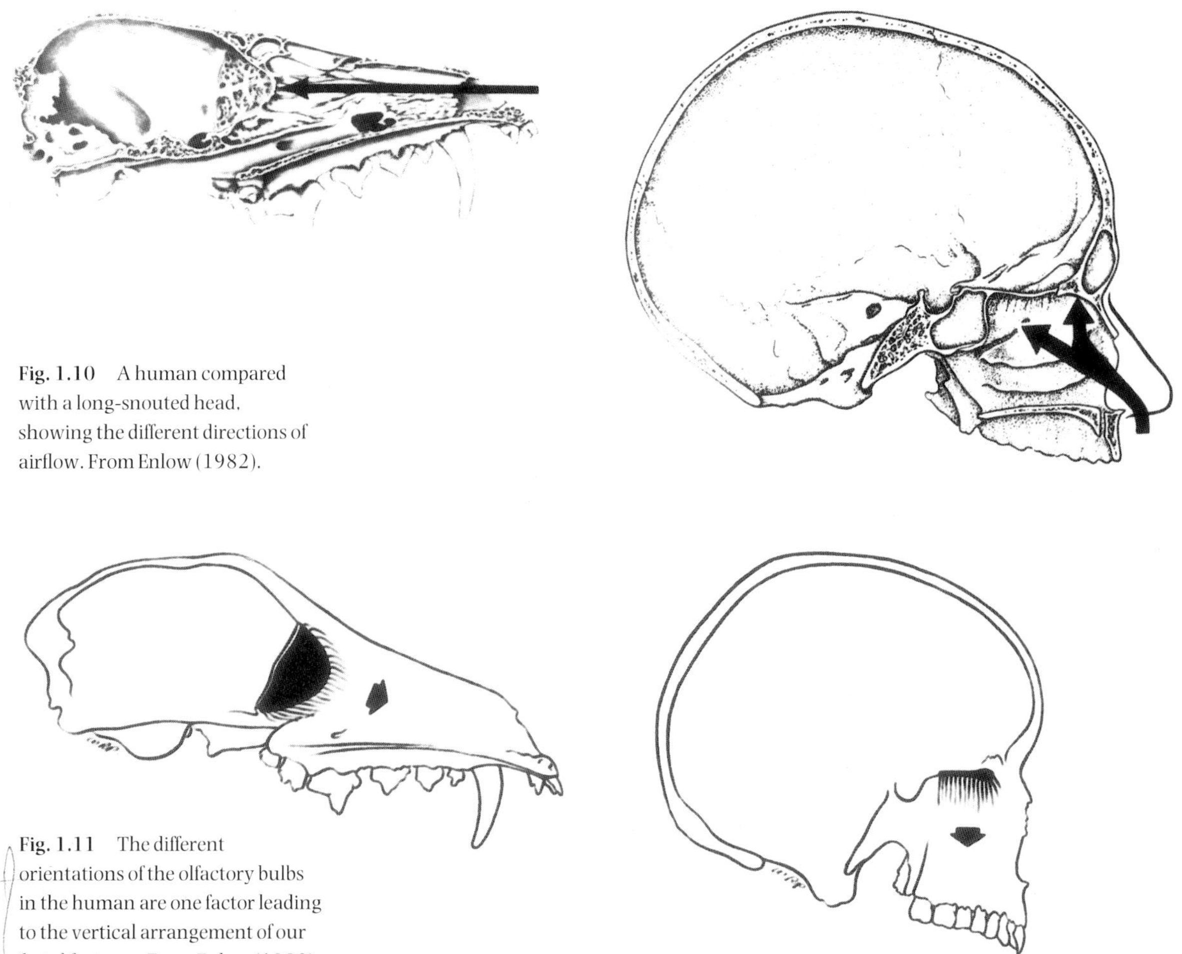

Fig. 1.10 A human compared with a long-snouted head, showing the different directions of airflow. From Enlow (1982).

Fig. 1.11 The different orientations of the olfactory bulbs in the human are one factor leading to the vertical arrangement of our facial features. From Enlow (1982). Used with permission.

mouth, lips, and jaws are used in the vast majority of mammals to catch and grasp food (some exceptions include primates who can use their upper limbs to collect food, and elephants who use their noses). The shapes of jaws and teeth reflect the kind of food an animal eats. Herbivorous mammals have jaws which move from side to side to break down vegetable matter, while carnivorous ones have powerful scissor-like jaws for catching and eating their prey. Primate jaws are relatively unspecialized and allow them to eat a wide variety of different foodstuffs. Within primates, there is much variation, and humans have particularly weak jaws compared with some of their closer relations among the great apes (see Fig. 1.12). The face of the modern human, *Homo sapiens*, may be compared with prehistoric forms of the human face seen in Fig. 1.13. *Homo erectus*, an early ancestral form of human which emerged from Africa some two million years ago, had a weak jaw rather like modern *Homo sapiens*, but a much smaller brain.

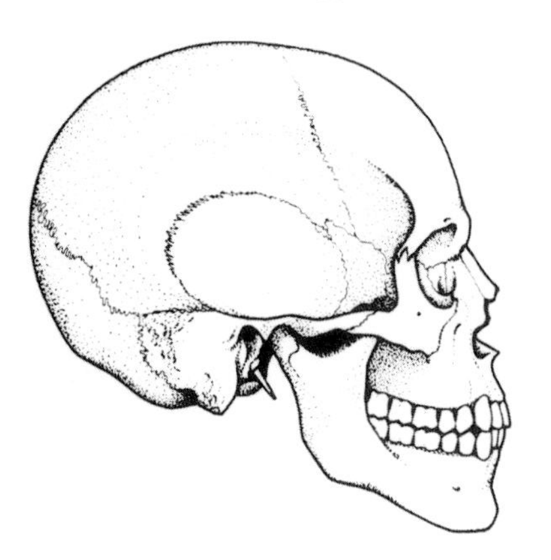

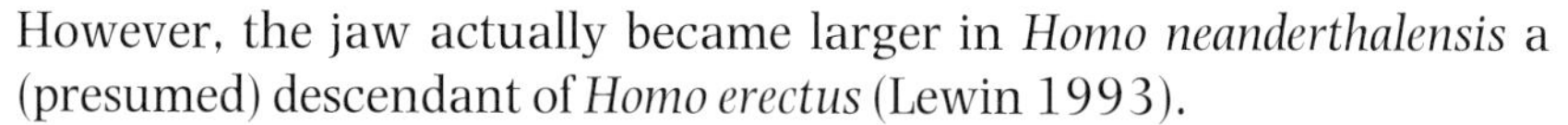

Fig. 1.12 The skulls of human and gorilla illustrate the comparatively small size of the jaw in humans. From Johnson and Moore (1989).

However, the jaw actually became larger in *Homo neanderthalensis* a (presumed) descendant of *Homo erectus* (Lewin 1993).

Once collected, food must be digested. Chewing or puncturing of food with the teeth enables the process of digestion to begin in the mouth, as enzymes can start to act on the food material. The nature and layout of teeth is a further factor which affects the appearance of an animal's face, and in the human the replacement of milk teeth with adult teeth, and associated growth of the jaw, has a significant effect on facial appearance in the developing child. In some animals teeth become elongated as ornamental features, as in elephant tusks.

Human mouths serve other functions than eating, since the tongue, teeth and lips are uniquely adapted to produce the variety of vocal signals upon which human language depends. Many animals produce elaborate vocal signals, and many show emotional states with movements of their faces (for example by movements of the ears). Many species of primates use their lips in expressive displays. But the human lips are particularly adapted to their role in helping to form speech sounds and these adaptations may allow them to be particularly expressive in our faces. This is a further example of the complexity of evolutionary adaptation, where a feature which emerges for one purpose (such as speaking) may prove useful for another (such as smiling).

Forehead

Human and other primate faces differ from other animal faces in a number of ways. Larger brains, particularly the frontal lobe regions, require more space, giving rise to the high brow area (compare the dog

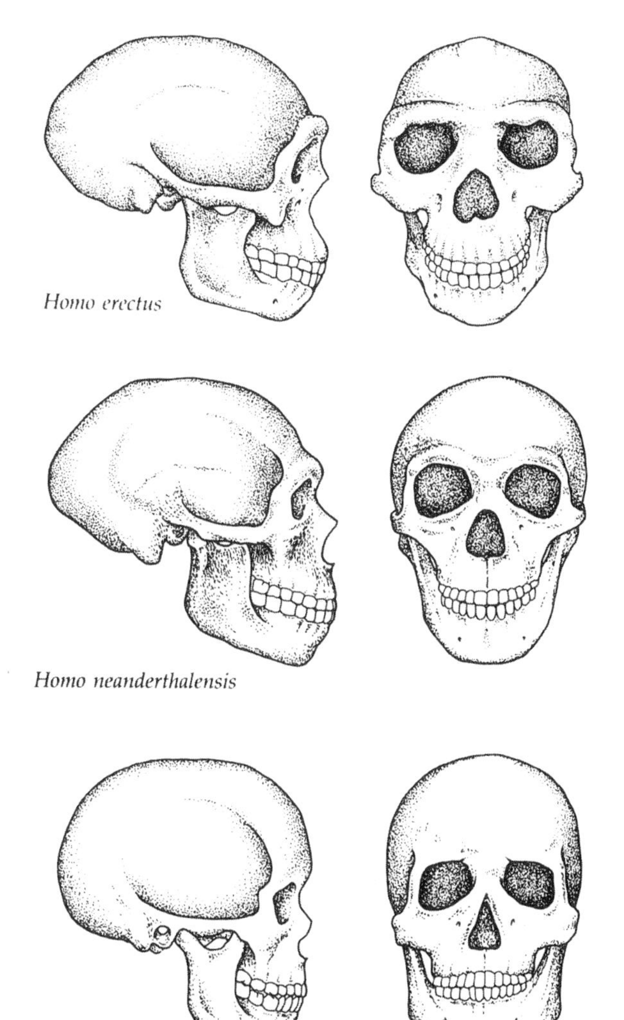

Fig. 1.13 Comparison of the skulls of *Homo sapiens* (modern human), *Homo neanderthalensis* and *Homo erectus*. From Lewin (1993).

and human faces in Fig. 1.4). The high brow plus forward-pointing eyes for stereoscopic vision result in a relatively flat rather than a pointed face (owls and cats also have flat faces). However, this itself is made possible because there is no pointed muzzle to get in the way of the overlapping fields of vision, and as we saw earlier, the shape of the muzzle is itself an adaptation to the growth of the brain. Thus the development of different facial layouts is interdependent (see Fig. 1.14). Moreover, these adaptations of the face and head also interact with other aspects of human shape. For example the upright, bipedal posture of the human frees the upper (front) limbs for tool manipulation, which capitalized on the three-dimensional vision made possible by the overlapping visual fields.

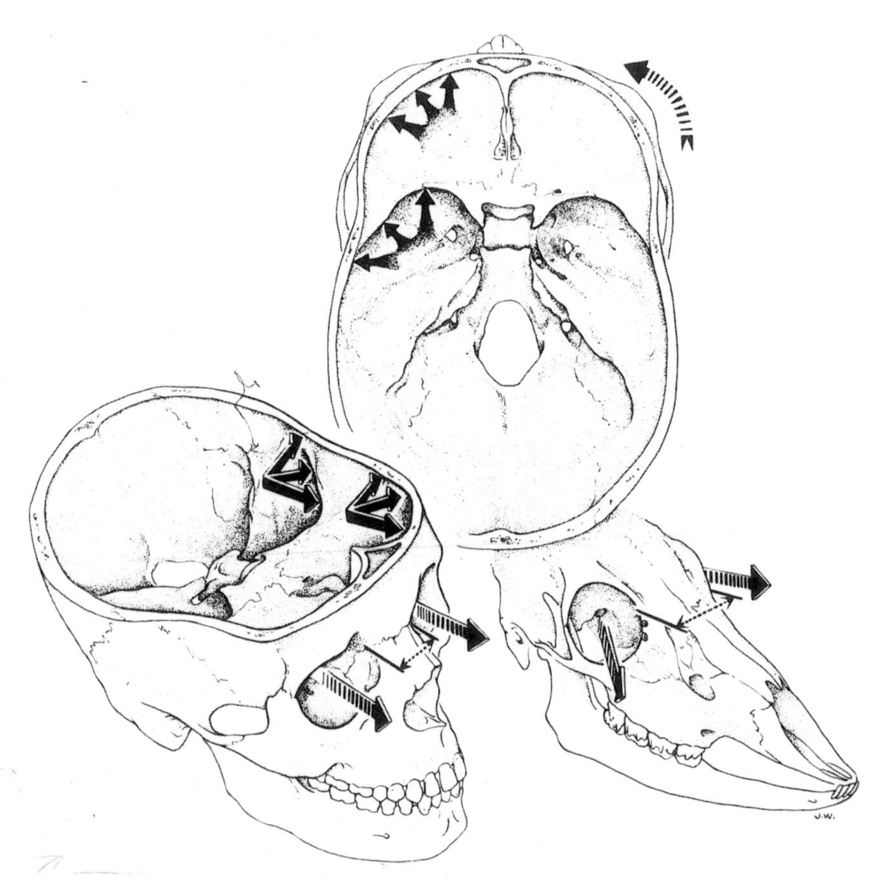

Fig. 1.14 The interdependence of different adaptations to the structure of the human face. The overlapping visual fields are made possible by the change in layout of the nose and muzzle, itself an adaptation to increased brain size. From Enlow (1982).

Distinctive features of human faces

Many species of animal have distinctive additional features of their faces, which play roles in sexual and/or aggressive displays, such as the antlers of deer or the horns of the rhinoceros. Human and other primate faces also have some unusual features. The distribution of hair on human and some primate faces differs from most other animals. In the mandrill monkey (Fig. 1.15) the absence of hair has allowed the evolution of the most extraordinary facial colouration. The red nose and blue cheeks mimic the red penis and blue scrotum which many species of monkey use in display to threaten males from rival groups.

Human female faces are almost hairless, with hair only on the eyebrows and head. As human bodies are generally hairless, we must ask what is the function of the remaining areas of hair? Head hair may well serve a function in preventing heat loss, but the function of eyebrows is more mysterious. One suggestion is that they serve to prevent sweat dripping into the eyes. However, another very important function of eyebrows is to exaggerate facial signals (see Fig. 1.16), and it may also be this role in interpersonal communication which has preserved hair in the eyebrow region.

Fig. 1.15 The face of the mandrill reproduces the colours of monkey genitalia, which are displayed to threaten rivals. From Fogden and Fogden (1974).

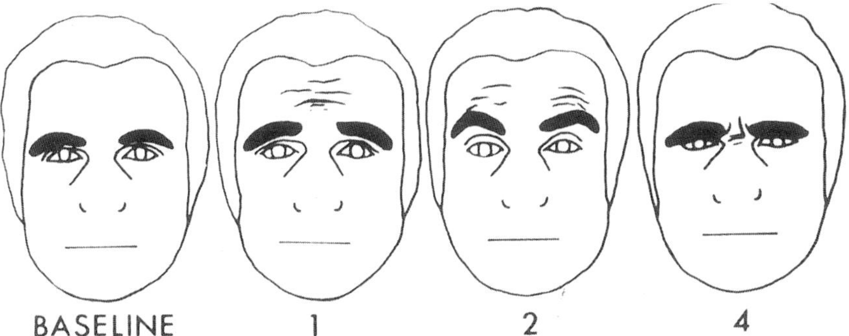

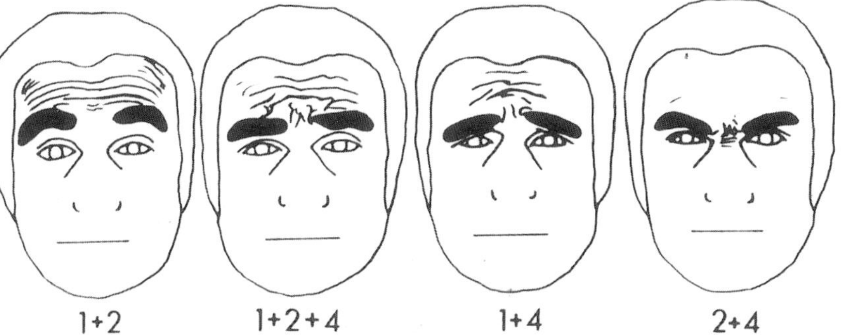

Fig. 1.16 Different eyebrow movements are involved in a range of distinct human expressions such as concern, surprise, and anger. The numbers signify the different groups of muscles which are active to produce each of the distinct eyebrow postures shown. Reproduced with permission from Ekman (1979).

A number of further differences between human and other animal faces can be seen in specific adaptations for speech. The social and structural pressures which led to the evolution of vocal speech are not clear, but one interesting account (Dunbar 1996) suggests that important factors were that humans can sweat, rather than pant, to regulate temperature, and have upright posture and flattened chests. This also happens to free the mouth, tongue and respiratory apparatus for communication, and enables humans to vary their respiration in the ways necessary to use expelled air for long periods in speech. Other adaptations for language through the shape and mobility of the jaws and tongue also affect the appearance of the human face (see Box 1A).

Box 1A: Adaptations of the human face which are important for speech

Many aspects of the internal and external appearance of the human face and air-passages enable us to articulate rapidly the different sounds which make up our languages. Whether these were specific adaptations for speech, or whether speech was structured by the shapes of our faces cannot be known, but there are certainly a number of unique features of our faces, not shared with other primates.

Human speech is made up of a series of sounds known as 'phonemes' which play a similar role in spoken language to that played by letters in written language, though in English there is only an irregular mapping between phonemes and letters. For example, the words 'pay' and 'they' differ by several letters but only by a single phoneme, 'p' compared with 'th'. If you say these words to yourself you will be able to notice that the 'p' sound involves the closure of the vocal tract at the lips, while in the 'th' sound the vocal tract is closed with the tongue against your teeth. All the different sounds of our language are produced by us modifying the shape of the vocal tract in different ways.

Figure 1A.1, from Fromkin and Rodman (1974) illustrates the major features of the human vocal tract, with the different places of articulation of speech sounds numbered from 1–8. Closure of the vocal tract at these different places results in different speech sounds. Try saying to yourself the following series of sounds—ba, va, tha, ta, ga—you will be able to feel how your lips and tongue close the vocal tract at different places in this series (see Chapter 6 for further discussion of speech sounds).

An additional and important adaptation that facilitates speech in the human is the lowering of the larynx in the human and lengthening of the pharynx compared with other animals including primates. One unfortunate consequence is that the human food and

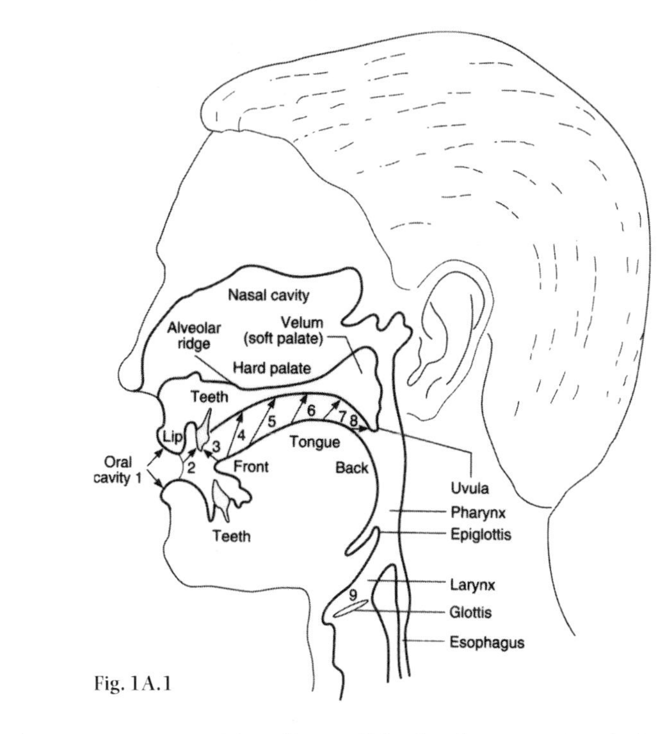

Fig. 1A.1

airways cross, making it possible for humans to choke on food, something which is virtually impossible in any other species.

Table A1 lists how different features of the human face allow us to make a range of speech sounds which commonly occur in different languages. Attempts to teach non-human primates to communicate with spoken language have all ended in failure, though there has been more success in teaching them to use words using sign language or via symbols. Such explorations of primate linguistic abilities have indicated that human vocal speech depends on a complex combination of neurological, cognitive, and facial adaptations.

Table A1 Biological foundations of language

Groups of widely occurring speech sounds	Some of the structures involved in their production
p, b, m	Muscles: muscular rim in lips; muscles of cheeks;
f, v, w	Teeth: vertical position of incisors; reduced canines
t, d, n	Position of teeth
k, g	Bulging of tongue with ability to raise back
l, r	Blade of tongue with facility for changing shape of its cross-section
vowels	Muscles in corner of mouth; small size of mouth

Adapted from *Biological foundations of language* (Lenneberg 1967)

Thus the basic face template is similar across the vast majority of mammals, with variations between species usually arising as specific evolutionary adaptations. The arrangement of features in the human face reflects particular sensory, dietary, and linguistic adaptations, and to perform these functions all human faces must be essentially identical in their underlying design.

Artistic violations of the basic face 'schema' are particularly striking and often disturbing (Fig. 1.17), and natural violations which arise

Fig. 1.17 Violations of the face schema are disturbing for us. Rene Magritte 1934, *Le Viol* (The Rape). Copyright ADAGP, Paris and DACS, London (1998).

from facial deformities and disfigurement can have profound effects on people's lives. Despite the similarity between all human faces, subtle differences in appearance convey information about group and individual identity, and movements of the face convey a variety of other social signals.

1.3 Alas, poor Yorick: the structure of the human face

The human face comprises underlying skeleton (skull) to which the hard and soft tissues of the face are attached (see Figs 1.18–1.20). The three-dimensional shape of a particular face results from the combination of the underlying skeletal structure, the hard tissues such as the cartilage in the nose, and the soft fatty tissues and skin. Differences in these tissues, plus variations in colouration and texture of skin, eyes, and hair, provide the basic information which is used to categorize the face (for example as a man or a woman). Because all faces must be identical in basic design, sensitivity to rather subtle differences is often needed to determine group membership and to identify individuals from their faces.

The orthodontist Enlow (1982) describes how the growth of faces and heads is constrained, by the interdependence between different aspects of their form, in ways which result in a small number of basic face types in different races. So, for example, long thin noses go with long narrow heads, and short, wide noses go with broad, wide heads. The two extreme forms which occur in humans are the 'dolichocephalic' head, which is long and narrow, and the 'brachycephalic' head, which is wide and short. These different types of head in turn give

Fig. 1.18 Sir Charles Bell's illustration of the skulls of adult female (*left*), male (*centre*), and infant (*right*) faces.

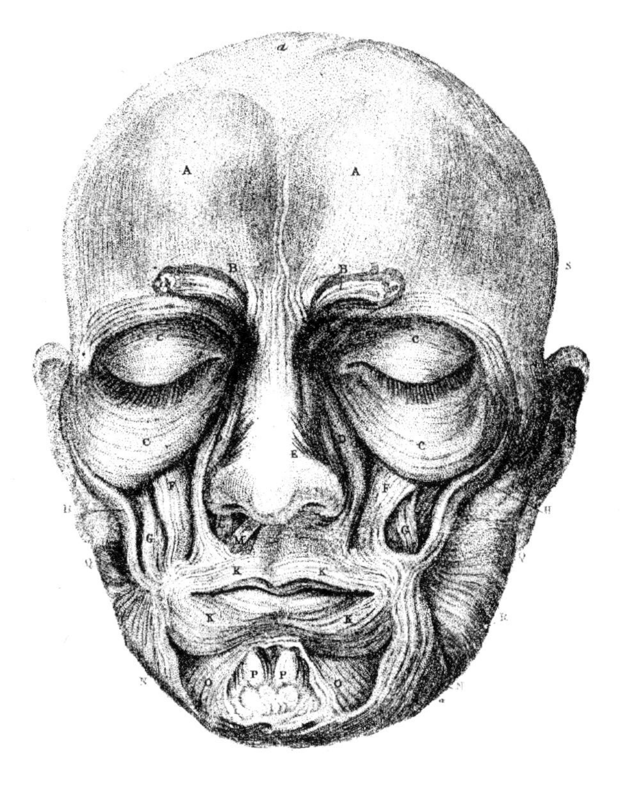

Fig. 1.19 Sir Charles Bell's illustration of the muscles of the face which are attached to the underlying bony structures of the skull.

rise to different types of face. The 'leptoprosopic' face (see Fig. 1.21) is long and narrow with protrusive features, typical of faces from southern Europe, for example. The 'euryprosopic' face is broad and wide with flatter features, typical of Asian faces.

Growth of the face during childhood produces a characteristic transformation in appearance as the nose and jaws grow (Fig. 1.22). In the infant, the eyes are relatively lower down the face, and relatively larger than in the adult, which gives rise to the characteristic 'cute' look of the baby and young child. The change in shape which occurs during growth was described geometrically by the Scottish naturalist D'Arcy Thompson, who argued that similar principles appear to apply to growth in a variety of biological forms.

> The form, then, of any portion of matter, whether it be living or dead, and the changes of form which are apparent in its movements and in its growth, may in all cases alike be described as due to the action of force. (Thompson 1917, p. 11)

Thompson was able to demonstrate that forms that were related by growth or evolution could be shown to be related one to another by a simple deformation induced by strain operating on the original structure. In Fig. 1.23 Thompson illustrates the deformation needed to map the shape of a human skull onto that of other primates. Indeed

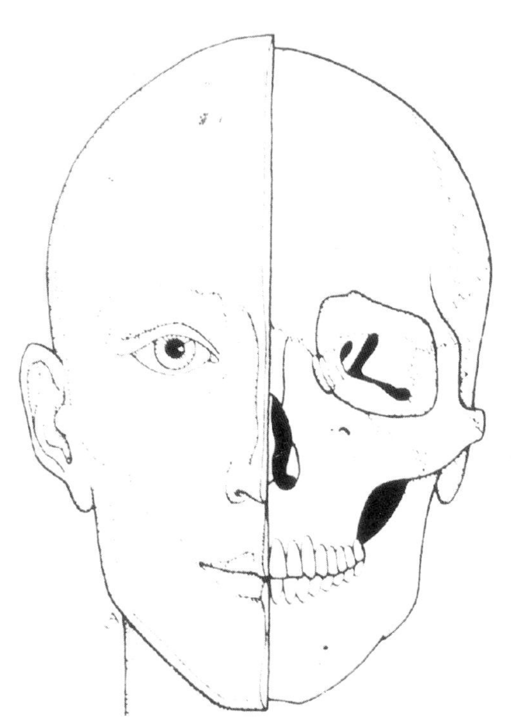

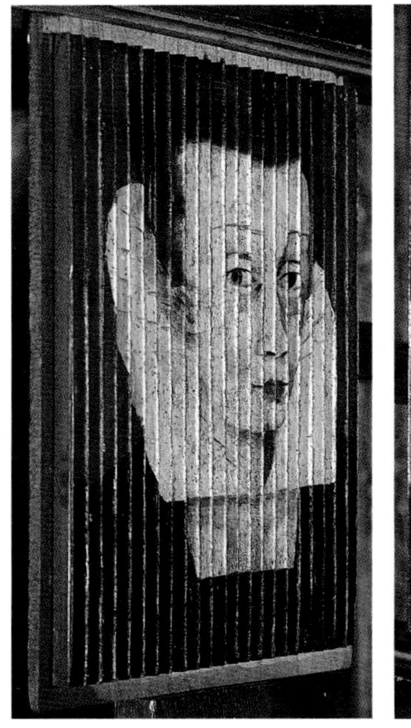

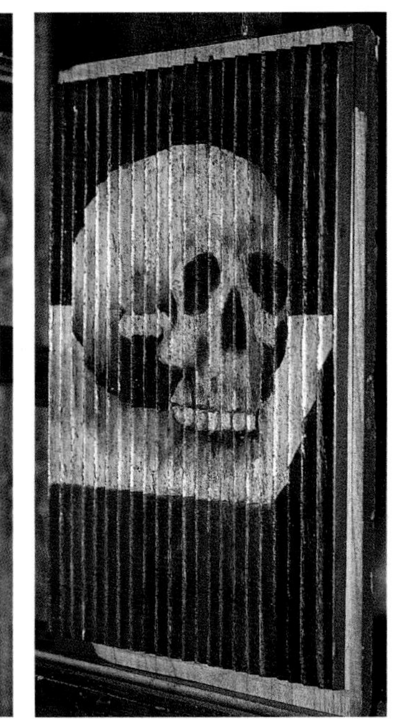

Fig. 1.20 *Left*: the skin surface of the face peeled back to reveal the skull beneath. From Johnson and Moore (1989). *Right*: an unknown artist played upon this theme to produce this Anamorphosis, called *Mary Queen of Scots*, in which the panel viewed from one side shows the face of the Queen, but viewed from the other side her ultimate fate is revealed by the skull. Courtesy of the Scottish National Portrait Gallery.

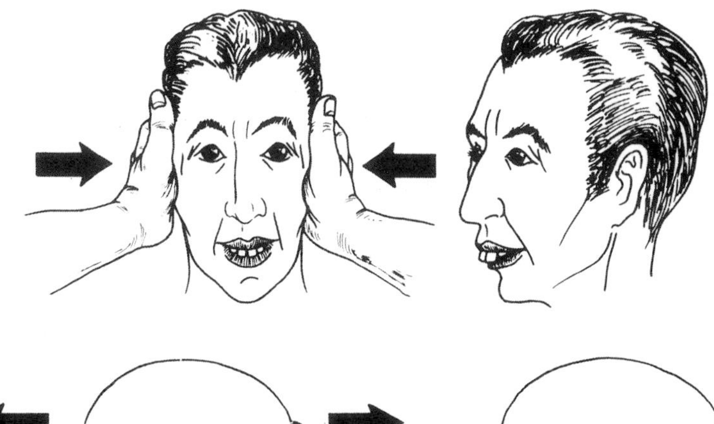

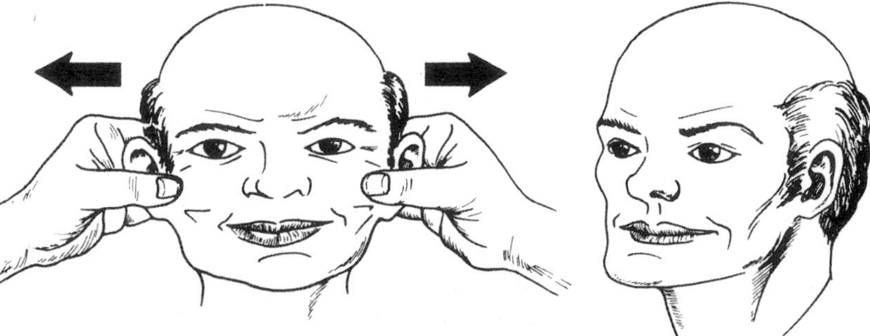

Fig. 1.21 The leptoprosopic (*upper*) and europrosopic (*lower*) faces described by Enlow (1982).

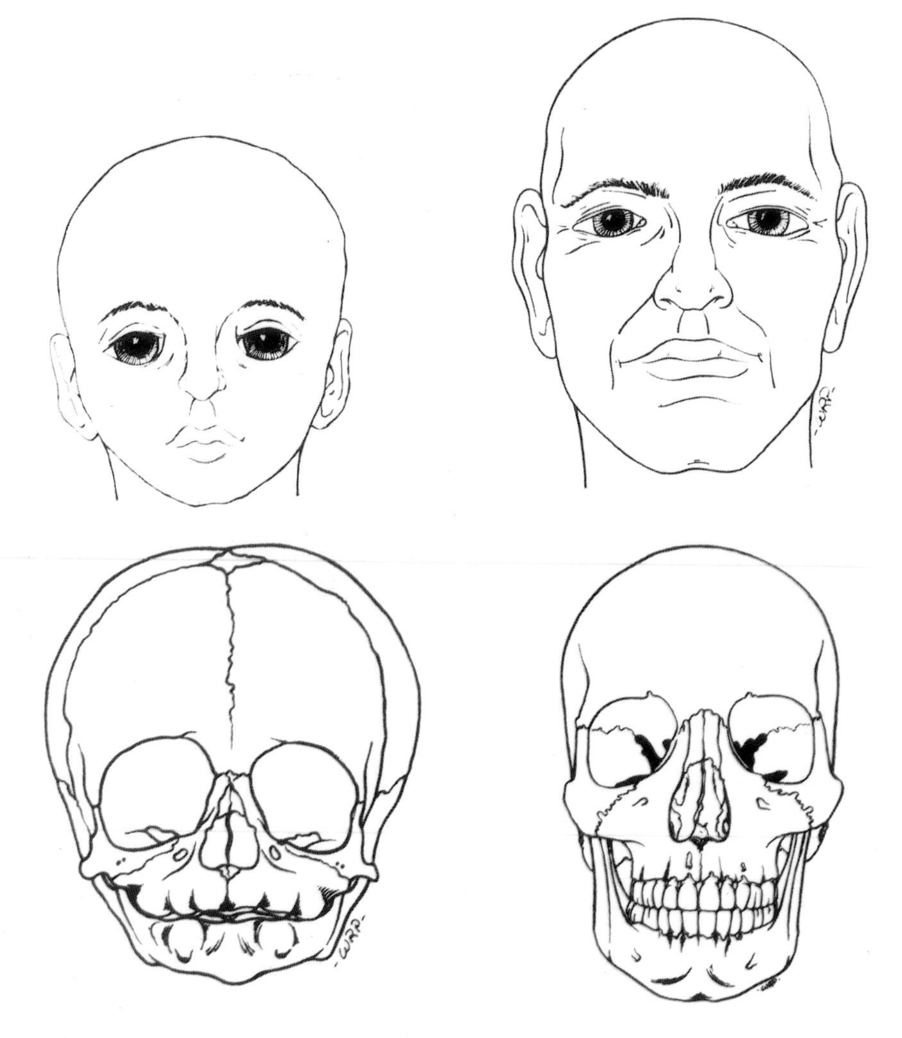

Fig. 1.22 Transformations in the shape of human skull and head between infancy and adulthood. From Enlow (1982).

some failures to produce simple mappings in this way led Thompson to draw conclusions about lines of descent:

> Mr Heilmann tells me that he has tried, but without success, to obtain a transitional series between the human skull and some pre-human, anthropoid type . . . which series should be found to contain other known types in direct linear sequence. It appears impossible, however, to obtain such a series, or to pass by successive and continuous gradations through such forms as Mesopithecus, Pithecanthropus, Homo neanderthalens, and the lower or higher races (*sic*) of modern man. The failure is not the fault of our method. It merely indicates that no one straight line of descent, or of consecutive transformation, exists; but on the contrary, that among human and anthropoid types, recent and extinct, we have to do with a complex problem of divergent, rather than of continuous variation. (Thompson 1917, p. 772)

This conclusion is in fact consistent with contemporary views of human evolution (see also Fig. 1.13 for an example of lack of continuity of evolutionary change).

Psychologists have built upon D'Arcy Thompson's pioneering

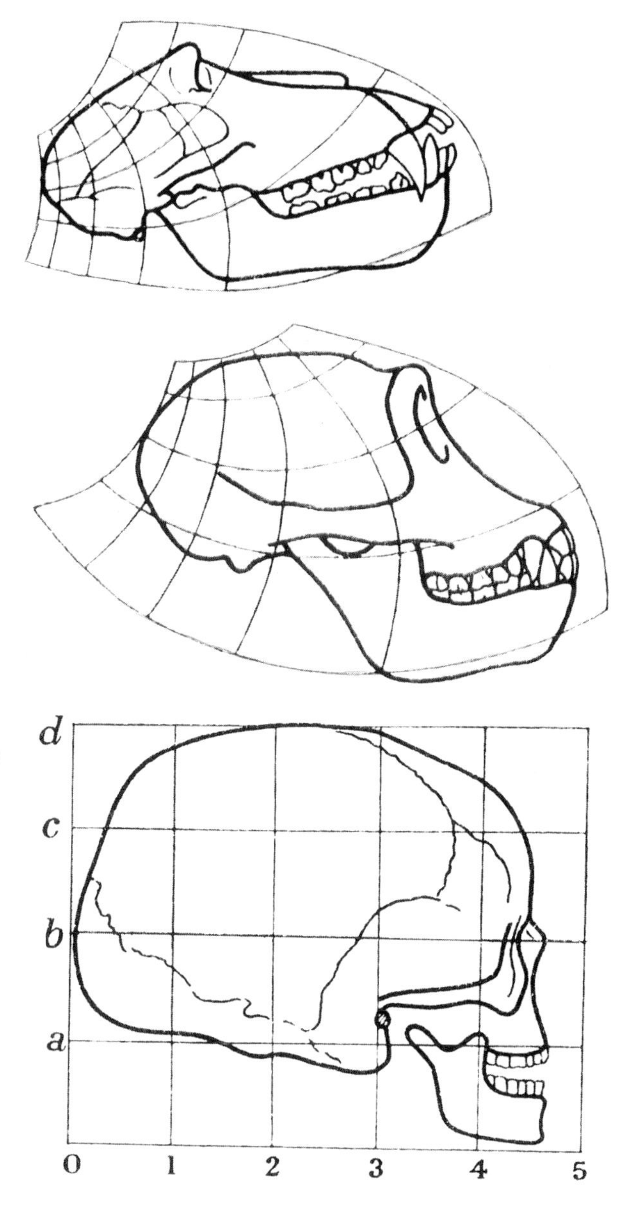

Fig. 1.23 Transformations in the shape of human skull (*bottom*) needed to map it to the shape of a chimpanzee (*centre*) and baboon (*top*). From Thompson, D. W. (1917). *On growth and form*.

demonstrations to characterize the shape change which occurs as the face and head grow from infancy to adulthood in terms of a mathematical transformation termed 'cardioidal strain', which we describe in more detail in Box 1B.

Cardioidal strain can only provide an approximate description of one of the global changes which arises through growth. Changes in facial hair, skin texture, etc. are also important as we will see in Chapter 3. During adolescence the facial appearance changes considerably with the emergence of secondary sexual characteristics: a prominent voice box and beard (or visible stubble) for men.

Box 1B: Growth of faces modelled by cardioidal strain

Pittenger and Shaw (1975) were the first to apply D'Arcy Thompson's work directly to the perception of human faces. They argued that as people are able to identify individuals despite changes produced through the ageing of their faces (though see Chapter 3), human vision must be able to decipher the geometry of this non-rigid transformation and extract information that remains invariant across such changes. Building upon Thompson's demonstration that strain characterizes growth processes in a range of biological forms, they compared the growth of the human skull with the growth of dicotyledonous plant and vegetable structures such as kidney beans

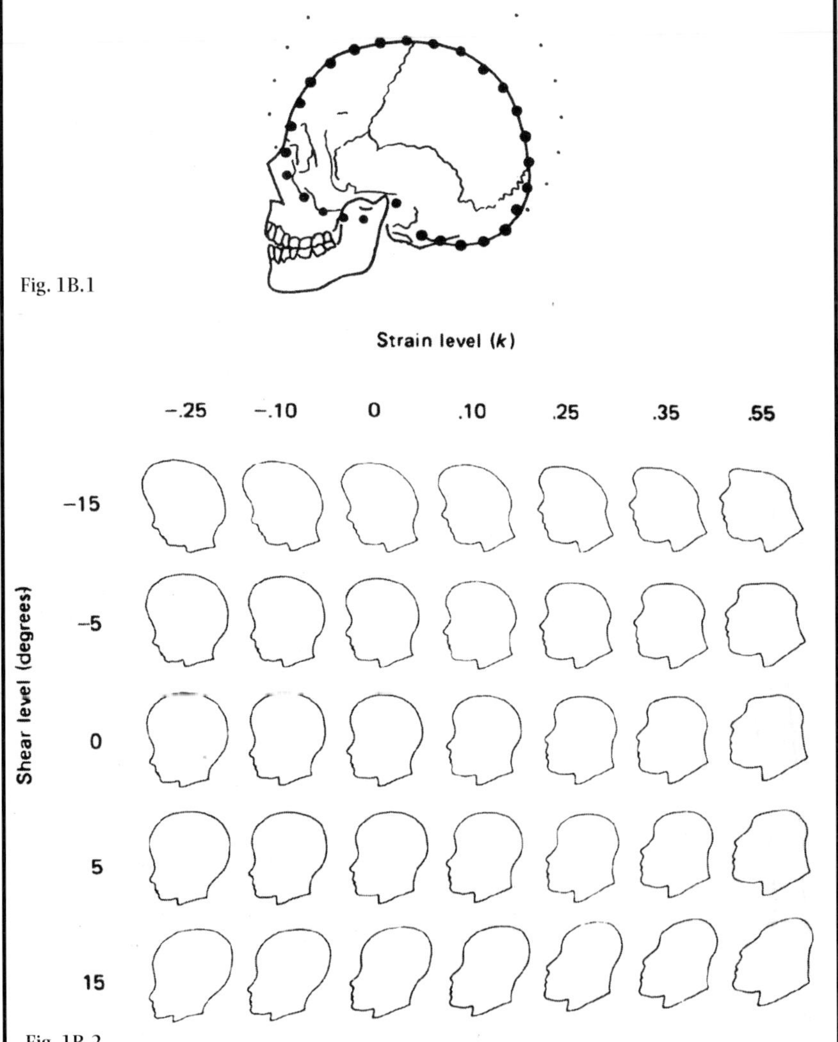

Fig. 1B.1

Fig. 1B.2

and apples, all of which expand symmetrically from a nodal point where growth is constrained (for example at the stem of the apple or the point of attachment of a bean to its pod). The growth of the human brain and skull appears very similar, and is also constrained around the nodal point created where the brainstem meets the spinal cord. Pittenger and Shaw showed that a strain transformation called cardioidal strain (from 'cardioid'—a heart shape) characterizes well the transformation of the human cranio-facial profile as it ages, as illustrated in Fig. 1B.1 (from Shaw *et al.*, 1974).

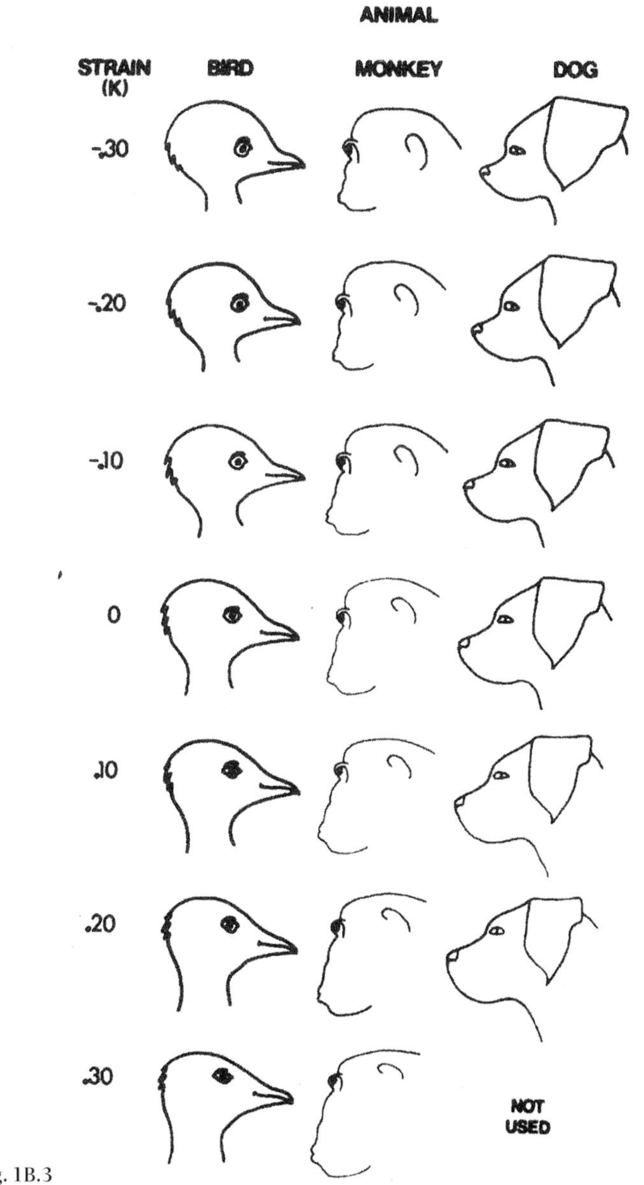

Fig. 1B.3

Box 1B: Continued

In addition to cardioidal strain, a 'shear' transform also models some of the changes due to human growth, in particular the change in the angle of the face between infancy and adulthood. Figure 1B.2 (from Pittenger and Shaw, 1975) shows a series of skull profiles which have been deformed to different levels of cardioidal strain (horizontally) and shear (vertically). Perceptual experiments have shown that observers are reasonably consistent at rank-ordering different skull profiles transformed in this way according to their apparent age, though cardioidal strain level has a much greater influence upon their judgements than does shear. This same transformation can be applied to the profiles of other animals to produce age-related transformations, as in the animal profiles illustrated in Fig. 1B.3 (from Pittenger *et al.*, 1979).

In an interesting extension of this work Mark and Todd (1983) obtained a database of three-dimensional measurements of a girl aged 15 years, which was used to produce a computer-sculpted bust. The cardioidal strain transformation was then applied (in reverse) to yield a transformed bust which observers judged as looking several years younger. In further work using this same technique, Bruce *et al.* (1989) applied this transformation to a database of 3D measurements of a face obtained using a laser range-finding device (see Box 1C) and displayed graphically. Such data can also be transformed in a way which should make the head look younger or older. This is quite successful in transforming the apparent age of these 3D images (see Fig. 1B.4 produced by Mike Burton, University of Glasgow), although the absence of hair in these images and the adult shape of other features such as the nose gives rise to some strange perceptual effects.

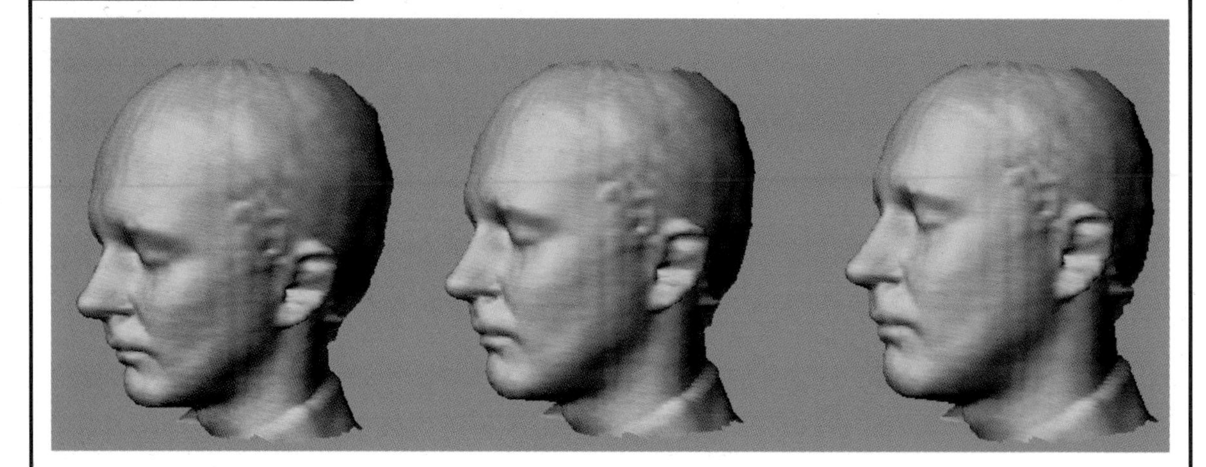

Fig. 1B.4

Fig. 1.24 Portraits of Bonnie Prince Charlie, courtesy of the Scottish National Portrait Gallery. *Top left*: infant, by unknown artist; *top right*: boy, by Antonio David; *bottom left*: young man, by unknown artist after Maurice Quentin de la Tour; *bottom right*: old man, by Hugh Douglas Hamilton.

Fig. 1.25 Morphing Bonnie Prince Charlie. The pictures on the diagonal running from top left to bottom right are portraits of the Prince as an infant, as a boy, and as a 65-year-old. Intermediate pictures have been produced by morphing between these images. (Compare the morph in the bottom left with the portrait of Charles as a young man seen in the bottom left of Fig. 1.24: The resemblance is quite reasonable). Courtesy of Peter Hancock, University of Stirling.

These age-related changes are clearly illustrated in Fig. 1.24 which shows a series of portraits of Bonnie Prince Charlie, produced at different ages and by different artists. Using the computer-technique called 'morphing' (described in more detail in Chapter 5) it is possible to take each of these portraits and produce intermediate stages between them to provide an idea of how the Prince's appearance changed gradually over the years (Fig. 1.25).

As people grow still older, the appearance of their faces changes further as the skin loses its elasticity, leading to wrinkles and sagging (Fig. 1.26). Figure 1.27 shows a bust of Albert Einstein, and Fig. 1.28 a portrait of the former Prime Minister, Sir Alec Douglas Home, in which these signs of ageing are evident.

Thus the appearance of the face is determined in part by the shape and growth of the underlying bone, and in part by other factors such as the age of skin, the distribution of fat, the texture and pigmentation of the skin and so forth. Because of the interplay between bone, hard and soft tissues of the face, and the important effects of facial and head hair

Fig. 1.26 Sir Charles Bell's illustration of the changes between infancy and old age.

Fig. 1.27 *Albert Einstein* by Sir Jacob Epstein. Courtesy of the Scottish National Gallery of Modern Art.

Fig. 1.28 *Lord Alec Douglas Home* by Avigdor Arikha. Courtesy of the Scottish National Portrait Gallery.

on appearance, it is only possible to make estimates of likely appearance from the skull alone. None the less it is often important, for historical or forensic reasons, to attempt such reconstructions.

Reconstructions of possible appearance are made by adding a face surface to the underlying skull. This may be added physically, using

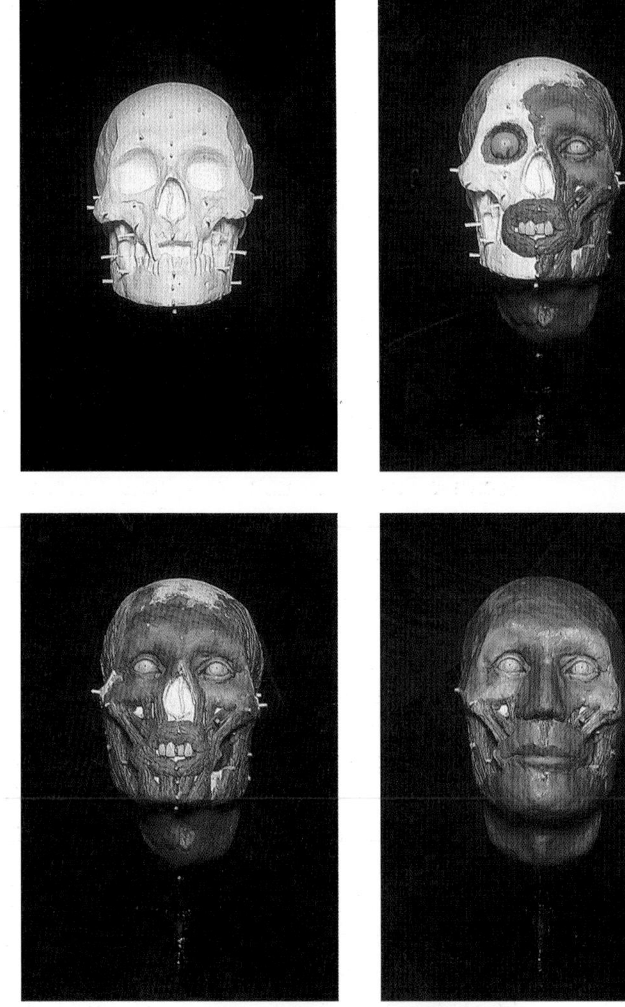

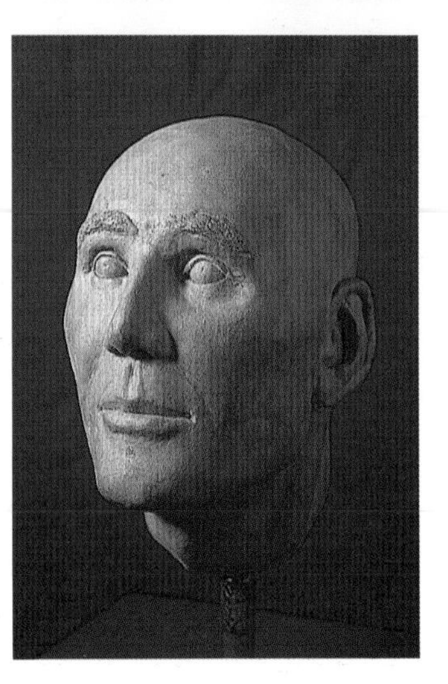

Fig. 1.29 Reconstruction of Robert the Bruce's face. Stages in the sequence from skull to reconstructed head. 1. A plaster replica skull showing 25 tissue depth pins *in situ*. 2. Eyes set in place and the facial musculature half finished. 3. Facial muscles almost completed. 4. Facial tissue is laid up upon hardened musculature. 5. The completed face. Courtesy Iain Macleod, Consultant in Oral Medicine, The Dental Hospital, Edinburgh (Project Director); Rosalind Marshall, Assistant Keeper, The Scottish National Portrait Gallery, Edinburgh; Brian Hill, Head of The Department of Medical Illustration, The Dental Hospital, Newcastle upon Tyne (Facial Reconstruction).

plasticine or other modellling materials, or, increasingly, it can be carried out electronically, using similar techniques to those used to simulate cranio-facial surgery (see Box 1C). A surface is built up based upon information about average thickness of the soft tissues in people of the same approximate age and sex as the target person. As there may occasionally be ambiguity even about the sex of a person whose skull is recovered it is clear that this process requires a good deal of guess-work. Guesses about the possible hair-style and colouration must also be made. One recent attempt was made by medical artists commissioned by the Scottish National Portrait Gallery to reconstruct the face of Robert the Bruce from his skull, as illustrated in Fig. 1.29.

1.4 Abnormal faces

Because faces are such complicated structures, a number of different things can affect their function and/or their appearance. Clearly, sensory deficits prevent the normal functioning of the afflicted organs, but this may also affect the appearance of the face. Blind people, for example, may show patterns of eye movement which are not typical of normally sighted individuals. Although deafness has no visible signs, the use of sign language rather than vocal language means that facial expressions are used differently in deaf signers, since facial expression has a linguistic function which it does not have for users of vocal language.

Malformation of parts of the head and face can arise as a result of genetic factors or accident. Cleft lip and palate affects approximately 1 baby born per 1000 births and is one of the commonest forms of facial abnormality. Its cause is not clear, though the slight increase in incidence in families where one child has already been born cleft suggests there may be some genetic component. Figure 1.30 shows the different forms and severity of cleft lip and palate.

As well as producing visible disfigurement, clefting affects speech development in the child, so early surgical correction is usually performed. However, following early correction of the cleft, later growth of the face may not be normal, and further problems for the cleft patient may arise in later life.

Although Leonardo da Vinci (Fig. 1.31) managed to create grotesque faces which he imbued with a strange beauty, the sad truth is that most people react badly or awkwardly in the initial stages of acquaintance with someone with a facial disfigurement. The difficulties reported by afflicted people have been confirmed in research by social psychologists. In one series of studies, Ray Bull and Nichola Rumsey secretly observed the reactions of passers-by to an actress who was made up to appear with a disfiguring birthmark, and compared the reactions given to the same actress in her normal appearance. Passers-by tended to approach the disfigured person less closely, and to avoid approaching her

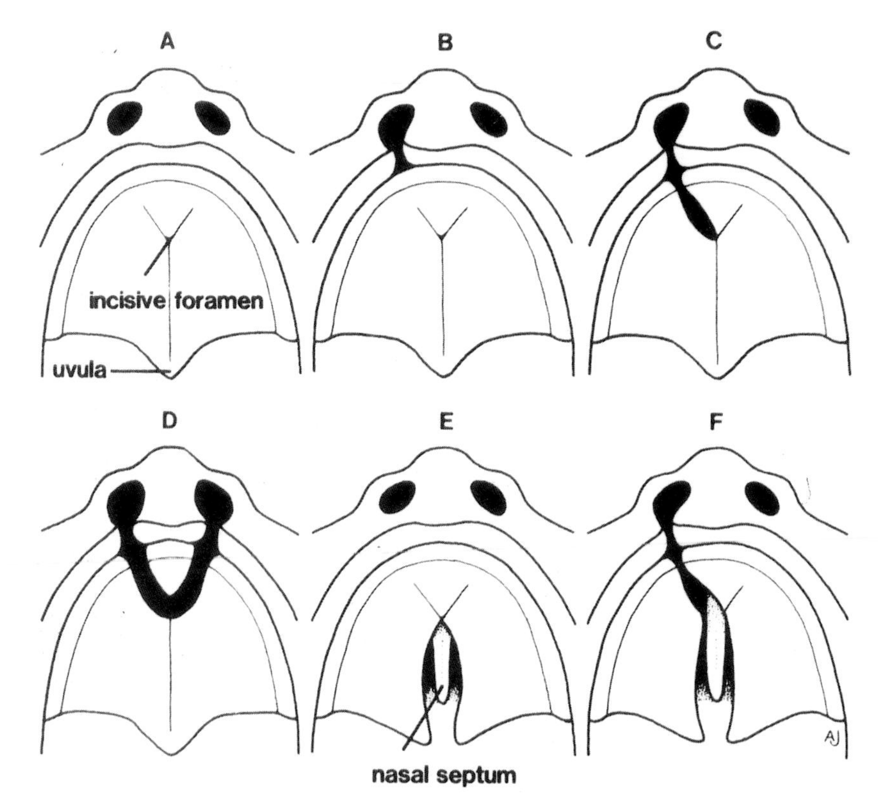

Fig. 1.30 Varieties of cleft lip and palate. A: normal growth; B: unilateral cleft lip; C: unilateral cleft lip and anterior cleft palate; D: bilateral cleft lip and anterior cleft palate; E: posterior cleft palate; F: unilateral cleft lip, anterior cleft palate and posterior cleft palate. From Johnson and Moore (1989).

disfigured side (Rumsey *et al.* 1982). In another study where the actress was collecting money for charity, fewer people donated when the actress appeared with the disfigurement (Rumsey and Bull 1986), though those that did donate money then gave larger sums of money.

Given such reactions, it is probably easier, and more satisfactory to the disfigured persons themselves, to intervene surgically to try to make faces more normal in appearance than it is to try to change the behaviour of bystanders. Certain surgical procedures, such as that for cleft palate, are remarkably successful and conducted routinely in most countries. However, correcting the kinds of damage which can occur to the face as a result of road traffic accident or fire remains a major challenge.

Simon Weston, a British soldier who suffered dreadful burns to his face and hands during the Falklands war, underwent numerous separate operations in order to restore to him a face which is still far from normal in appearance (Fig. 1.32).

> I can joke about my looks now, but there was a time when every glimpse of myself in the mirror was enough to send shivers down me. I was frightened I looked so terrible that no one would ever look at me or touch me again. (Simon Weston, *Going back*, p. 16)

Simon Weston became very famous in the UK for his bravery, and his story was told in TV documentaries and in his own books. His celebrity continues through his charity work, and the whole country delighted

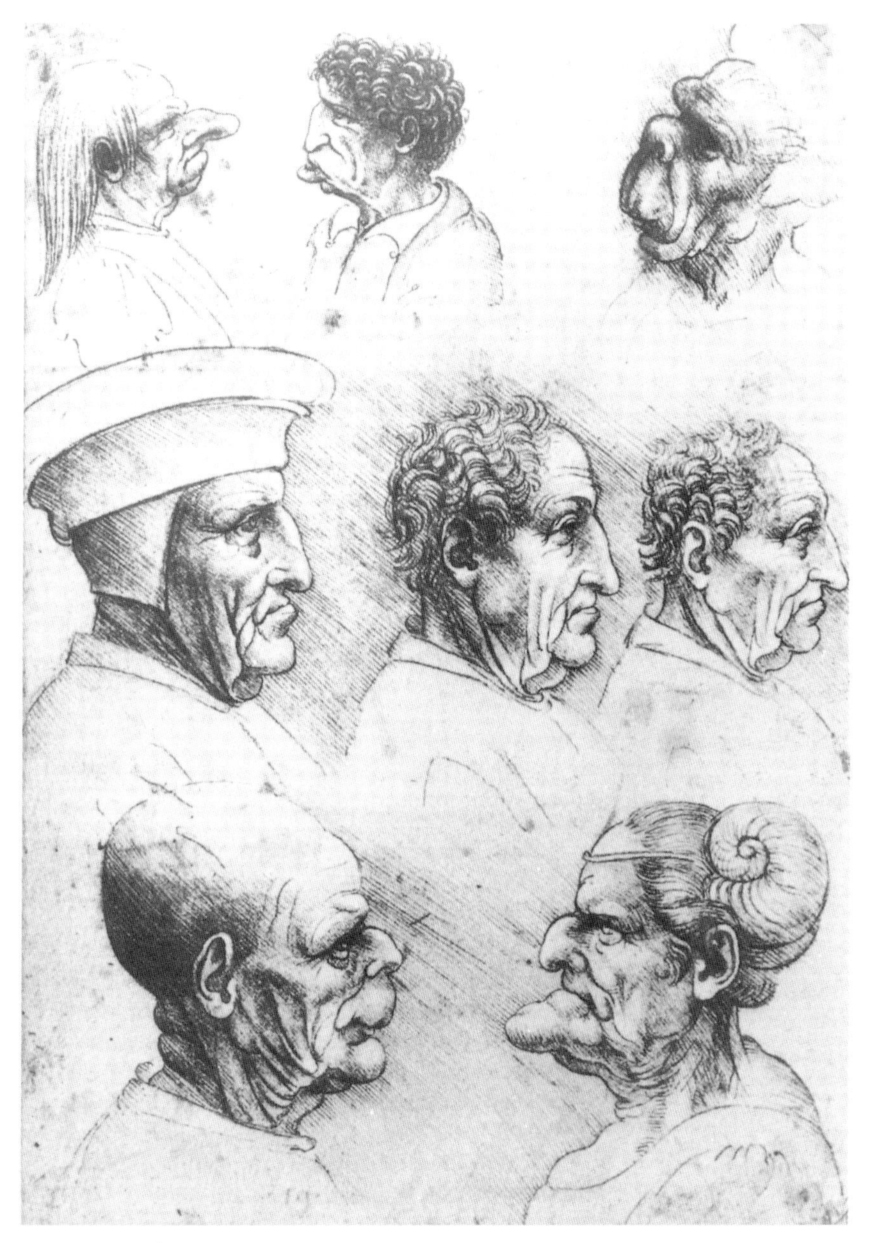

Fig. 1.31 Grotesque faces drawn by Leonardo da Vinci. The Royal Collection © Her Majesty Queen Elizabeth II.

in his marriage in 1990 and the birth of his child in 1991. His case provided a high profile illustration of the profound effects that facial disfigurement can have on an individual's life, and the importance of developing improved techniques for facial surgery and skin grafting. 'Face and hands—these are the bits that really define you as an individual in the eyes of other people, the bits that everyone notices, that can't be covered up...' (Simon Weston *Going back*, p. 4).

Many of the standards which have been used to plan and assess surgical interventions to the face were based on two-dimensional

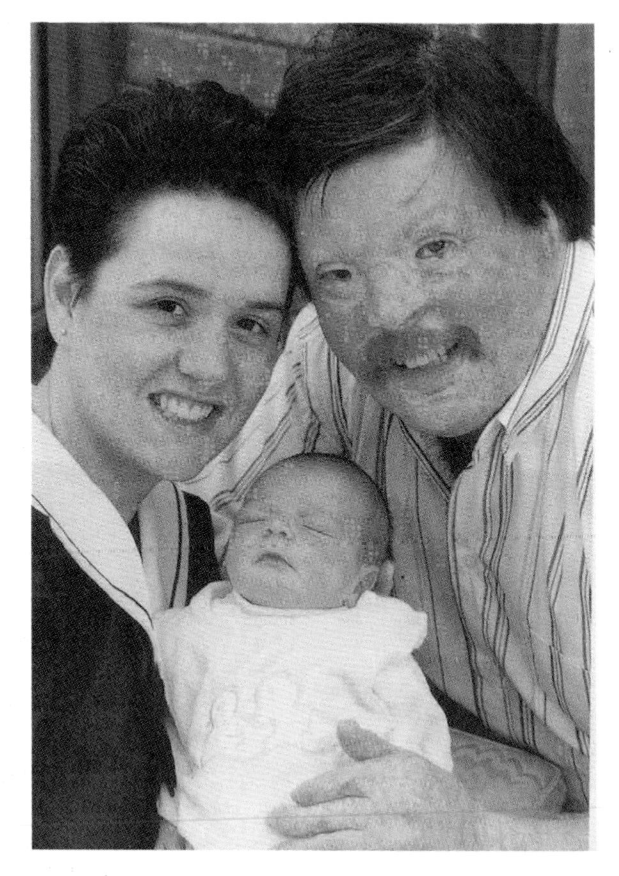

Fig. 1.32 Falklands war veteran Simon Weston, with his new face and new family.

assessments of and measurements to the cranio-facial profile. However, the human face is a complex three-dimensional surface, and operations must be conducted on complicated 3D bone structures. Improvements in surgical techniques are arising through better ways of understanding and describing three-dimensional shape.

Modern X-ray and other imaging techniques, combined with developments in computer graphics have made it possible for surgeons to simulate facial surgery in advance of actual operations. A three-dimensional graphical representation of the bones can be obtained, and surgeons can examine the effects of making changes to this underlying bone structure, examining the results of different possible interventions in three dimensions. An image of the facial surface (skin) can be obtained by measuring its shape with a laser, and the resulting surface image can be superimposed on that of the bones. The likely appearance of the face before and after surgery can then be inspected from different angles. The same techniques can be used to reconstruct a face surface from a skull. Box 1C describes this process in more detail.

Successful facial surgery involves more than three-dimensional remodelling. For example for burns victims much of the repair requires grafting skin from other areas of the body on to damaged areas of the

Box 1C: Electronic manipulation of three-dimensional images of faces

Over the past ten years, there has been remarkable progress at producing representations of bone and skin surfaces in 3D interactive computer graphics, enabling surgeons to plan their operations systematically to produce the most effective (in terms of minimizing patient trauma and medical costs) means of producing a particular outcome.

3D images of the facial bones can be obtained using a number of modern imaging techniques, and surface images of the skin surface can be obtained by measuring the undulations of the face with a laser. Typically, the subject sits in a chair which rotates in front of a laser beam. The ins and outs of the beam as it is reflected from different regions of the face are recorded, so that the face is then represented as a very large number of laser 'profiles'. Figure 1C.1 shows the 3D profile as measured by the laser down the midline of the human face. The points on different profiles can then be joined together to form a wire mesh model of the face. This model can be displayed as a smooth surface by showing how a model with all these facets would look if illuminated by light from a particular direction. Figure 1C.1 illustrates the positioning of the laser profile on a surface image constructed in this way.

A database of surface measurements obtained by laser can be pulled over a skull to show how the surface of the face would be affected by some underlying restructuring of the bones. One of many advantages

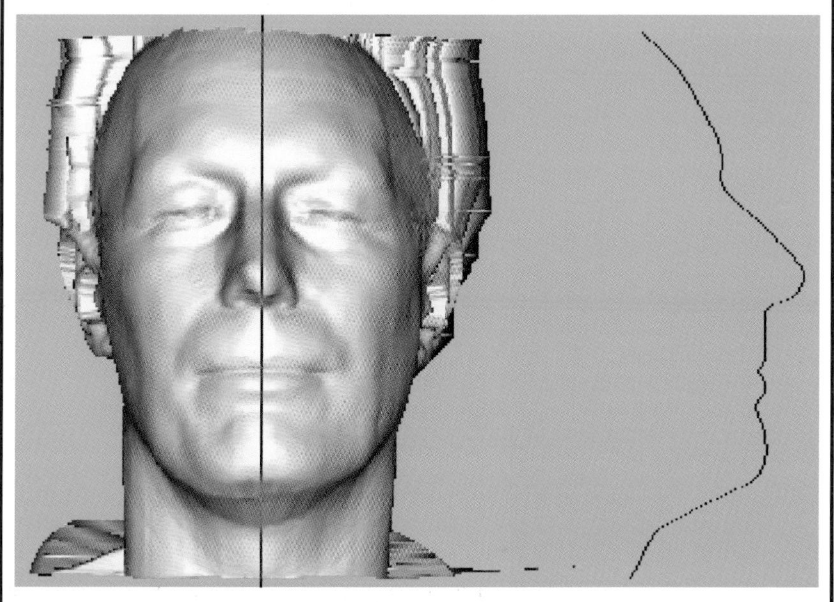

Fig. 1C.1

Box 1C: Continued

of these techniques is that it means that surgeons can 'try out' different operations electronically, and then redisplay the surface of the face to view the effects of a hypothetical operation. Moreover, the face can be viewed from any angle to see the effect on appearance (Linney *et al.* 1989).

This same technique of pulling facial surface measurements onto a skull can be used to try to reconstruct the actual appearance of a face from recovered bones (Vanezis *et al.* 1989). Figure 1C.2 shows an image of a skull, and Fig. 1C.3 a surface image to be fitted over it in order to try to predict the appearance of the actual face that went with this skull. The spider's web patterns drawn on the face and skull are used to bring appropriate points on the two images into correspondence. Figures 1C.4 and 1C.5 show front and side views of

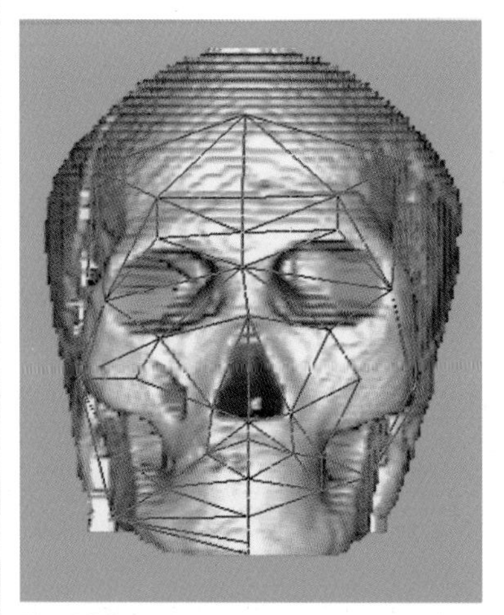

Fig. 1C.2

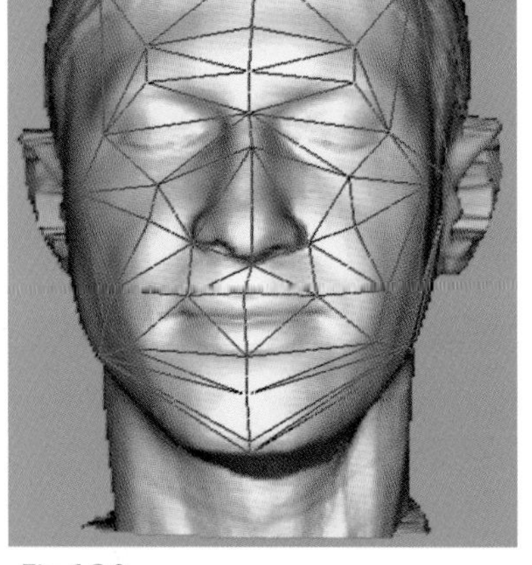

Fig. 1C.3

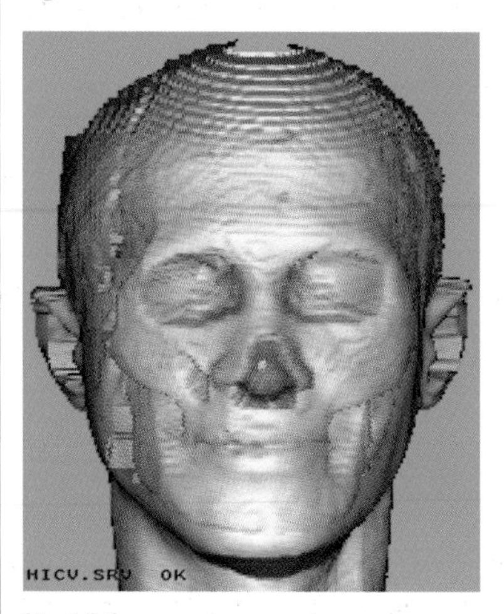

Fig. 1C.4

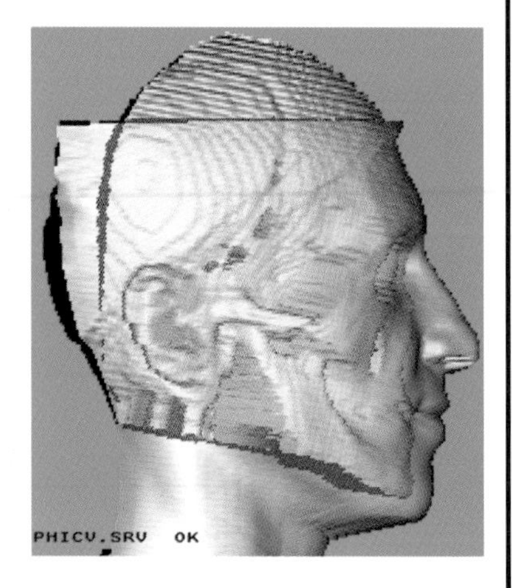

Fig. 1C.5

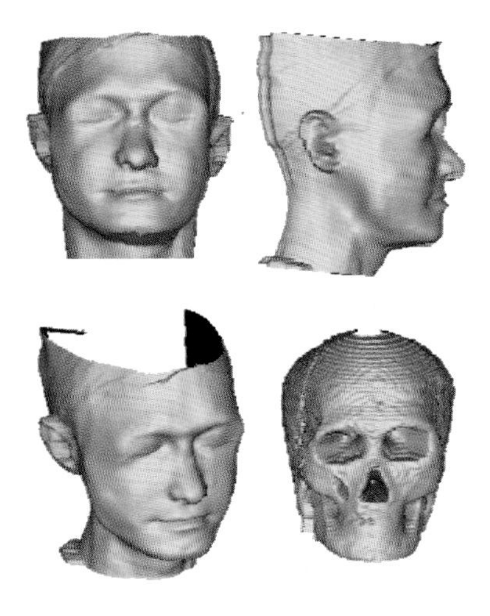

Fig. 1C.6

the surface image superimposed upon the skull, and Fig. 1C.6 shows three different viewpoints of the reconstructed head alongside the skull used to produce it.

All the images in this box were produced for us by Alf Linney at University College London, whose team has developed many of these pioneering interactive graphical techniques for three-dimensional measurement and display of face images (e.g. Coombes *et al.* 1991; Fright and Linney 1993).

One problem with these ways of displaying faces is that until recently these facial surfaces have lacked pigmented 'features'. Research has shown that such featureless surfaces can be very difficult to identify (see Chapter 5). New techniques, described in more detail in Chapter 3, allow the texture as well as the three-dimensional shape to be measured and displayed, opening up new possibilities for future surgical prediction. However, any attempt to add such pigmented features onto the reconstructed faces from skulls will still involve a lot of guesswork, and could be misleading if the wrong choices are made.

face. Differences in texture and appearance of grafted skin can make a repaired face look strange even when its underlying shape has been restored. 'Natural' variations in the skin surface such as port-wine stains ('birthmarks') can also have considerable impact on people's social interactions, as we described earlier.

1.5 The muscles of the face

While the hard and soft tissues of the face produce the individual variations in appearance which are important for categorization and identification, it is movements of the face which are responsible for its ability to transmit a range of other social signals. Expressive movements provide information about emotional states, eye and head movements provide information about the direction of attention, and movements of lips, tongue, and jaws provide information which aids speech perception. All these different kinds of movements are controlled by a bewildering variety of muscles (see Fig. 1.33).

The different functions of the human face—looking, eating, breathing, and sending social signals—all require muscular movements. For example muscles are needed to control lips, tongue, and jaws during speech, to chew or expel food, and to adjust the posture of external sensory organs. Darwin and others have argued that emotional expressions of the face build upon these other kinds of activity rather than involving specific muscles which have developed solely for expression; 'there are no grounds, as far as I can discover, for believing that any muscle has been developed or even modified exclusively for the sake of expression' (Darwin 1872, p. 355). According to Darwin and others since, our specific human expressive movements are explicable as remnants of behavioural responses to emotionally arousing events.

Understanding how such movements are produced and controlled has helped to improve the accurate depiction of facial movements in paintings. The artistic depiction of facial expressions was influenced strongly by advances in basic anatomy. The Scots artist Sir David Wilkie (see *Distraining for rent* Fig. 6.14, Chapter 6) is acclaimed for the realistic portrayal of character and narrative in his paintings. Wilkie was a student of Charles (later Sir Charles) Bell, the anatomist. Bell's *Essay on the anatomy of expression in painting* examined in detail how different emotional states resulted in changes to facial musculature and Wilkie incorporated these findings into his artistic repertoire.

In Chapter 6 we describe the painstaking research by Duchenne and more recently by Ekman to understand the effects of movement of different muscle groups on the expression of different emotions (e.g. see Fig. 6.2 and elsewhere in Chapter 6). Applying this knowledge, computer graphics artists have been able to animate facial expressions by applying a model of human muscle action to distort the skin surface of the human face as represented by a mesh of 3D data points (see Figs 1.34 and 1.35). The surface representations used in such animations are very similar to those used to reconstruct faces electronically (Box 1C).

Sir Charles Bell's careful anatomical study of expressions also influenced Charles Darwin's thinking on the origin of emotional expressions through evolution. However, while Darwin was a great admirer of Bell's work, setting it apart from that of the physiognomists (see Chapter 4) which he disparaged, he was also critical that Bell did not attempt to explain why certain muscle movements characterized the different expressions; 'why, for instance, the inner ends of the eyebrows are raised, and the corners of the mouth depressed, by a

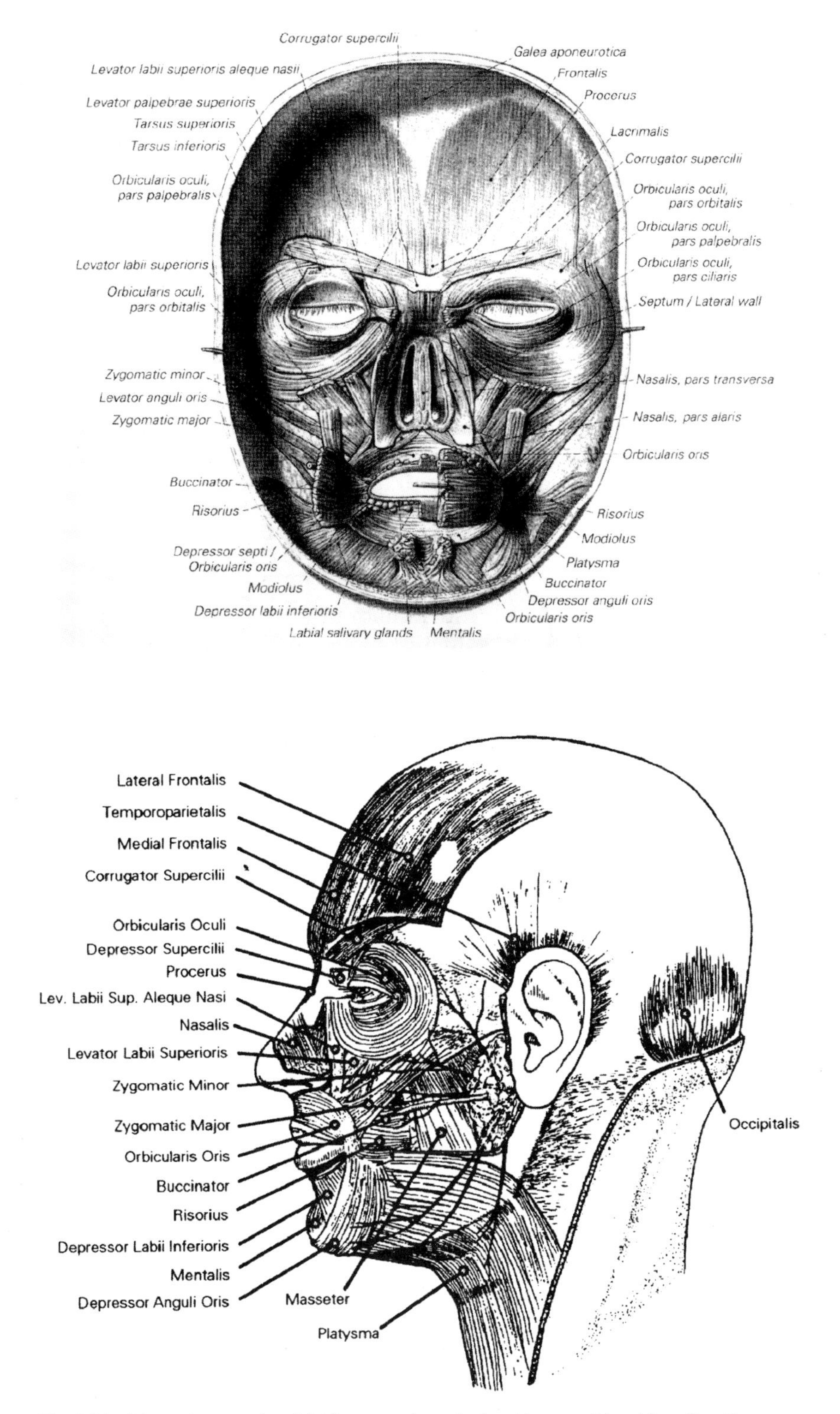

Fig. 1.33 The major muscles of the face, seen from the front (as a mask) and from the side, as adapted by Fridlund (1994) from Pernkopf (1963) and from Hiatt and Gartner (1982).

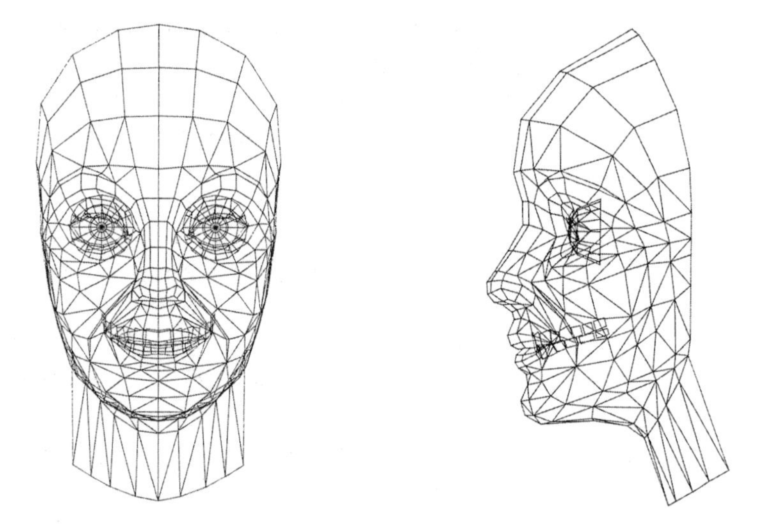

Fig. 1.34 A wire-frame model of a head used as the basis for animation of facial muscle models from Parke (1982). Copyright © IEEE.

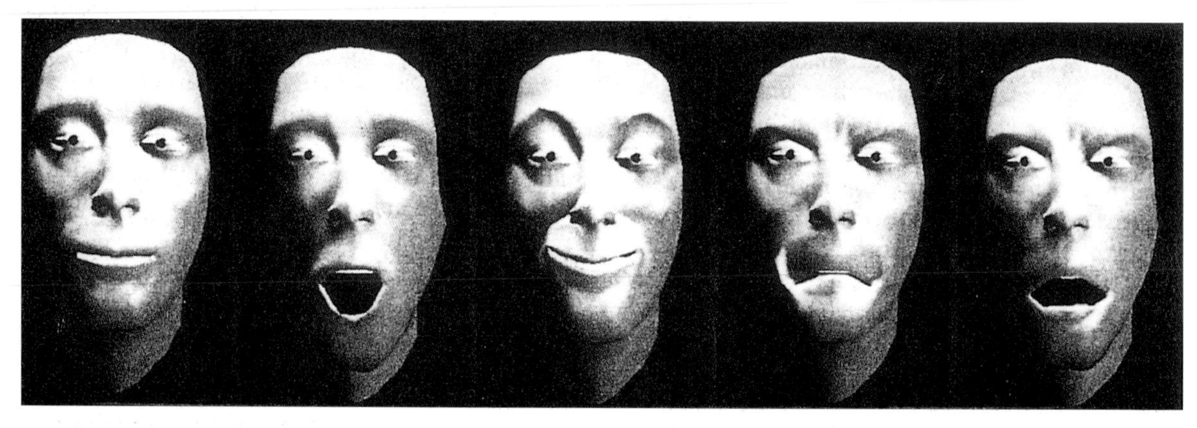

Fig. 1.35 Computer animation of facial expressions. Surface texture is 'mapped' from a photograph of a person's face onto a three-dimensional model (see Fig. 1.34) and deformed by simulating the effects of muscle movements on this surface. (From Waters and Terzopoulos (1992).)

person suffering from grief or anxiety?' (Darwin 1872, p. 3). It was this question which formed the basis of Darwin's own examination of the nature and origins of expressions in man and other animals.

1.6 Evolution of facial expressions

In *The expression of the emotions in man and animals* Charles Darwin produced an early and thorough analysis of the origins and nature of facial expressions. Darwin disagreed with the prevailing view that each individual species was uniquely created with a specific repertoire of behaviours, emphasizing the continuity of expressive behaviours across species, and the origin of human facial expression in more primitive evolutionary stages; 'some expressions, such as the bristling of the hair under the influence of extreme terror, or the uncovering of the teeth under that of furious rage, can hardly be understood, except on the belief that man once existed in a much lower and animal-like condition' (1872, p. 12).

In his treatise, Darwin claimed three important fundamental principles of expressive acts, which he deduced from observations of expressions in a number of species. The first principle is that of *service*. An action which accompanies some biological act (such as grimacing with pain, or ejection of foul-tasting food) becomes habitually associated with the accompanying emotions. This does not mean that how to make each expression must be learned by each individual during their lifetime. Darwin was clear that such expressive 'habits' are inherited in man just as other complex behaviours are instinctive in other species.

Darwin's second principle was that of *antithesis*. When habitual and physiological tendencies give rise to one set of expressive acts to accompany one emotion, there is a tendency for opposite expressive acts to accompany an emotion of opposite kind. Emotional expressions are contrastive, so that the expressions which accompany friendliness, for example, are distinctively different from those accompanying hostility. This principle, while arising for clear physiological reasons, ensures that expressive acts can act as clear social signals, since there will be maximum visible contrasts between opposite emotions.

The final principle is that *direct action of the nervous system* can produce expressive actions which do not arise through habit or antithesis, but simply from as yet unexplained physiological responses. Examples given by Darwin include trembling with intense emotion or blushing with shame.

There have been many valiant attempts to describe the origins of human facial expressions in the expressive actions of primate relatives (see Figs 1.36 and 1.37). Not all such comparisons are convincing, and this is perhaps not surprising given that there is considerable diversity within non-human primates. Chimpanzees (see Fig. 1.36(b) in the left panel) are thought to be most similar in their facial musculature and expressions to humans.

The commonest expressions among primates are threat faces, grimaces, the play-face and lip-smacking. The threat face, grimace, and the play-face are all rather similar one to another, yet they have very different functions. The primate play-face is similar in structure and function to the human laugh. What is more controversial is whether the human smile is also linked to the play-face in some diminutive way or, as some would argue, more closely related to monkey threat and grimace gestures.

It may be oversimple to seek a single origin for the human smile, since there are many different varieties of smile. Ekman and Friesen (1978) go so far as to claim that there are over 180 different smiles which are distinguishable anatomically and visually. In particular,

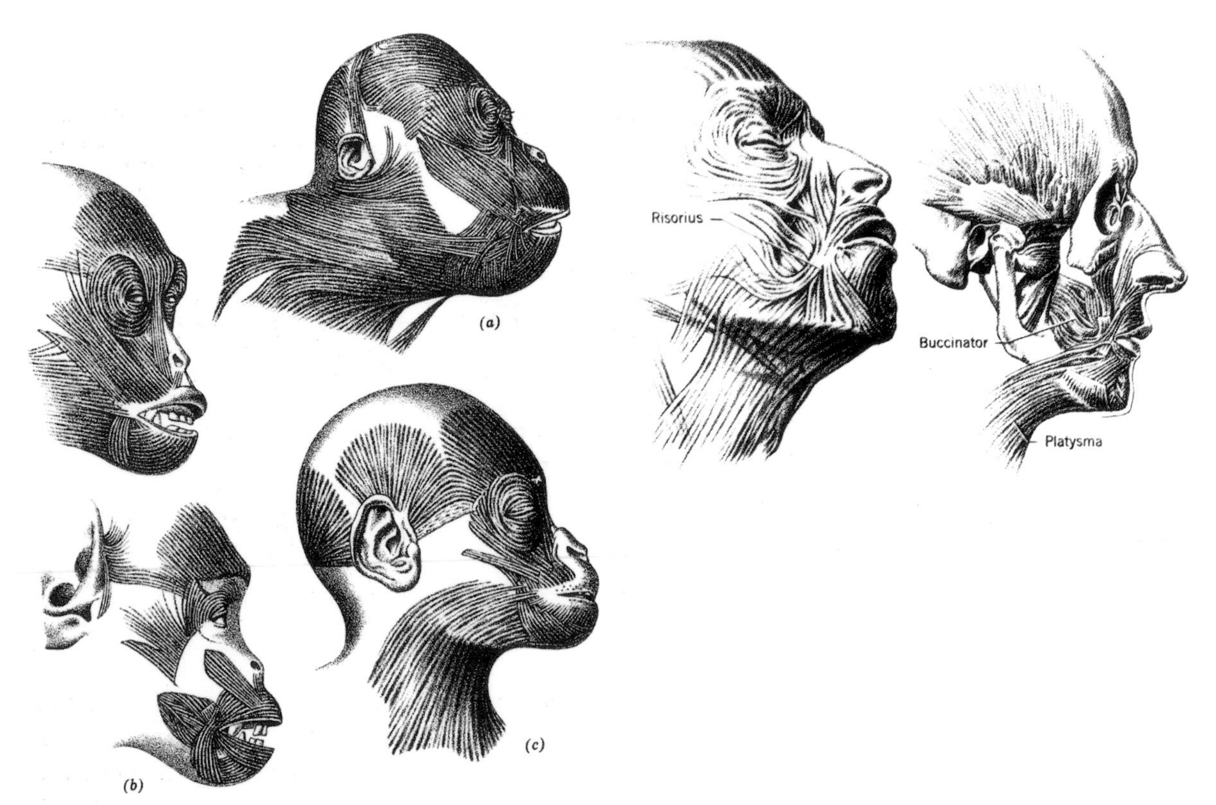

Fig. 1.36 Similarities in the musculature of human and primate faces allow comparisons to be made about the different expressions used by different primate and human species. The left panel shows facial musculature in (a) orang-utan; (b) chimpanzee; and (c) gorilla infant. From Lenneberg (1967).

Duchenne and later Ekman draw important distinctions between genuine smiles and false smiles used to hide other feelings, and there is no doubt that smiling can be used in a nervous or threatening way by humans. Interestingly, Darwin devotes some discussion to 'grinning' in dogs, 'A pleasurable and excited state of mind, associated with affection, is exhibited by some dogs in a very peculiar manner; namely, by grinning' (1872, p. 120), and notes that Sir Walter Scott's famous greyhound Maida was much prone to this (see Fig. 1.4). This was not mere fancy on Darwin's part, since the same was noted by Sir Charles Bell, 'Dogs, in their expression of fondness, have a slight aversion of the lips, and grin and sniff amidst their gambols, in a way that resembles laughter' (1844, p. 140).

Such observations may suggest that there are more fundamental linkages between pleasurable arousal and movements of the lips across a range of species—or else that we read more into dogs' expressions than we should!

Human facial displays of anger also differ. Where teeth are visible the anger face may reflect a relic of pre-human ancestral fighting with teeth. Other comparisons may be made between the closed mouth human angry face and the tense mouth display in primates.

The human fear face resembles the primate grimace. Similar muscles

Fig. 1.37 A range of chimpanzee expressions with human expressions for comparison. From Chevalier-Skolnikoff (1973).

are involved in each, and these are quite different from the smile musculature. However, neither sadness nor disgust are so easy to compare with other primate gestures. The disgust face may have evolved from the movements which physically accompany the ejection of foul-tasting substances from the mouth, and a wrinkled up nose to prevent the inhalation of foul-smelling air. Sadness as well as disgust involve a turned-down mouth which is not evident in other primate signals. The expression of adult sadness clearly originates from the more dramatic facial movements which accompany crying, but why these particular movements occur is not so clear.

Other recognizable facial 'expressions' include surprise and interest, but these do not have the same emotional content as others. Some have argued that the particular facial actions which signal these states may reflect the orientation of the sense organs—eyebrows raised and eyes wide in surprise as though to take in as much external information as possible, and eyes narrowed in interest in order to focus on the object of attention.

Whatever the specific reasons for the repertoire of facial expressions exhibited by humans, there is some remarkable consistency in both the display and 'universal' understanding of these different signals across different cultures. Later in this book (Chapter 6) we will discuss the extent and possible significance of the universality of facial expressions, and consider the relationship between the perception of expression and the experience of emotion.

This chapter has reviewed the many different biological influences on the form and function of the human face, which explain why all faces are so similar to one another, and which constrain the nature of the social signals that faces can transmit. In Chapter 2, we turn to consider physiological and psychological processes of human visual perception which are involved in deciphering these signals.

2 Light, colour, and shape: the science of vision

2.1 Overview

The human visual system uses various features of a visual image to interpret the three-dimensional shape of the objects viewed. Abrupt changes in brightness across the image signal 'edges', boundaries between one object or part of an object and another, and patterns of shading are used to interpret the undulations of the surface. Humans can also use colour vision in daylight, and variations in colouration of faces can affect perception in subtle ways. This chapter illustrates how the visual system deciphers patterns of light and colour, sometimes making use of implicit assumptions about likely possibilities in the scene being viewed, and considers the consequences for the way faces are perceived. We also consider the important role of movement for vision, and consider what may be lost from the absence of movements in photographs and portraits.

2.2 Physiology of the visual system

Vision is a miraculous achievement of the eye and brain, working together to reconstruct the world from patterns of reflected light. Each eye is a little like a camera, focusing light onto a photosensitive surface, the retina, which captures an 'image' of the scene being viewed. However, the analogy between eye and camera quickly breaks down when we consider how the focused image is processed by the eye and brain, since this processing involves both enhancement and interpretation of the retinal image.

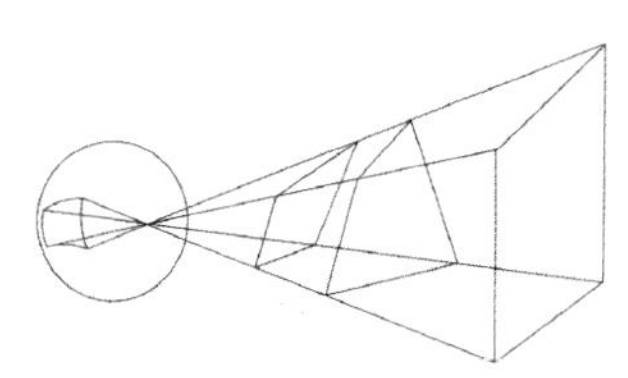

Fig. 2.1 An infinite number of different shapes project the same shape onto the retinal image. From Wade (1990).

At any particular instant in time, the retinal image is ambiguous, since many different scenes could result in the same image (see Fig. 2.1). Artists such as M.C. Escher and Salvador Dali have exploited the ambiguous and uncertain nature of seeing by producing images with multiple interpretations. Our usual, stable perceptions arise because assumptions and knowledge about the world can be used to help decipher retinal images (Fig. 2.2).

Fig. 2.2 Ambiguous images using faces. *Top panel*: How many face profiles are there? The black ones are easier to see than the white ones. Turn the image upside-down and a new set emerges. From Wade (1990). *Right panel*: Salvador Dali, *Slave Market with the Disappearing Bust of Voltaire* (1940), oil on canvas, 18¼ × 25⅜ inches. Collection of The Salvador Dali Museum, St. Petersburg, Florida. © ADAGP, Paris and DACS, London, 1998. *Bottom panel*: drawing by G. M. Rand.

The first stage in vision is a chemical reaction within the photoreceptors of the eye, resulting from the absorption of light by the pigments contained in these receptors. This changes the electrical potential of the cell, and when there is sufficient change a signal can be transmitted to the next layer of neural cells in the retina (Fig. 2.3). Receptor signals are collected by a number of different kinds of cell and pooled at the ganglion cells, whose axons leave the retina to form the optic nerve.

Different photo-pigments are contained within different types of receptor, rods and cones, distributed within the retina of the human eye. Rods are more sensitive to low levels of illumination, and at night we rely mainly on rod vision. Cones allow us to see finer detail and colours during daylight. Colour vision is possible because most human eyes contain three different types of cone, with pigments responsive to slightly different wavelengths of light. Because cones are not as sensitive to light, colour vision is possible only when overall illumination levels are high—in daylight or high levels of artificial light. A small percentage of the population have congenital conditions which result in the absence of one or more of the different types of cone. These people see colours differently from the majority, and in the most common form of 'colour blindness' (as the condition is popularly called) people confuse red with green.

The visual system is built in a way which responds to change, and

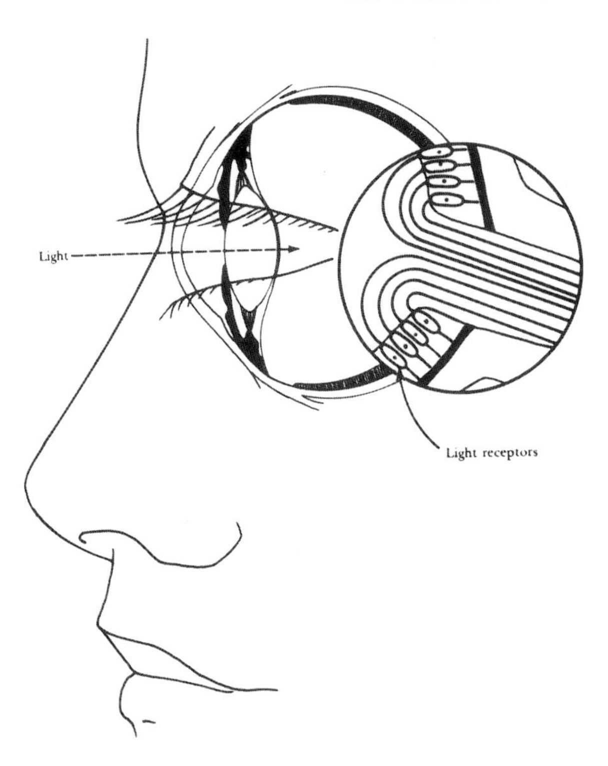

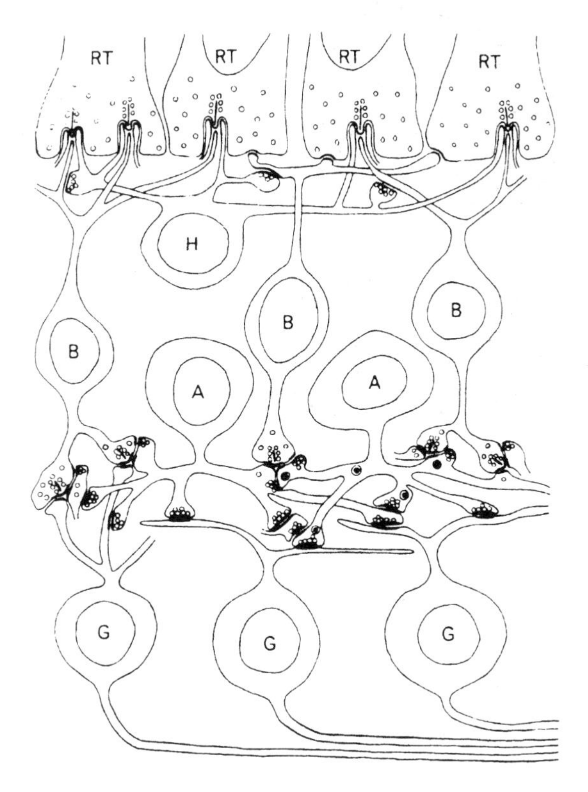

Fig. 2.3 *Left*: The light-sensitive retina is the first layer of neurones on the pathway to the brain. From Lindsay and Norman (1977), used with permission. *Right*: The structure of the vertebrate retina, showing receptor terminals (RT), horizontal cells (H), bipolar cells (B), amacrine cells (A), and ganglion cells (G). Reproduced from Dowling (1968) with permission of the Royal Society.

hence to informative aspects of visual input. Information about boundaries between objects in the world, and the shapes of objects, is carried by the spatial pattern in the image (differences in intensity across the image) and information about events in the world is conveyed by changes over time—temporal pattern. In fact, cells early in the visual system will adapt to continuous unvarying stimulation in a way which makes them less responsive after a few minutes exposure. For example, cells in the visual system respond with a burst of activity when a light comes on, but then their rate of firing diminishes and if the pattern remains unchanged will eventually drop away as the cell fatigues. You can witness such adaptation effects yourself by seeing the after-effects which arise as a result of selective fatigue of cells in some areas of the visual field.

For example if you stare at a bright light or patch for a minute or so, and then look at a uniform field, you will see a dark patch or 'after-image' which corresponds to the area of the retina which was stimulated by the original bright part. A rather striking example of this is illustrated in Fig. 2.4. Stare at the left-hand panel for at least a minute. Try not to move your eyes at all by fixating the central dot.

Fig. 2.4 A face illusion seen in a visual after-image. See text for viewing instructions. The after-image is easier to see if you blink your eyes a few times when looking at the blank region.

After a minute, look at the blank panel on the right. You will see a famous face! This dramatic effect is quite easy to explain. You should be able to understand why the famous face is seen once you are told that the upper panel contains a negative image of the face, with light and dark areas reversed from the original image. Selective fatigue of receptors stimulated by the bright region means that when gaze is transferred to the uniform field, a positive version of the original face will be seen in the pattern of the after-image. As we will describe later in this chapter, this 'illusion' relies on the fact that faces are extremely difficult to recognize in photographic negative images (and hence the initial identity of the image is concealed in Fig. 2.4), but readily identified from black on white images provided these contain appropriate regions of light and dark (see pp. 67–8).

You can establish that this effect arises as a result of the fatigue of cells early on in the visual system by repeating the demonstration, but this time closing one eye when you stare at the adapting figure. After one minute, transfer your gaze and now compare the after-effects seen by the adapted and unadapted eye. Only the adapted eye sees the after-effect, confirming that the effect arises because of activity in cells prior to the combination of information from the two eyes.

This general technique of selective adaptation and the study of after-effects has proved remarkably informative about the processes of visual perception in humans. For example colour after-effects demonstrate how signals arising from the different types of cone receptor are combined in opposition; we code colours by their positioning on blue-yellow and red-green dimensions. Stare at a red patch for a minute or so, and the resulting after-effect will be tinged with green. A yellow patch gives rise to a blue after-effect.

After-effects of this type do not usually arise in everyday life, as the eyes do not normally remain still long enough for selective adaptation ('fatigue') to occur. The eyes are in constant motion, darting from one part of the visual scene to another (through large scale movements called 'saccades'), smoothly pursuing moving objects, and also making continuous tiny flicks and tremors. All these movements mean that different receptors are rarely exposed to constant stimulation. However, after-effects of over-stimulation may be seen when you accidentally look at a bright light such as a light bulb, and then look at a white wall—you will see a grey patch which only slowly fades. An interesting motion after-effect may be experienced if you stare too long at something which moves in one direction (such as a waterfall). A stationary pattern (e.g. the river bank) viewed soon afterwards will appear to have an illusory motion in the opposite direction. We consider the perception of motion again in Section 2.7.

2.3 Edge detection

The signals which arise from the photoreceptors in the retina are combined at retinal ganglion cells in a way that enhances responses to changes in light intensity (for example at an edge between a bright and dark object) and is relatively unresponsive to uniform areas. The enhanced response to changes in illumination arises from a mechanism termed *lateral inhibition*. Cells which are active tend to inhibit the responses of other cells close by. The consequence is that when neighbouring cells are looking at a uniform field, they will mutually inhibit each other and the net response to such uniformity will be quite low. However, lateral inhibition serves to sharpen responses to spatial pattern. For example a cell responding to an isolated bright spot will experience little inhibition from neighbouring cells looking at dark areas, since they will not be activated.

The mechanism of lateral inhibition was first discovered in the eye of the horse-shoe crab (Hartline *et al.* 1956). Much of the evidence for similar mechanisms in humans arises from careful investigation of a range of illusory perceptual effects. One apparent consequence of lateral inhibition in the human retina is the enhancement of relative brightness and darkness which may be seen at boundaries between dark and light regions. Figure 2.5 illustrates this 'Mach band' effect, named after the German physicist Ernst Mach who first described it. The pattern you look at shows a dark area to the left and a light area to

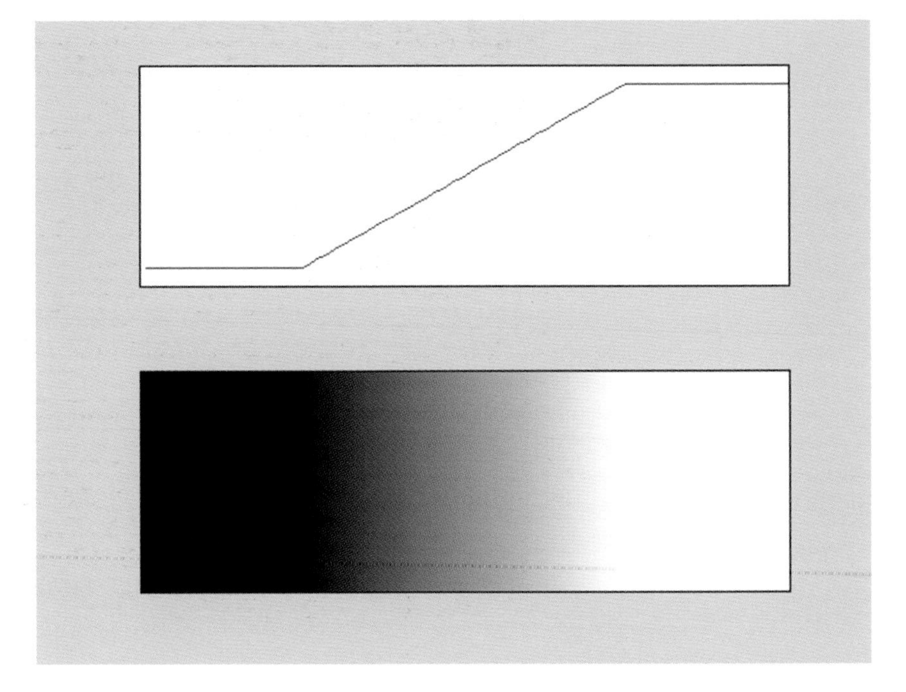

Fig. 2.5 Mach bands. A strange extra dark and extra light stripe will be seen near to the ramp in intensity. The upper panel shows a plot of the light intensity shown in the pattern below, to illustrate that these super-dark and super-light strips are a product of the visual system looking at the pattern.

the right, with a uniform smooth 'ramp' of intensity between these two regions, as plotted in the graph above. However, when you look at this simple pattern you will see what appears to be a super-dark and super-light strip at either side of the ramp: these 'super-strips' are an illusion—a creation of your visual system.

To explain why you see these super-strips we must consider the balance of inhibition which will be experienced by cells which are 'looking' at the different areas of the pattern. Consider a cell looking at the even dark area towards the left of the pattern. It will be activated very little by the dark patch, but will also receive very little inhibition from its neighbours who also view dark areas. In contrast, a cell 'looking' at a region nearer to the ramp—the part corresponding to the super-dark strip—will have some neighbours who are looking at the lighter strip to the right. These neighbours will be more active than the neighbours in the dark area, and because they are more active, they will pass more inhibition to the cell centred on the super-strip. The result of this will be that the net activation of such cells will be lower than the activation of those cells further from the border area. A similar argument applies to cells looking at the light side of the pattern. Those cells looking at a region over on the far right will receive a large amount of inhibition from their neighbours, which will reduce their overall firing patterns. A cell looking at the light area close to the border will receive less inhibition in total, since some of it will come from its neighbours looking at the darker area to the left. A cell which

Fig. 2.6 Nicholas Wade's portrait of *Ernst Mach* (1838–1916), based upon a photograph in D. G. Runes Pictorial History of Philosophy. © 1992, Nicholas Wade.

receives less inhibition will end up with a higher overall level of activation—thus signalling a brighter strip.

Figure 2.6 shows an ingenious display by the psychologist-artist Nicholas Wade, in which a portrait of Ernst Mach is concealed in a pattern of stripes which themselves further illustrate Mach banding in the super-dark and super-light tinges around the edges of the face. The portrait is superimposed by the addition of very tiny variations in the overall tapering stripes, and is conveyed only by very coarse-scale variation (see next section). You will see the portrait more easily if you view the figure from a distance of a metre or so, and you may find blurring your eyes helps by filtering out some of the interfering information from the stripes. The face is shown in three-quarter view, looking towards your right (the subject's left) with the right ear clearly visible. In the next section we will describe how different spatial scales are differentially important in conveying information important for face perception.

Interestingly, some painters use the technique of 'irradiation' to sharpen edges and contours in a way which mimics the natural mechanisms of vision, but was discovered independently by artists much earlier. Latto (1995) describes how Seurat made use of this technique to sharpen borders in a way which was much admired and copied by others including Picasso (see Fig. 2.7).

A more general consequence of lateral inhibition may be its effect on perceived contrast and brightness between different parts of the visual image. A grey blob surrounded by white appears darker than the same

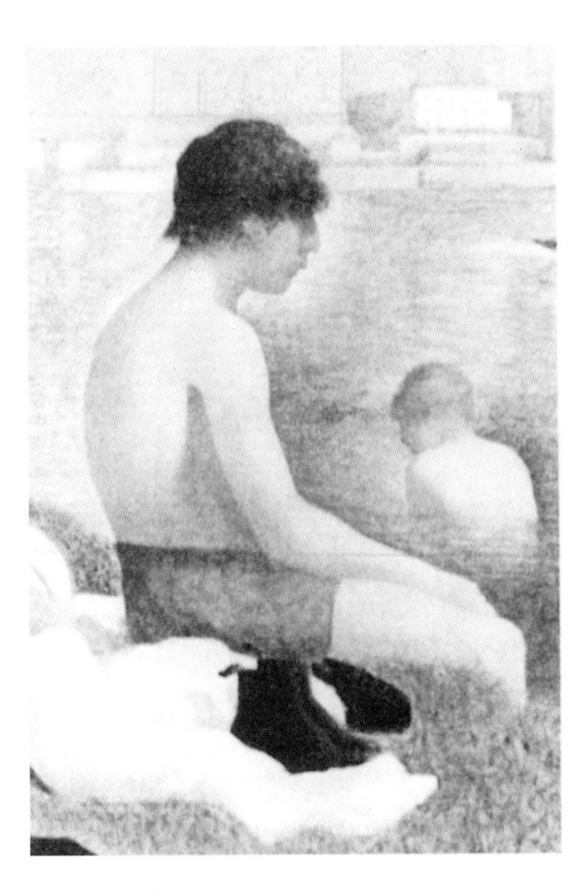

Fig. 2.7 Use of irradiation by George Seurat, 1883–4, *Bathing at Asnieres* (detail). Courtesy of the Trustees of the National Gallery, London.

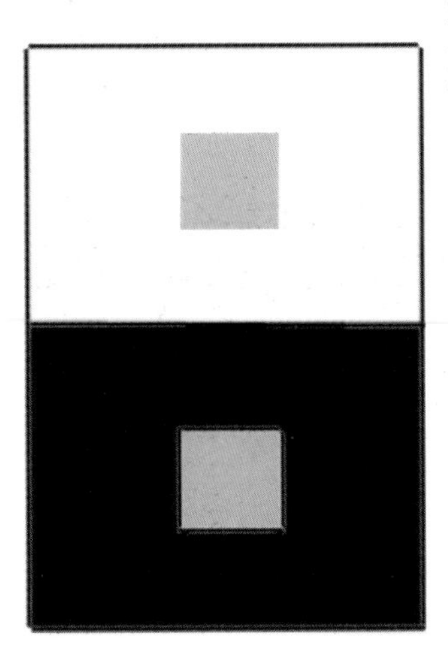

Fig. 2.8 Brightness contrast. The central shape has the same brightness in each half of this image.

grey blob surrounded by black (Fig. 2.8). This could be explained in the following way. The cells which look at the grey blob at the left get much more inhibition from their more active neighbours looking at the bright surround than do the cells looking at the grey blob on the right. Although we can explain these effects in terms of direct lateral interactions between receptor cells, as has been demonstrated in the retinas of some species, another way to understand this effect is to consider how the activity from receptors looking at the central (grey) blob and the surround (white or black) is pooled at the next neural stage in the visual system.

This same principle of balancing excitation and inhibition is found in the way that cells later on in the visual system pool activity from the earlier stages of photoreception. The activity at each ganglion cell is influenced by activity feeding in from photoreceptors covering a particular area of the retinal image, known as the 'receptive field' of the ganglion cell (see Fig. 2.9). Ganglion cells in the retinas of mammals and primates generally have receptive fields within which there is a central zone where there is excitation from their input and a broader surrounding area from which inhibition is obtained. In some ganglion

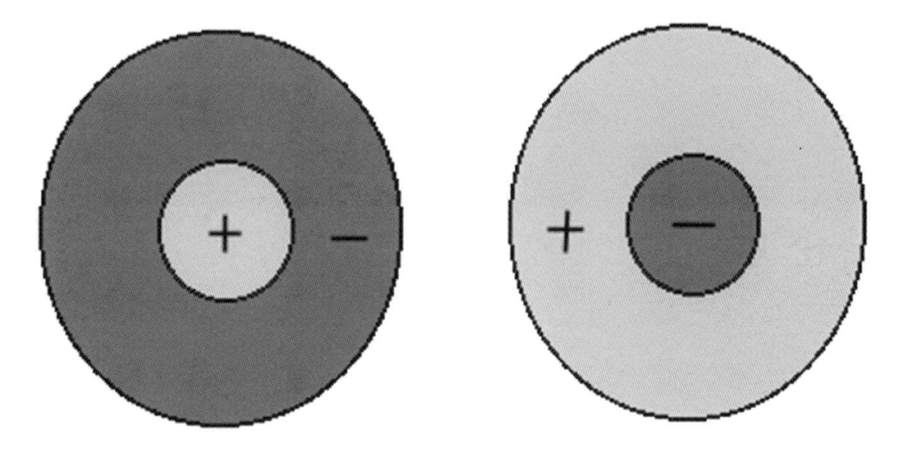

Fig. 2.9 Schematic illustration of receptive fields of ganglion cells showing inhibitory (dark) and excitatory (light) regions.

cells this pattern is reversed, with central inhibition and peripheral excitation. The result of this pooling is again that responses to change will be sharpened and those to uniform fields will be less strong.

Different ganglion cells have receptive fields of different sizes. Those near the edge of the retina (the periphery) have relatively large receptive fields, pooling information from receptors sampling a wide area. Those in the centre of the retina have much smaller receptive fields, and there is much less pooling, making this central area—the fovea—especially responsive to fine detail.

Figure 2.10 demonstrates the effects of this pooling of responses by retinal ganglion cells of different receptive field sizes. When you look at the grid, you will experience a faint grey spot at the white intersections at every location except the one which you are directly fixating. Moving your eyes to try to look at the grey patches is frustrating: wherever you look, there is no grey patch, but you are aware of them at other intersections. Figure 2.10 illustrates how these grey patches

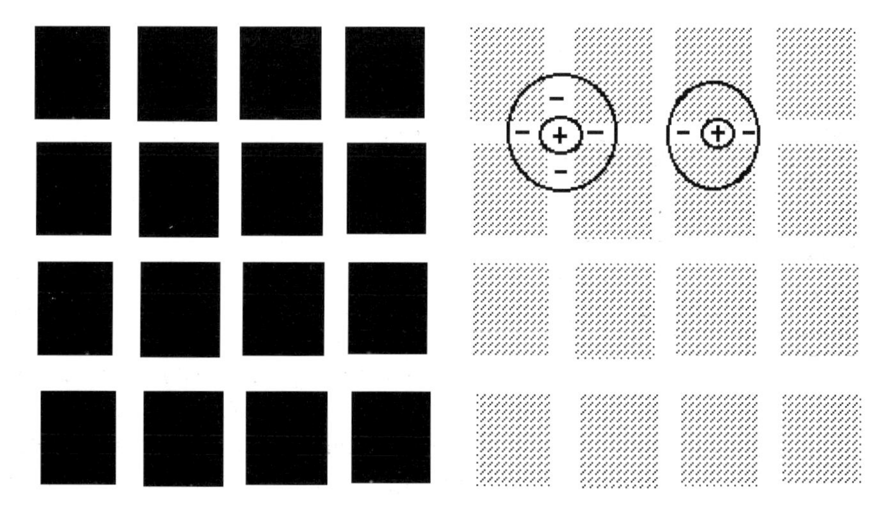

Fig. 2.10 *Left*: the Hering grid. *Right*: an explanation in terms of relative amounts of inhibition. The ganglion cell which 'looks' at a grid intersection gets four doses of inhibition from the four areas marked with minus signs. A ganglion cell positioned further along will only get two doses of inhibition. The one at the intersection will therefore have its activity reduced by more, leading to the apparent darker patches at these places.

could arise from the extra inhibition experienced by a ganglion cell with a receptive field centred on the intersection compared with one off-centre. The ganglion cells at the fovea (your point of fixation) have smaller receptive fields and thus the inhibition from broader areas of light is not evident.

Figure 2.11 shows another of Nicholas Wade's cunningly concealed portraits, this time of Hering, whose image is superimposed on the grid which bears his name. If you view Fig. 2.11 close up, you will experience the same kind of grid illusion as before, but this time as extra 'light' patches at the intersections of the dark bars everywhere except where you are currently fixating. Can you apply the same principles discussed for Fig. 2.10 to understand why these brighter patches are seen?

If you hold the page further away you should be able to see the faint impression of Hering's face which has been portrayed by subtle variations in the thickness of the black lines in the grid. (The face shown has a receding hairline and moustache, and is nearly full-face but angled slightly to the right.)

Fig. 2.11 Nicholas Wade's portrait of *Karl Ewald Konstantin Hering* (1834–1918), based upon a photograph in S. Duke Elder (1968) *System of Ophthalmology Vol. IV. The physiology of the eye and of vision*. Henry Kimpton, London. © Nicholas Wade.

The mechanisms of lateral inhibition mean that discontinuities in brightness between regions of the visual image are enhanced. The advantage of this is that abrupt changes normally signal the edges of objects in the world. This mechanism of edge enhancement can thus help the visual system construct a 'sketch' of the objects in view, rather as a human artist might produce a line drawing. There is now a good deal of evidence that a simple 'sketch' of the main boundaries and parts of an object provides the right kind of information to allow an object to be categorized as a particular type. For example in Fig. 2.12 you can readily recognize each of the distinct types of object shown. However, to recognize specific individual objects within a particular category, for example to recognize your own car in the car park or your own suitcase at the baggage return at an airport, more information about the surface may be needed. This is important here because face recognition involves discrimination within a category of objects (the face) whose members all have a similar overall shape. To do this, a simple 'sketch' of major features does not seem to be enough, and the relative lightness and darkness of different regions on either side of these edges, and gradual changes in brightness from shading, are also essential to our proper perception of faces.

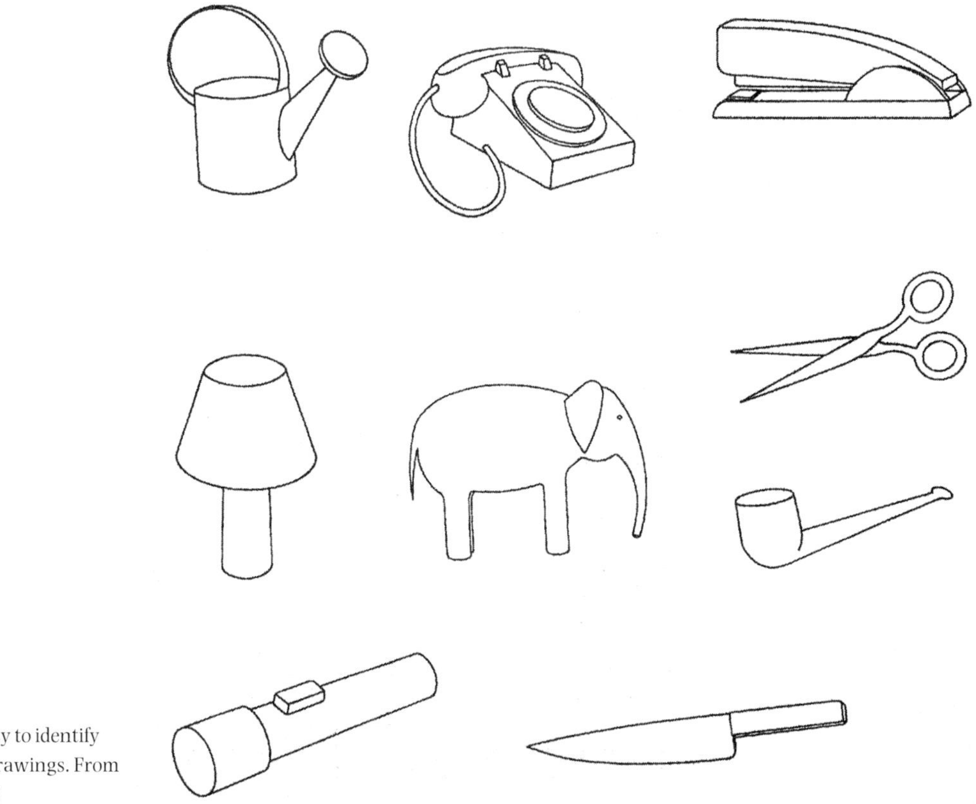

Fig. 2.12 It is easy to identify objects from line drawings. From Biederman (1987)

2.4 Information at different scales

Retinal ganglion cells pool information across their receptive fields. When receptive field sizes are relatively large, fine spatial detail is lost and a set of ganglion cells with large receptive fields effectively 'blurs' the image, removing all fine detail but preserving the coarse-scale features of the image. When receptive field sizes are small, then finer spatial detail is preserved but coarser-scale features—such as how the intensity changes across a region of the image—are lost. The retina is just one of a number of sites within the visual system where such spatial filtering operations take place. There is interesting evidence that the visual system processes the image through a small set of different channels, each of which analyses a different range of spatial scales (see Box 2A).

Box 2A: Spatial filtering by eye and brain

The French mathematician Jean Baptiste Fourier showed that any complex waveform, such as the sound waveform produced when someone is speaking, or a pattern of light intensities on the retina, can be described as though it were constructed from the sum of a set of simpler patterns known as 'sine waves'. A visual sine-wave pattern is one whose brightness varies in the way shown in the top two rows of Fig. 2A.1. The graphs on the left describe the sine-wave undulations in

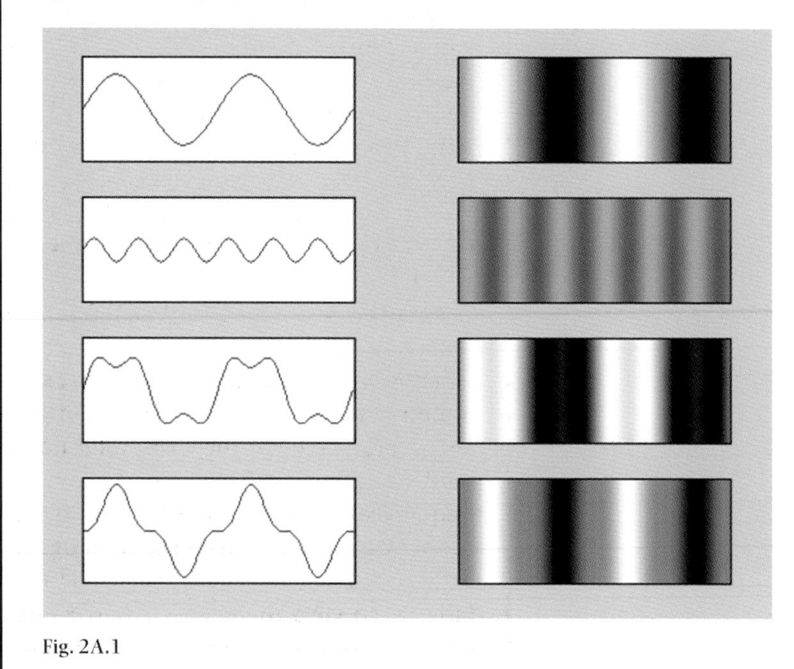

Fig. 2A.1

the pattern of intensity which are drawn out in the images on the right. The image on the top right has a frequency of one-third that of the pattern below it, but an amplitude (the 'contrast' between the darkest and lightest parts) three times as great as the one below. Because these are patterns whose frequency varies across space, we can refer to them as sine waves with different 'spatial frequencies'. These two sine-wave patterns can be added together to form more complex patterns, as shown in the bottom two rows. These differ in terms of their 'phases'. At the bottom, the sine waves have been added so that their peaks and troughs coincide (they are 'in phase') to provide an apparently higher overall contrast between the darkest and brightest parts. In the row above, however, the two patterns have been added so that their peaks do not coincide (they are 'out of phase').

Since any complex pattern can be decomposed into sine-wave patterns in this way, one way of transmitting a complex signal would be to decompose it into its constituent frequencies. There is some rather strong evidence that the human visual system filters the images it receives through a number of distinct spatial frequency channels, which seem to have some degree of independence from each other.

This evidence has been obtained in experiments in which the contrast in patterns of stripes is varied to measure how much contrast is needed for human observers just to be able to detect that stripes are present. Fergus Campbell and John Robson (1968) found that the amount of contrast needed to see a complex pattern of stripes (like those in the bottom two rows of Fig. 2A.1) was almost the same as that required for the detection of each of the individual sine-wave components presented individually, suggesting that the complex pattern was detected by combining the outputs of channels which selectively responded to distinct component frequencies. Consistent with this, Graham and Nachmias (1971) showed that the combination of the two frequencies shown in Fig. 2A.1 with peaks added was no more detectable than their combinations with peaks subtracted, despite the apparently greater contrast in the peaks-added pattern.

Another series of experiments by Colin Blakemore and Fergus Campbell (1969) showed that observers can selectively adapt to different frequencies of patterns. For example if an observer stares for a minute or so at a high-contrast pattern of stripes at a frequency of six stripes per inch, they will then find it much harder to detect a faint pattern of stripes at this same frequency, but not at other frequencies. It is as though cells responding to a specific band of spatial frequencies have become fatigued by the adapting stimulus, rather in the way that cells looking at a bright light may become fatigued to yield the after-effects discussed in Section 2.2. However, the effects of spatial frequency adaptation, unlike those of after-effects

Box 2A: Continued

of brightness or colour, do not depend on the eye being kept fixated on the adapting image.

It therefore appears that the human visual system filters its input in a way that keeps different spatial scales separate. This filtering

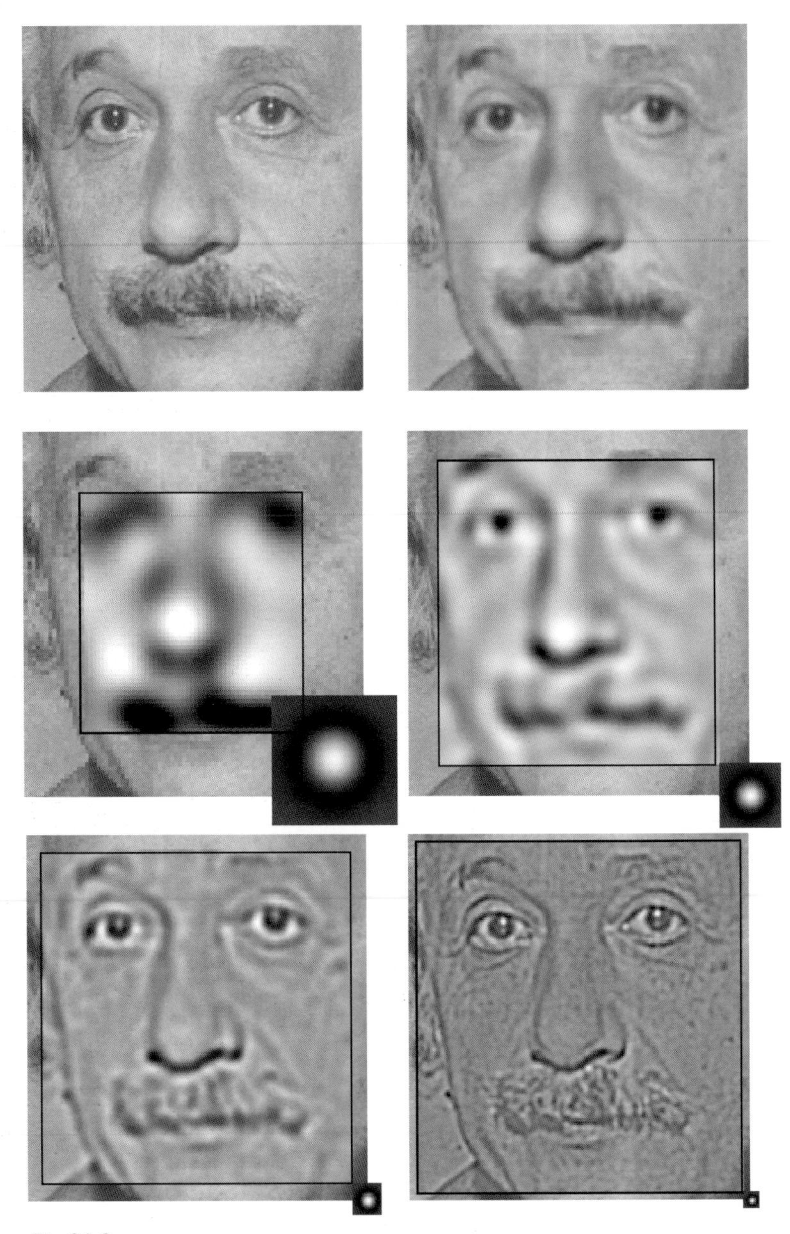

Fig. 2A.2

operation could be achieved if the outputs of ganglion cells with similar receptive field sizes were pooled together. Figure 2A.2 has been produced by Mark Georgeson (from Bruce *et al.* 1996) to show how the image of Albert Einstein's face might be seen by four distinct spatial frequency filters. The receptive field sizes and properties (excitatory inner portion, inhibitory outer portion) of the ganglion cells that would be pooled to produce these filters are shown at the bottom right of each image. The separate outputs of these four frequency channels can transmit virtually all the information from the original image, as seen by the reconstruction of the original from the four filter outputs shown in the top right of Fig. 2A.2.

Figure 2A.3 shows Nicholas Wade's picture 'Contrast sensitivity' in which a photograph of Fergus Campbell has been superimposed on a sine-wave pattern or grating. As you move away from the page, the stripes become less visible and Fergus Campbell's image becomes more prominent. The sine wave grating acts as 'noise' to conceal the image of Campbell shown in lower-spatial frequencies, similar to the demonstrations elsewhere in Figs 2.13 and 2.15.

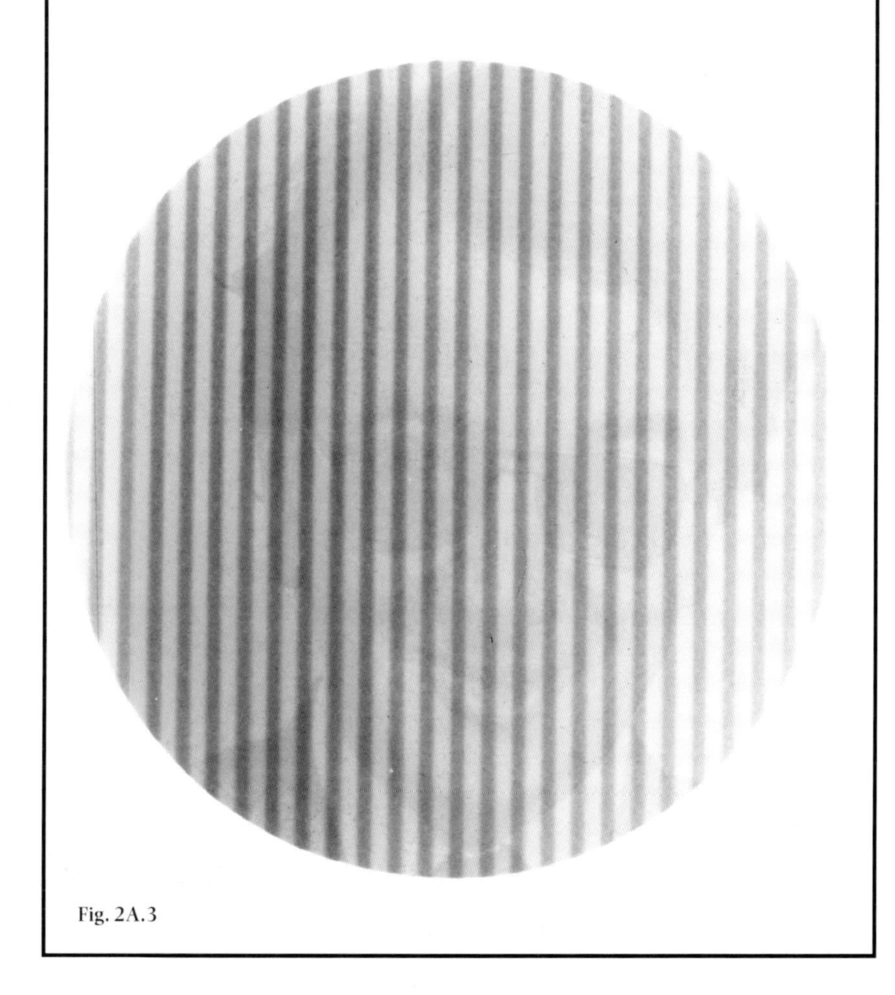

Fig. 2A.3

One function of this decomposition of the information in the image into separate spatial channels is as an aid to edge localization, since coarse-scale channels 'see' blurred edges and finer-scaled channels see sharp edges. However, it is also the case that the pattern of intensity variation within the different filter outputs conveys other information important for face perception.

If face images are filtered, as in Fig. 2A.3, to produce outputs that preserve information at different spatial scales, it can be shown that much information that is important for face perception is captured by the very coarse-scaled (blurred) representation (for example the centre left panel from Fig. 2A.3). A striking demonstration of this was provided by Leon Harmon (Harmon 1973; Harmon and Julesz 1973) using a portrait of Abraham Lincoln (see Figs 2.13 and 2.14). To explain how this image was produced, we need to describe how computers display pictures.

A picture is displayed on a computer as a grid of 'picture elements' or pixels, each showing a specific intensity or 'grey level' (for monochrome pictures). The more pixels are used to display a picture, the better will be the reproduction of the fine-scale information from the original. As the number of pixels used to display a picture is reduced, so the finer-scale details of the original are lost. Harmon

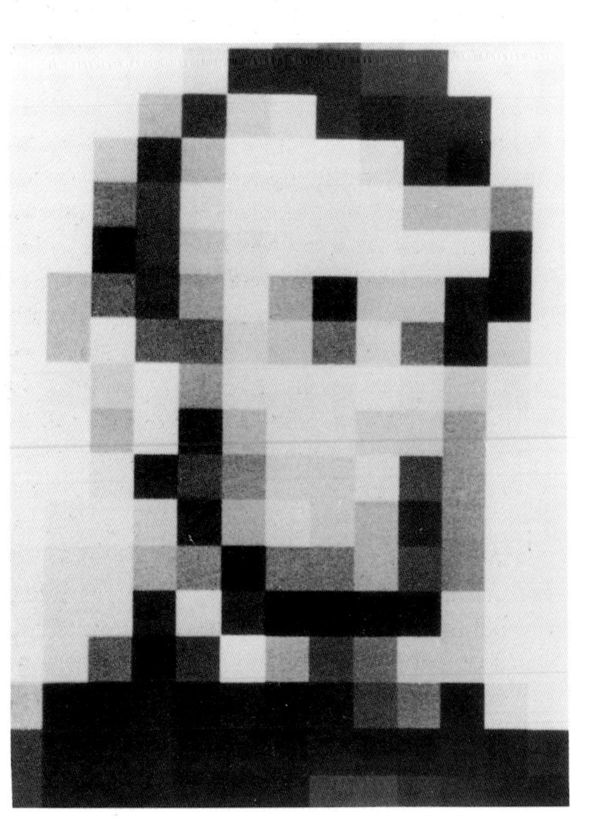

Fig. 2.13 Leon Harmon, *Abraham Lincoln*

Fig. 2.14 Salvador Dali, 1976 *Gala looking at the Mediterranean*. ©DACs, 1998.

produced an extremely coarse-scale version of a portrait of Abraham Lincoln by radically reducing the number of pixels used to display the image, replacing the specific intensity within each of the original pixels with the average intensity over a larger block. With block sizes set large then the image is severely degraded compared with the original, and this 'coarse pixellation' technique effectively blurs out the fine-scale information from the original image. However, unlike the smooth filtering that can be obtained with retinal ganglion filters (see Box 2A), the resulting pixellated images introduce sharp edges between each of the blocks, which seem to conceal the identity of the face. These sharp edges are due to irrelevant fine scale detail, introduced by the process of dividing the image into the large blocks, and these sharp edges are superimposed on the coarse-scale information. If you look at Figs 2.13 and 2.14 you may find that you are able to recognize the original face behind the pixellated disguise by blurring your eyes (try also the bottom right image in Fig. 2.24). Blurring your eyes filters out the fine-scale block edges, and leaves the coarser-scale information unaffected. This demonstration shows how this coarse-scale information alone can contain information sufficient to identify a face. Careful experiments have shown that relatively coarse-scale information can allow

Fig. 2.15 Thatcher/Blair composite produced by Philippe Schyns and Aude Oliva at the University of Glasgow. © Schyns and Oliva 1997

faces to be categorized in a number of important ways, for example according to their age and sex.

Figure 2.15 shows an intriguing illustration produced by Philippe Schyns and Aude Oliva to show how different information can be contained in different spatial scales. Look at Fig. 2.15 close up. What or who do you see? Now place the book against a wall and look at it from a distance of a couple of metres. What or who do you see now? At close distance, your visual system is able to see the high spatial frequencies in which Tony Blair's face is depicted. At a distance, this fine detail cannot be seen and the low spatial frequencies of Margaret Thatcher's face dominate.

Harmon's original studies and considerable research since (see Costen *et al.* 1996 for one recent study) made use of static images of faces, but the technique has been applied to moving images in many television documentaries where there is a need to conceal the identities of people being interviewed about sensitive subjects. The pixellation used for concealment alters rapidly between frames, producing a bewildering array of intensities. More recently, television production editors have started to use defocusing of the faces rather than pixellation, as it became apparent that identities could be seen if people squinted their eyes when looking at the pixellated images, just as you can see Lincoln's face in Fig. 2.13. In current research in our laboratory in Stirling, we are exploring systematically the kinds of image degradation which are successful in making faces hard to recognize, and the effects that motion may have on identification performance (Lander *et al.*, submitted).

It is also possible to filter faces so that only the fine-scale information from the original image is preserved (see Box 2A) and the result is like a line sketch of the original. As we see later, simple line drawings of faces are not good for their recognition.

Why is it that the coarser-scale information is especially useful for face perception and recognition? There are a number of possible reasons. One is that coarse-scale information provides information about shading which may be important in building up a three-dimensional representation of the face. Another is that it is the coarsest-scale information that is available over the longest distances. Try moving away from the page with Fig. 2.13 and you will find that it eventually becomes recognizable without squinting your eyes; this is because the coarser but not the fine-scale information can be perceived at the longer distance. It may be that the way we process faces is in part adapted to our biological need to know quickly who is approaching and what kind of greeting or emotional signal they are displaying!

Moving Fig. 2.13 away from you also shows a further interesting point about our use of information at different spatial scales, which is how flexible we are. As the image is moved away from you, the coarse-scale details gradually shift to becoming finer and finer, because of the changing size of the image relative to the retina. Yet the information about the hidden figure is picked up pretty instantaneously as soon as the masking information is lost. One way of thinking about how this might be achieved is that the visual system may be registering information by coding it relative to the face, as this would be unaffected by changing distance. To do this, though, we would have to be able to find face patterns quickly and effectively. An intriguing finding of Roger Watt and Steve Dakin has suggested that the coarse-scale visual channels may deliver a characteristic 'signature' about the face in the pattern of light and dark stripes (see Figs 2.16 and 2.17). Watt (1994) suggested that this pattern was like a bar-code; identifying a face-like

Fig. 2.16 The pattern shows a raw face image, and one that has been filtered with horizontally oriented filters. The broad sweep of the face results in a simple alternation of light and dark stripes, corresponding to the hairline (dark), forehead (light), eyes (dark), and so forth. Such a distinctive pattern may be easy for the brain to spot, forming a 'bar-code' for the human face. Courtesy of Roger Watt, University of Stirling.

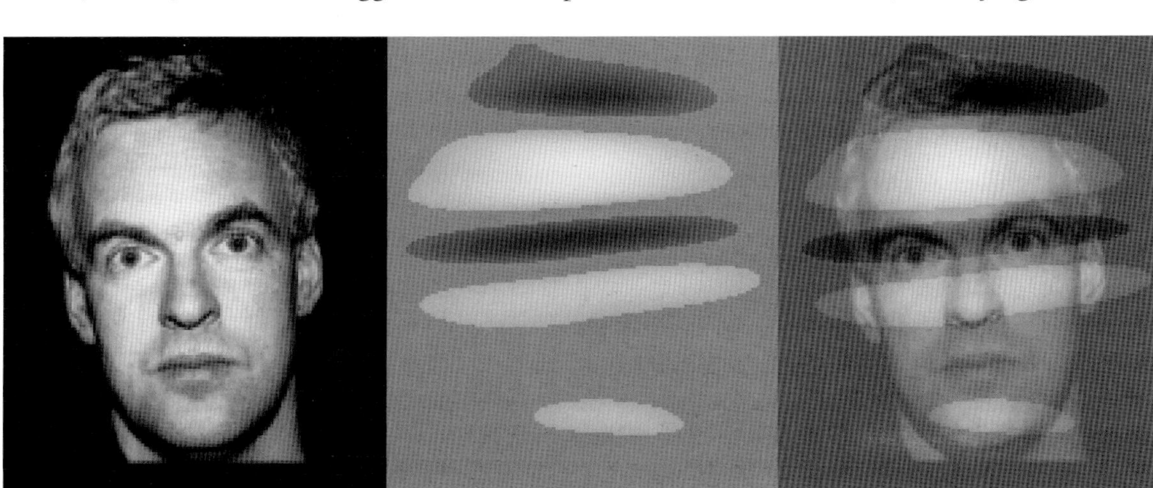

Fig. 2.17 The characteristic bar-code signature for the human face is shown here for each of the faces in the portrait of Bill Forsyth. Courtesy of Roger Watt, University of Stirling.

object and allowing faces to be quickly detected in visual scenes. These ideas are currently being tested.

2.5 Patterns of brightness and colour

A lump of coal looks dark even if illuminated by a bright light, and a piece of writing paper looks white even in a gloomy corner of the office. Our perception of brightness appears to be based upon the relative rather than absolute amounts of light reflected by different surfaces in the scene. Similarly, when we see colours, a banana will look yellow and a leaf look green under a range of different illuminations which may considerably change the actual wavelengths present in the reflected light. The mechanisms by which relatively bright things look bright, and yellow things appear yellow under varying illuminations are called brightness and colour constancies.

Our perception of faces relies upon information contained in patterns of relative brightness, but preservation of exact levels seems unimportant. For example, we are reasonably good at recognizing images of faces portrayed simply as arrangements of black on white (for example see Fig. 2.4) and, as we will see, we are remarkably tolerant of certain variations in colour.

However, line drawings which do not preserve some information about regions of light and dark are poor representations of faces. Graham Davies and colleagues (Davies *et al.* 1978) illustrated this very neatly in an experiment on the recognition of famous faces. They carefully traced around the features of the faces in photographs, and showed these detailed line drawings to observers. Their volunteers were able to identify only 47 per cent of line drawings compared with 90 per cent of the original photographs. Now if our memory for individual faces was based upon a set of feature measurements or descriptions (for example length of nose or width of eyebrows) it is difficult to understand why these accurate line drawings are so difficult to recognize.

Line sketches of faces by human artists usually contain shading and texture information to convey impressions of shape and colouration, and other studies inspired by engineering rather than artistic considerations have shown the benefits of capturing such information in black-and-white images of faces.

During the 1980s, Don Pearson and his colleagues at the University of Essex explored ways to compress the information in moving images

Fig. 2.18 Building a computer-drawn 'cartoon' from separate threshold (*left*) and line (*centre*) components. Their combination (*right*) produces an extremely recognizable 'cartoon' of Michael Caine. From Bruce *et al.* (1992*b*).

Fig. 2.19 The same image of the same face as drawn by human artist (*left*) and computer (*right*). From Pearson *et al.* (1990).

of faces and hands. Their aim was to provide video-telecommunication for deaf people using sign language. A full monochrome image of a face is very 'expensive' in terms of information, since each of its many thousands of pixels can contain 256 different brightness values. However, if the image of the face is converted to black-on-white then the image has been compressed to only one 'bit' of information per pixel, making it possible to send more frames of a moving sequence along a telephone line with limited information capacity. If a recognizable image could be sent in this way, then sign language or even lip-reading via the telephone system might be possible. (Since the late 1980s developments in larger capacity digital telephone lines, and new compression techniques, mean that it is now possible to use richer images of faces in video-telephony than could be achieved a decade earlier.)

Pearson and Robinson (1985) produced a way of automatically compressing moving face images into such black-on-white images. As well as providing a 'sketch' of the layout of the major face features, the procedure they developed also preserved areas of light and dark by setting to black all regions whose average brightness was above a certain threshold. Experimental evaluation of this kind of computer 'cartoon' showed that this 'thresholding' of the images provided information essential for their recognition. Cartoons which just showed the edges were very poor likenesses. With the addition of the threshold component, the resulting images were almost as recognizable as the originals. Figure 2.18 shows how the two separate components of the computer algorithm contribute to building up an image of Michael Caine's face which is readily recognizable. Figure 2.19 shows a comparison between a drawing made by the algorithm of a picture of a female face, and the same face drawn in black on white by a human artist. The similarity is striking.

Fig. 2.20 Original portrait of *Sir Walter Scott* by Andrew Geddes. Pencil drawing based upon this portrait by Andrew Geddes, 1823. Both courtesy of the Scottish National Portrait Gallery. Computer-drawing produced using Pearson and Robinson's (1985) algorithm, by Mike Burton, University of Glasgow.

Figure 2.20 shows an original portrait of Sir Walter Scott, and two different line-drawings based upon this portrait. The first was drawn from the original by a human artist. The second was drawn automatically from the image by computer, using the Pearson and Robinson algorithm. Figures 2.21–2.23 show a series of further line-drawn portraits by humans, illustrating clearly how human artists use shading and cross-hatching to preserve 'mass' and contribute to a three-dimensional impression of the face.

Fig. 2.21 Pencil and chalk drawing of *Sir Walter Scott* by Sir William Allen. Courtesy of the Scottish National Portrait Gallery.

Fig. 2.22 *Isobel Wylie Hutchison.* by David Foggie, 1935, Courtesy of the Scottish National Portrait Gallery.

Fig. 2.23 *5th Earl of Rosebery* by unknown artist. Courtesy of the Scottish National Portrait Gallery.

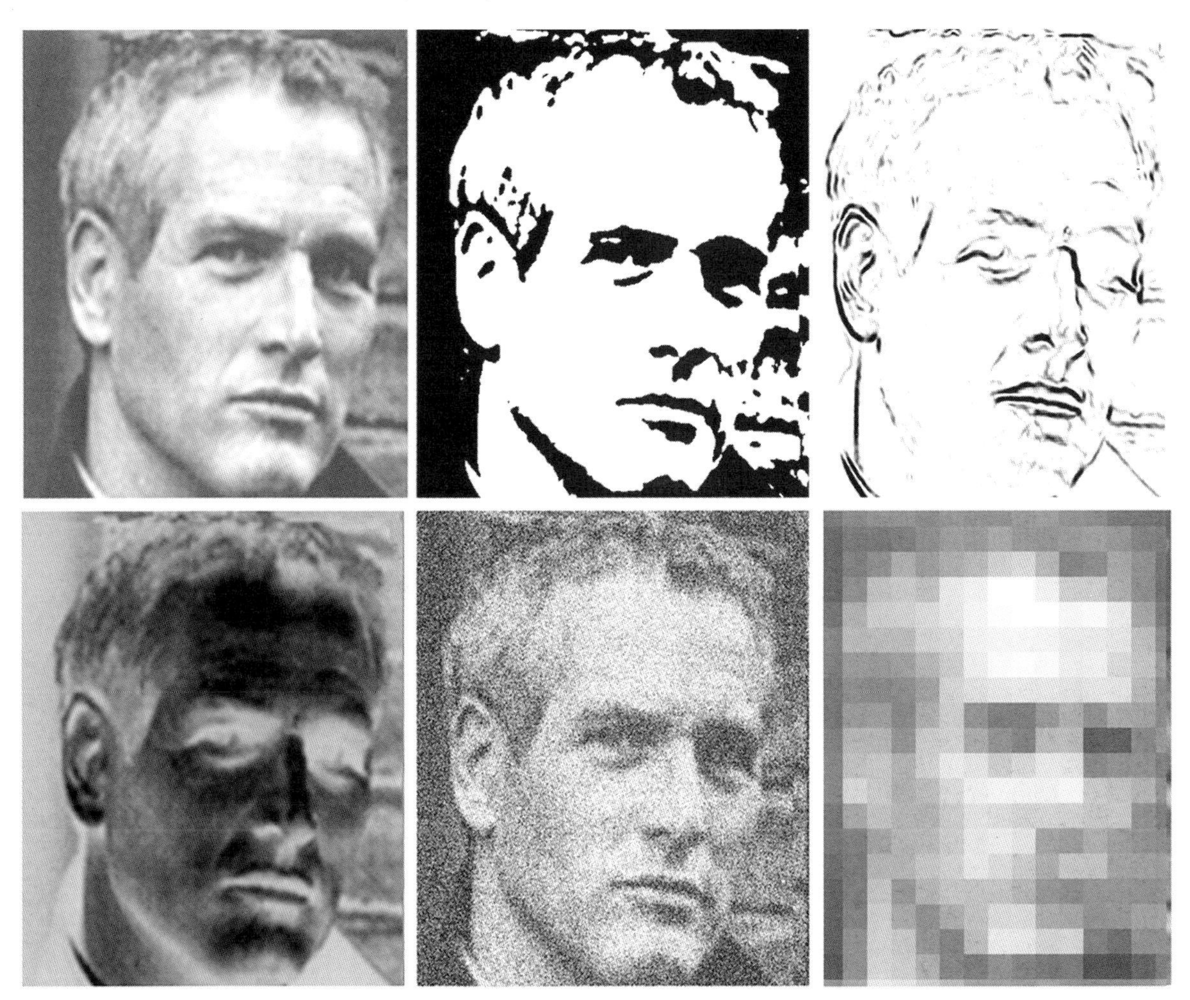

Fig. 2.24 Computer-produced studies of Paul Newman's face which mimic some styles used in modern art. Identity is well preserved in the lithograph (*top centre*) and noise-neoimpressionism (*bottom centre*), but less-well preserved in the outline drawing (*top right*) and negative images (*bottom left*) which do not preserve areas of light and dark. The mosaic image (*bottom right*) uses the same technique as in Fig. 2.13. Identity is revealed when eyes are blurred. Produced by Helmut Leder, University of Fribourg.

This research shows that human recognition of faces uses more than the information contained in a sketch of the main face features and their layout. There is important information in the pattern of light and dark itself. This is consistent with our earlier observation of the relative importance of coarse-scale rather than fine-scale information for face perception.

This conclusion is also consistent with the observation that photographic negatives, which maintain the range of brightnesses in the original but reverse all the brightness values, are extremely difficult to recognize. Face recognition seems to require the correct maintenance of relative brightness. This may be because light and dark areas of the face are informative about the pigmentation of hair and skin (a negative image of a Caucasian face shows the skin darker than the hair), or because negating an image inverts shading patterns. Notice how the areas of dark provide an impression of depth in the computer-drawn cartoon of Michael Caine's face in Fig. 2.18. Reversing light and

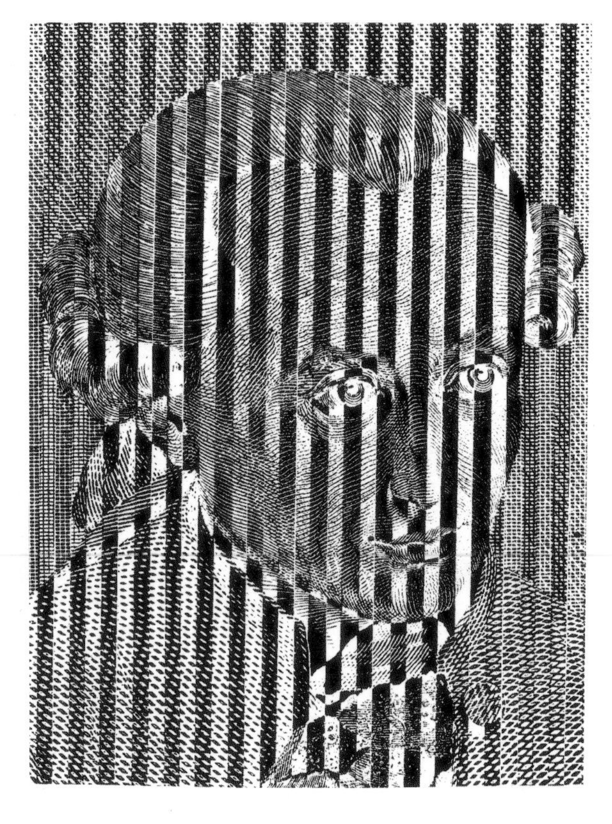

Fig. 2.25 Apperception. Portrait of *Immanuel Kant* by Nicholas Wade. Portrait after frontispiece engraving in: G. Hartenstein (ed.) (1853). *Immanual Kant's Kritik der reinen Vernunft*, Leopold Voss, Leipzig. © 1992 by Nicholas Wade.

dark inverts such information about shadows, and in a photograph with a range of brightnesses it will impair the interpretation of these shading patterns, which we discuss in the following section. Later in this chapter, Box 2B describes some clever recent research by Richard Kemp, Graham Pike, and colleagues which examines different possible reasons for the effects of photographic negation.

Figure 2.24 summarizes some of the conclusions we have reached in this chapter about the importance of relative brightness and coarse-scale information for face recognition, by showing the same face depicted in a number of different formats, which differently reveal or conceal the identity. Each of these formats has been used artistically in order to create particular impressions, but each also reveals something about the processes involved in human vision and face recognition.

Human vision is immensely adaptable and under certain circumstances we can cope despite the disruption of patterns of relative brightness. In Fig. 2.25 we are able to see the face in Nicholas Wade's portrait of the philosopher Immanuel Kant even though this requires that we group together alternating blocks of black and white.

While effects of relative brightness on face perception are clear, those of colour are not. People may become difficult to recognize following a dramatic change in hair colour, but it is not known whether such effects arise because of changes in resulting lightness or hue. The

Fig. 2.26 John Bratby, *Sir David Steel*. In a private Scottish collection.

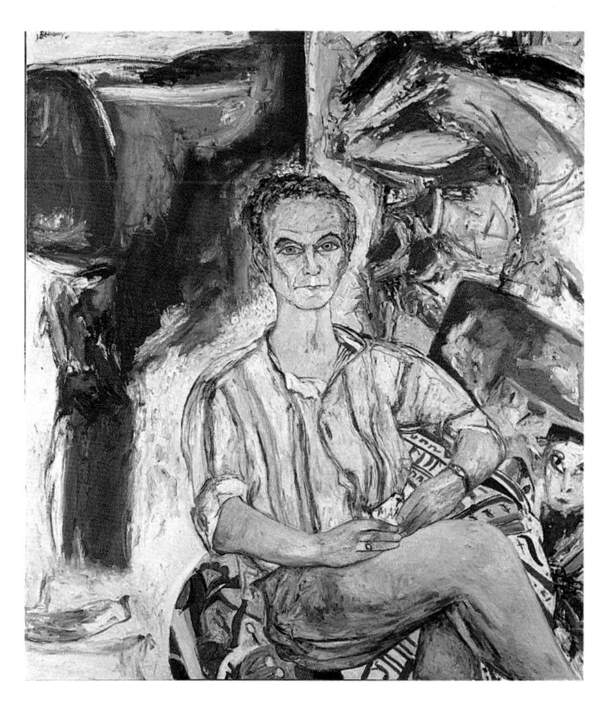

Fig. 2.27 John Bellany, *Sir Peter Maxwell Davies*. Courtesy of the Scottish National Portrait Gallery.

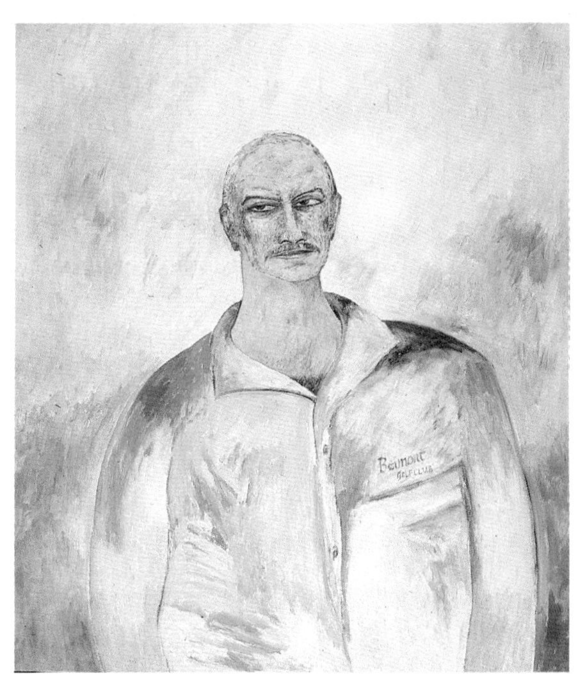

Fig. 2.28 John Bellany, *Sean Connery*. Courtesy of the Scottish National Portrait Gallery

modern portraits shown in Figs 2.26–2.28 show that we can readily recognize likeness despite the bizarre colourations employed. Richard Kemp and colleagues (1996) have shown that face recognition is virtually unaffected by a transformation which reverses hue but maintains lightness (see Box 2B). None the less, we may respond to subtle changes in people's colouration to detect signs of ill-health or embarrassment.

Box 2B: Colour, shading, and negation

Richard Kemp, Graham Pike, and their colleagues at the University of Westminster have analysed and illustrated the separate impacts on face perception of hue and brightness changes (Kemp *et al.* 1996).

What we normally refer to as the colour of an object has three distinct components, the brightness or 'luminance', (essentially the amount of light reflected from the surface of the object), the hue (the wavelength of the light reflected from the surface of the object) and the saturation (the purity of the colour—unsaturated colours appear washed out). Artists often think of shades of colour arranged around a circle, where the hue is represented round the circumference of the circle and saturation is represented by the distance form the centre of the circle (i.e. along the radius). Luminance is the third dimension—bright colours would be above the plane of the circle and dark colours below the plane.

A very simple colour circle is illustrated at the top left of Fig. 2B.1. Note how the different colours are arranged around the circle so that complementary colours are opposite each other. The colours on the outer edge of the circle are fully saturated, those on the inner circle are at 50 per cent saturation (and therefore appear more washed out).

Because hue and luminance are coded separately in a colour image it is possible to manipulate them separately. In a normal colour negative (which Kemp *et al.* (1996) referred to as a 'full-negative') both the hue and the luminance channels are reversed. The strip of

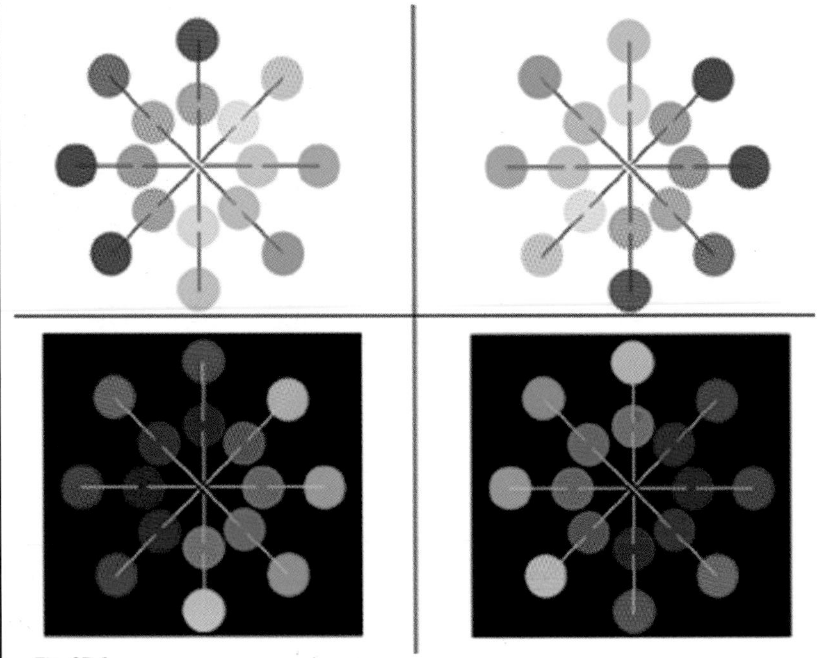

Fig. 2B.1

photographic negatives which you receive when you send a colour film for processing shows this effect which is represented by the bottom right of Fig. 2B.1. Note that blue/green patches now appear yellow/red (and vice versa) and that the black lines of the figure appear white and the white background appears black.

In the top right panel of Fig. 2B.1 Kemp and colleagues show the effect of reversing only the hues. This has the effect of rotating the colour wheel through 180° so that each original colour is now replaced by its complementary colour. The brightness has not been changed by this manipulation (so the black lines and white background are unchanged). We will refer to this manipulation as 'hue-negation'.

At the bottom left of Fig. 2B.1 we show the effect of reversing the luminance channel without altering hue information. The arrangement of the hues round the wheel is unaltered, but the black of the lines has become white and the white background has become black. We will refer to this manipulation as 'luminance-negation'. Figure 2B.2 illustrates the manipulations of hue negation, luminance negation and full negation on an image of a human face.

Hue-negated and luminance-negated images of faces were used by Kemp *et al.* (1996) to demonstrate the relative unimportance of hue for the recognition of faces. As well as being of artistic interest, this has

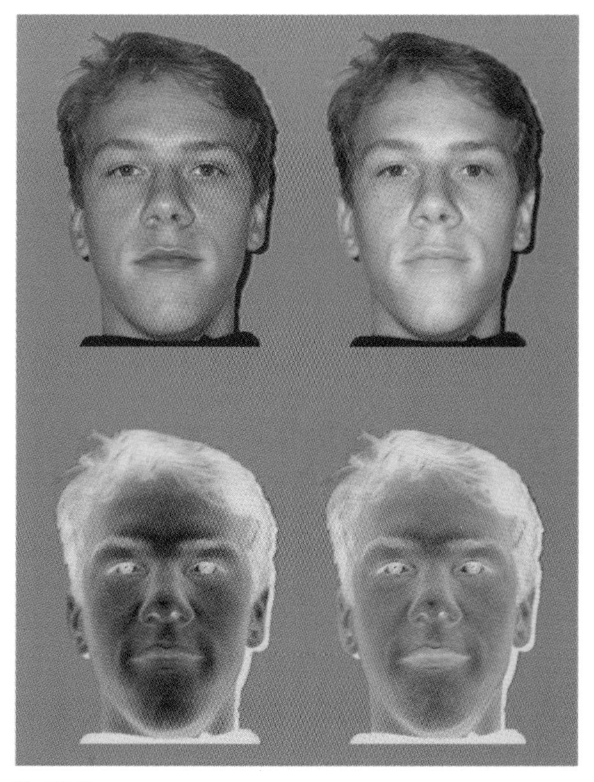

Fig. 2B.2

Box 2B: Continued

implications for our understanding of the physiological processing of faces, and the reasons why photographic negative images of faces are difficult to recognize.

Using normal negation (either colour or black and white) it is very difficult to separate different explanations of the negation effect, as the negated image both suggests a change in pigmentation and disrupts the relative brightness of different regions, possibly affecting the processing of three-dimensional shape-from-shading processes. However, Cavanagh and Leclerc (1989) demonstrated that the shape-from-shading processes are 'colour blind' as the visual system is insensitive to the hue of the shadows and the regions casting the shadows. It appears that artists have been aware of this fact for far longer than psychologists! Artists often make use of unusual and unexpected hues in their work, but the image is still perfectly recognizable because they take care to preserve the relative luminance of the colours they use. Thus in painting a shadow region on a face an artist can use orange, red, or even green pigments provided that the

Fig. 2B.3

chosen pigment is darker than the surrounding areas which are not in shadow. Several portraits in this chapter demonstrate this imaginative use of colour (for example Figs 2.26–2.28).

The reader is invited to inspect Fig. 2B.3 which shows a portrait of Lord Steel presented in its original colour (top left), in hue-negative (top right), luminance negative (bottom left), and full negative (bottom right). Although the hue-negated image does not show the colours as the artist intended, it has retained the relative lightness and darkness of the various parts of the image, and thus the shape-from-shading processes are still able to interpret the pattern of shadows and shading. For example the darker shadows that appear under the eyes of the original are still apparent in the hue-negated image, and the darker areas of shadow on the left side of the face are still apparent in the hue-negated image. In contrast, in the luminance- and full-negative images (bottom two patterns) these darker areas of shadow appear brighter than the surrounding areas of the face. Thus the hue-negative image preserves the artist's use of light and shadow and only alters the hues. This would seem to support the shape-from-shading explanation of the negation effect and further reinforce the notion that face recognition is mediated by the formation of a three-dimensional representation of the face.

Kemp and colleagues (1996) conducted a series of experiments using different types of negation on more conventional face images such as those shown in Fig. 2B.2. They were able to show that people were able to process hue-negated face images quite normally when

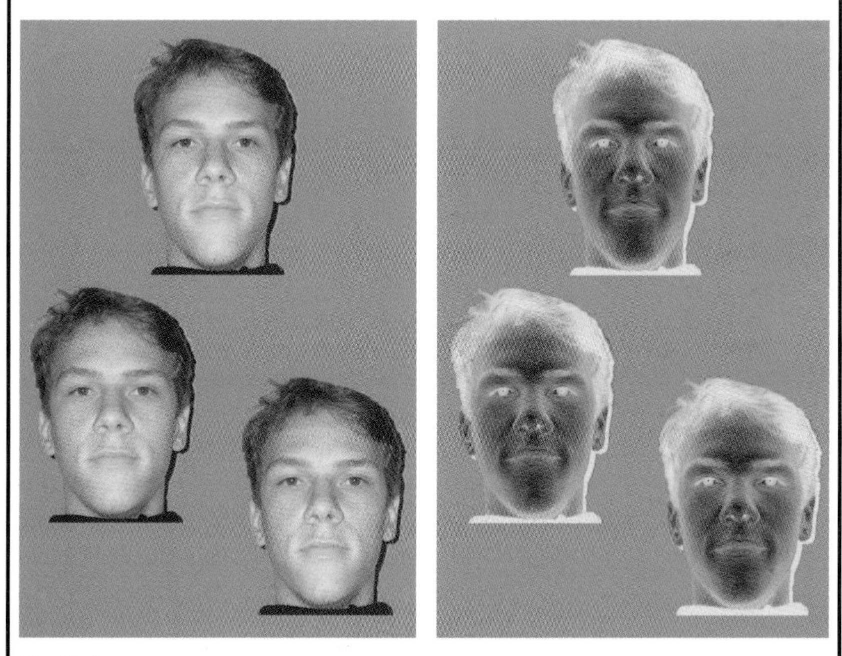

Fig. 2B.4

Box 2B: Continued

their task was to recognize familiar faces, or to spot tiny displacements made to the features of faces, but were very disrupted with luminance-negated or full negative images. In Fig. 2B.4 we illustrate some of the conditions used in their experiment. Your task is to decide which of the lower two faces is identical in every detail to the one at the top. You should find this easier in the hue-negated version (left) than in the luminance-negated version (right). (Answer: in each panel, the face at the bottom right has its eyes moved further apart by four pixels and so does not match the one at the top.)

These results suggest that photographic negatives may be difficult to recognize because of the reversal of brightnesses, perhaps because of the disruption of three-dimensional shape-from-shading, but that hue values are not important.

There was one task which Kemp *et al.* (1996) found was disrupted by hue-negation—the task of recognizing previously unfamiliar faces. This is probably because unfamiliar face recognition is strongly influenced by the precise details of the picture which is shown at study and test, and this experiment suggests that picture memory may be sensitive to hue, even though familiar face memory is not. This sensitivity of pictorial memory to colour also explains why we may be extremely sensitive to the faithfulness with which colours of faces are reproduced in copies of photographs or paintings.

2.6 Shape from shading

The human face is a complex three-dimensional surface, and variation in the specific shapes and relationships of protruberances and concavities helps specify the age, race, and sex of the face (see Chapter 3). Moreover, as we saw in Chapter 1, understanding and manipulating the three-dimensional facial surface is important for effective surgical interventions to correct facial abnormalities.

Human vision makes use of a variety of methods to recover information about the third dimension, which is missing from the two-dimensional image projected on each retina. The cues to three-dimensional shape which are most relevant for face perception include stereopsis, motion, and shape-from-shading.

Stereopsis is the process by which the slightly different images of the world seen at the two eyes are combined in the brains visual cortex to provide a unified image with information about relative depth. You can demonstrate for yourself that each of your eyes sees a slightly different image of the world by alternately opening and closing each eye. The objects that you are viewing will 'jump' from side to side in each eye's view. To view pictures in three dimensions means that some way of reproducing the effects of these different perspectives on objects must be found. A stereoscope is an instrument that allows three-dimensional pictures to be seen. Stereoscopes work by showing separately to each eye the different images of a three-dimensional object which have been taken from the point of view that each human eye would see. The invention of the stereoscope in the 1830s was claimed by both the English physicist Sir Charles Wheatstone and the Scottish scientist Sir David Brewster (see Fig. 2.29), who also invented a variety of other optical instruments including the kaleidoscope, and Wade (1983) reprints the lively correspondence between the two on this point.

Motion cues to depth arise if an object or the observer moves, since the images of different parts of the scene will travel across the retina at different rates depending on their distance. Both stereopsis and motion are important in our perception of live faces, but we cope extremely well when deprived of both these sources, as when looking at pictures. There are a number of cues to depth which are present in pictures, the most important for face perception arising from shading patterns.

The undulations of the facial surface moderate the amount of light reflected from different regions. When lighting comes from above, for example from the sun or from an overhead room light, parts of the

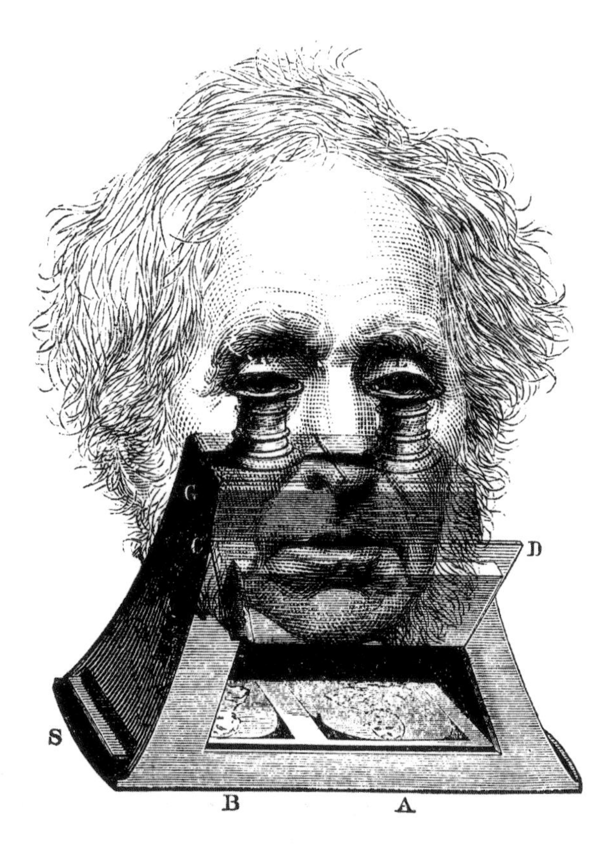

Fig. 2.29 *Brewster's stereoscope.* © Nicholas Wade.

surface higher up will reflect more light than parts lower down. Where one part of a surface blocks light, it will cast a shadow. The patterns of shading and shadow across a surface are therefore informative about the three-dimensional shape of a surface, and the visual system can use this information to retrieve a description of surface shape provided that some information about the direction of the light source is known. The direction of the light source needs to be known because any particular pattern of shading is ambiguous.

For example, consider the simple shapes shown in Fig. 2.30. This pattern of shading could arise if domed shapes were lit from above, or if hollow shapes were lit from below. Unless light direction is either known or assumed, the pattern cannot be interpreted. Moreover, it is often possible for the interpretation of such an image to shift, either spontaneously, or as a result of some other change of conditions, such as a change in the apparent direction from which the surface is lit.

Sir David Brewster was well aware of this ambiguity, and wrote at length about early observations reported to the Royal Society of London by Dr Gmelin in 1744. Gmelin had noticed that when objects were viewed through compound microscopes, whose lenses inverted the pattern of shading, these objects often appeared to reverse in depth, so that a concave surface ('intaglio') appeared convex ('cameo') or vice versa.

> It cannot therefore be doubted, that the optical illusion of the conversion of a cameo into an intaglio, and of an intaglio into a cameo, by inverting an eye-piece, is the result of the operation of our own minds, whereby we judge of the forms of bodies by the knowledge we have acquired of light and shadow. (Brewster 1826; in Wade 1983, p. 58)

In this quote Brewster clearly illustrates that he understands that human vision is the product of both the received image and its interpretation by the human mind.

In more recent times, V.S. Ramachandran (for example Ramachandran 1995) has made use of the perception of simple displays such as those illustrated here to demonstrate a number of simple assumptions made by the visual system when interpreting patterns of shading, and therefore to systematize the rules employed by the mind when interpreting such patterns.

The first principle that seems to guide human vision in interpreting

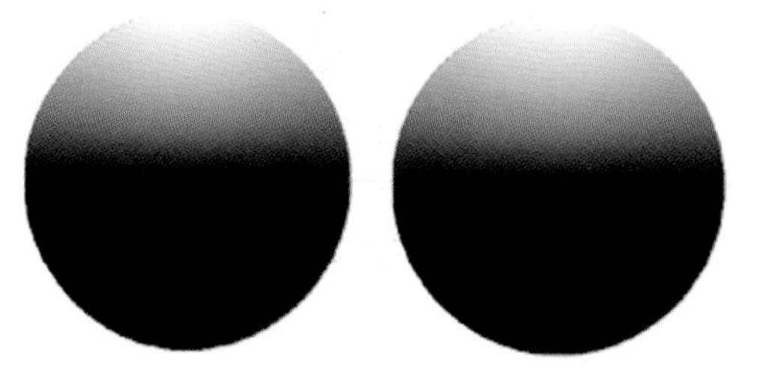

Fig. 2.30 Humps or dents? See text.

patterns of shading is that lighting is assumed to come from above. This is why most people will tend to see Fig. 2.30 as domed (convex) shapes. However, if you turn the page upside-down and thus invert the shading pattern, then hollow shapes will often be seen, consistent with the shapes that would give this pattern of shading if lighting remained from above. It makes biological sense that the visual system should assume lighting comes from above, since this reflects the predominant direction of light from the sun in the natural world in which vision evolved.

The second preference that human vision seems to exert is to interpret patterns of shading as though there was only a single light source illuminating the scene. Thus, Fig. 2.31 will usually be seen as a collection of humps and dents, rather than a collection of humps lit from different directions. Again, the assumption of a single light source is a sensible one for a visual system which evolved before even fire was used as an additional source of light.

The third assumption that seems to be made is that, other things being equal, shapes are convex rather than concave. In Fig. 2.32 a single light source from the side does not favour humps or dents, yet most people would tend to say 'humps'. This time, if you invert the page, you will still tend to see humps. Again, the assumption of convexity reflects the natural world where most objects are convex rather than concave (indeed the human ear is a rather interesting exception to this).

While these three principles are usually sufficient to explain our interpretation of patterns of shading, exceptions are made when other assumptions about likely events in the world are stronger. If we are presented with a hollow facial mask, then provided that we view this at a distance too great for stereo vision to work, we will see this as a real face,

Fig. 2.31 Humps and dents. See text.

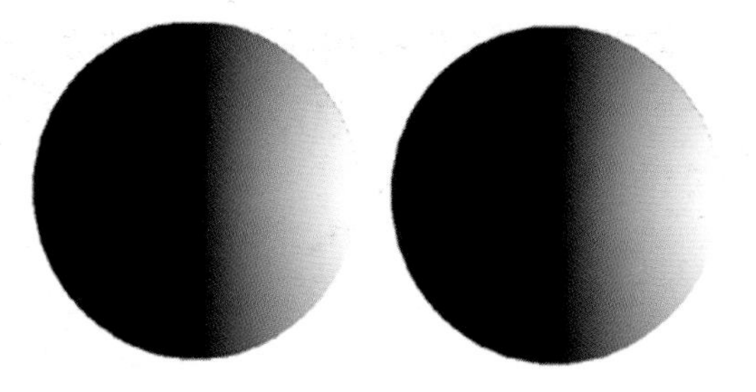

Fig. 2.32 Humps or dents? See text.

with its nose sticking out towards us, rather than as the mask it actually is. This hollow face illusion, popularized by psychologist Richard Gregory, appears first to have been described by Brewster (1826):

> The last species of illusion of this nature, and perhaps the most remarkable of all of them, may be produced by a continued effort of the mind to deceive itself. If we take one of the intaglio moulds used for making . . .bas-reliefs. . . and direct the eye to it steadily without noticing surrounding objects, we may coax ourselves into the belief that the intaglio is actually a bas-relief. It is difficult at first to produce the deception, but a little practice never fails to accomplish it.
>
> We have succeeded in carrying this deception so far, as to be able, by the eye alone, to raise a complete hollow mask of the human face into a projecting head. In order to do this, we must exclude the vision of other objects, and also the margin or thickness of the cast. This experiment cannot fail to produce a very great degree of surprise in those who succeed in it; and it will no doubt be regarded by the sculptor who can use it as a great auxiliary in his art. (Brewster 1826; quoted by Wade 1983)

Brewster seems to give the impression that this 'hollow face' illusion requires some training and will-power to observe, suggesting that he was observing the mask at a rather close distance. At greater distances it can be more or less impossible to see the true hollow shape. Moreover, the illusion shows that our preference for seeing faces is so strong that it can override the usual assumption that light comes from above. The illusion can be illustrated with a figure (Fig. 2.33) but it is much more dramatic in real life, since when the illusory perception of the face is seen, this has knock-on effects on the way in which the information is interpreted if the observer moves their head. Unlike a real face, the hollow face will appear to move with the observer as he or she moves their head from side to side. Some sense of this is seen in the way in which the angled shots of the hollow and solid face appear differently in Fig. 2.33.

If an observer walks towards the hollow face, there comes a point where the true hollowness of the shape becomes visible because of information from stereo-vision and cues to depth which arise from the observer's own motion. Harold Hill and Vicki Bruce (1993, 1994) explored how different variables affected the strength of the illusion. To do this, they used the distance from the mask at which it was seen as hollow as a measure of illusion strength. Using this method, Hill and

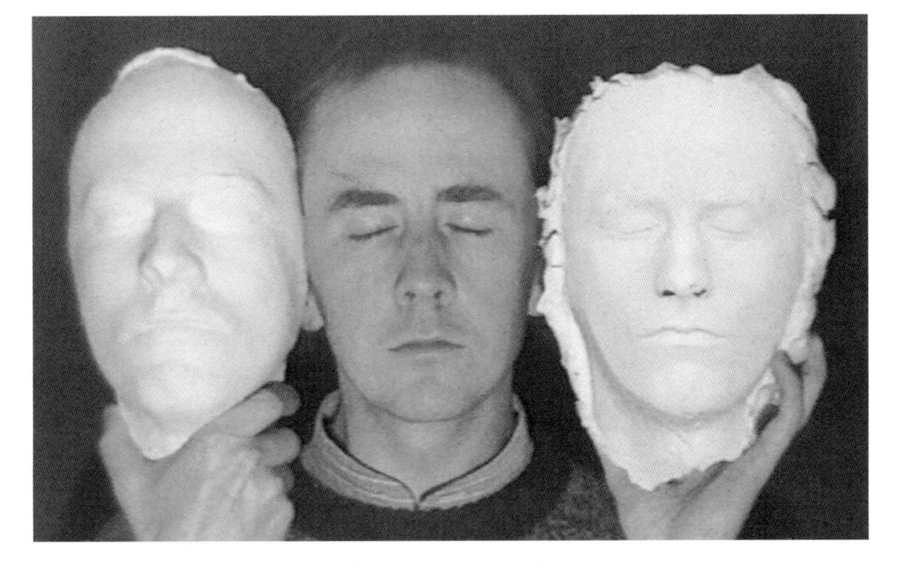

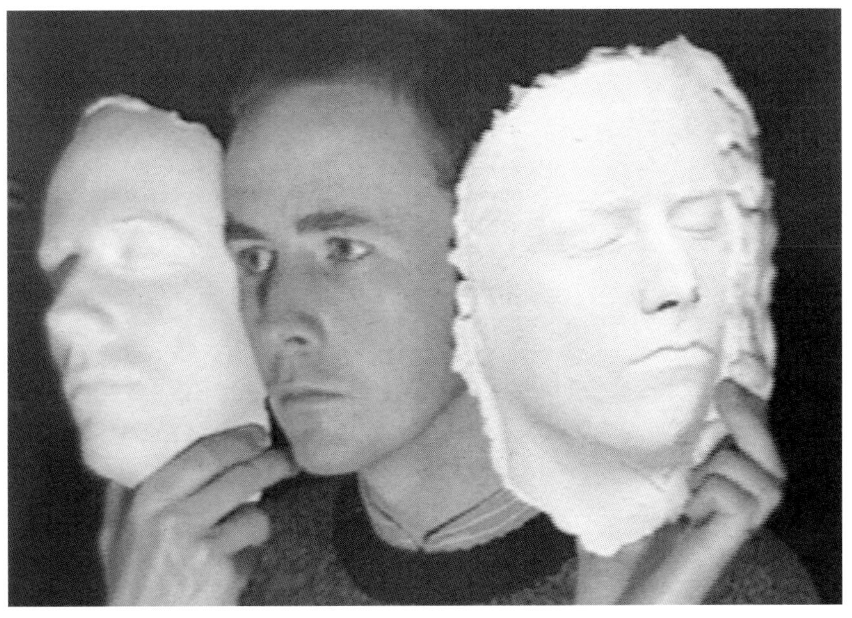

Fig. 2.33 The hollow face illusion. Ben (*centre*) holds a plaster cast of his face (*left of picture*) and a hollow mold of his face (*right of picture*). Both appear to look like faces, though apparently lit from different directions. As the camera turns (*lower panel*) the hollow face apparently turns with it, but still appears as a solid face. Pictures courtesy of Ben Craven and Harold Hill, University of Stirling.

Bruce showed independent influences on the strength of the illusion from lighting direction, stereoscopic information, and object familiarity (see results in Fig. 2.34). The illusion is stronger when the hollow mask is illuminated in a way which results in the illusory face being seen lit from above. The illusion is also stronger when the mask is shown upright than when upside-down, suggesting that one contribution to the illusion is indeed from our acquired expertise with faces. However, illusory percepts are also given, albeit more weakly, with the upside-down face, consistent with an additional factor involving a weaker preference for convex shapes. Further studies confirmed that the

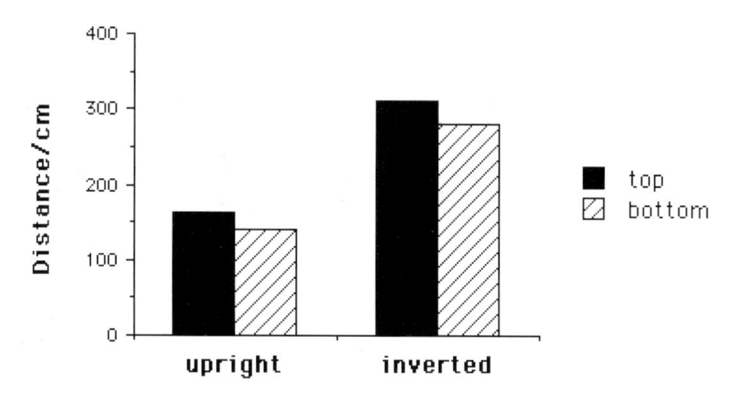

Fig. 2.34 Results of experiments by Hill and Bruce (1993). The figure shows the average distance at which the face perception flipped from illusory to correct (*hollow*) percept. Smaller distances (shorter bars) indicate a stronger illusion. The results show the illusion is stronger when the face is upright compared with inverted, and when the resulting face appears lit from the top rather than the bottom.

illusion is also stronger when observers view the face with one eye rather than two, since stereoscopic information signals the true three-dimensional shape when two eyes are used.

In a follow-up experiment, Hill and Bruce compared the strength of the illusion obtained with a hollow face, upright and inverted, and with a non-face shape ('hollow potato'), and were able to show that the strength of the illusion given by the inverted face was approximately the same as that of an unfamiliar, non-face shape.

Although human vision is quite tolerant of changes in lighting, face perception can be strongly affected by these changes. It is difficult to recognize even very familiar faces when these are lit from below, a fact well-known to children playing with underlighting from torches to make their faces look grotesque, and it is hard to decide that two faces belong to the same individual when they are lit from very different directions. This was demonstrated by Hill and Bruce (1996), who asked observers to decide whether two different images belonged to the same or different faces. When the two faces were lit from different directions, it was much more difficult to match different images of the same person's face, even if they were shown in identical viewpoints (see Fig. 2.35).

Because the human face is three-dimensional, its shape is seen better when the face is shown at an angle than when shown in a frontal view which foreshortens the features. Most portraits show an angled view (often described as 'three-quarter' though the viewpoint chosen is rarely exactly 45° from frontal). The portrait of the psychiatrist R.D. Laing (Fig. 2.36) is relatively unusual in showing a full face view, but notice how the strong side lighting from the window works to enhance the three-dimensional effect of this portrait.

2.7 Movement

One striking difference between the perception of most works of art and our usual perception of the world is that the latter is filled with

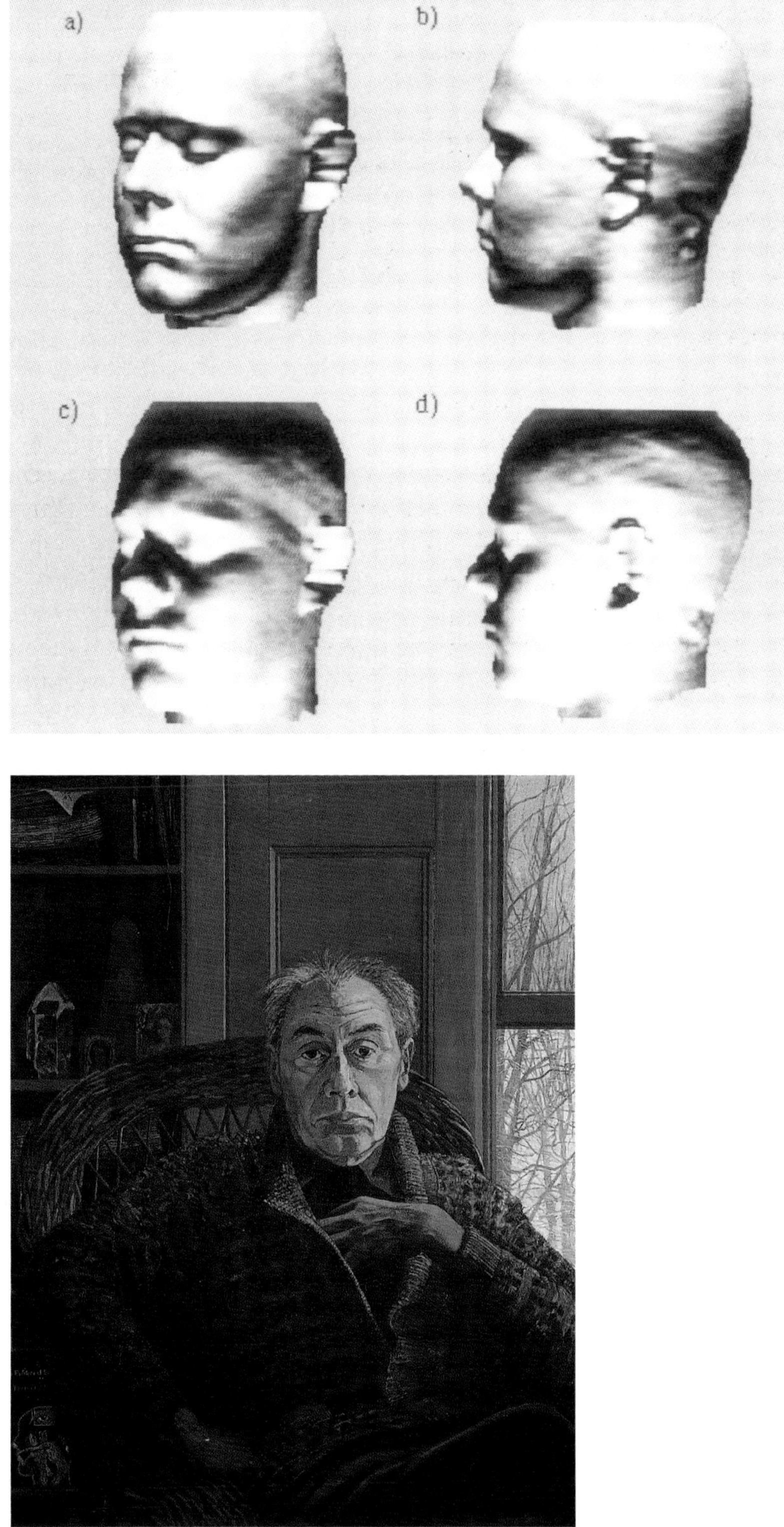

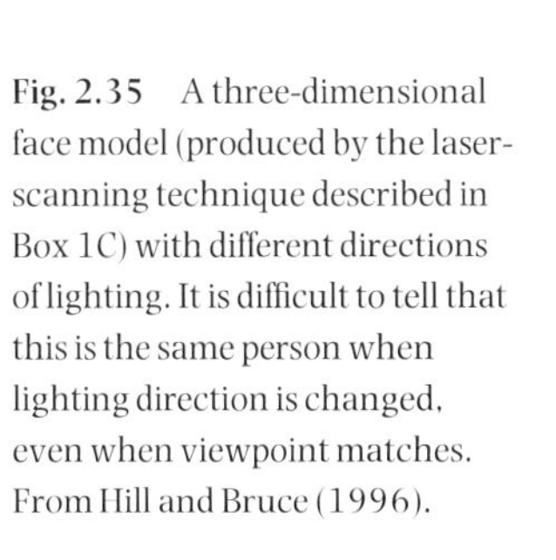

Fig. 2.35 A three-dimensional face model (produced by the laser-scanning technique described in Box 1C) with different directions of lighting. It is difficult to tell that this is the same person when lighting direction is changed, even when viewpoint matches. From Hill and Bruce (1996).

Fig. 2.36 Victoria Crowe, 1984, *Ronald Laing*. Courtesy of the Scottish National Portrait Gallery.

movement while most art conveys static images only. Braddick (1995) notes that few artists have attempted to depict movement in their works, though there are some occasional, dramatic successes as in Fig. 2.37. Portrait artists capture a likeness of their subject in a still pose, usually with little by way of expression depicted. The observation that photographs and paintings convey likeness so well might suggest that movement is unimportant for face perception. However, this conclusion would be premature.

One remarkable way to illustrate the important information conveyed by motion is to present displays where motion is just about the only information present. A technique to do this was introduced by the Swedish psychologist, Gunnar Johannson (1975), using 'point-light' displays of human motion. In these displays, points at the major joints in the body are illuminated; for example lights are placed on the shoulders, elbows, wrists, hips, knees, and ankle joints. It is possible to film this display as the actor moves in the dark, so that all the viewer can see is a moving set of these twelve points. A still frame from such a film looks completely unrecognizable as a human figure—a Christmas tree, perhaps. Yet, once the film is animated, the perception of a moving human figure is instantaneous. Observers can judge whether the person is male or female, what movements they are making, and what weights they are carrying or throwing (even if the weights themselves are invisible).

A similar technique was applied to faces by Bassili (1978), who

Fig. 2.37 Giacomo Balla, 1912, *Dynamism of a dog on a leash*. Courtesy of the Albright-Knox Art Gallery, Buffalo, New York.

scattered a large number of small bright spots over the surface of a human face, and then filmed the face as it moved in the dark, with the camera set to pick up just these brightest spots. Bassili was able to show that, when seen moving, people were highly accurate at recognizing what facial expression was displayed from this moving point-light display. In follow-up experiments, Bruce and Valentine (1988) made point-light displays of a small number of their colleagues and asked other department members to try to identify these individuals. When the tapes were seen moving, there was some modest ability to recognize individuals from these point-light displays, though performance was highly inaccurate overall. People also showed some ability to tell if the displays were male or female. These results can be explained if motion allows the human brain to build a better representation of the facial surface than can be obtained from static individual frames.

On theoretical grounds, we might expect a role for motion in face processing. This is because many of the characteristics of face perception—the importance of coarse-scale information containing patterns of overall lightness, and the unimportance of hue (colour)—are characteristics of a major visual pathway (termed the 'magno-cellular' pathway) in the brain which also codes information about movement. Moreover, research has already revealed that there may be important information in the fine timing of expressions which helps us to decipher such things as whether an expression is genuine or posed (see Chapter 6). You may be able to convince yourself of this by remembering how a 'forced' or 'insincere' smile may be over-quick or too slow in comparison with a genuine smile of pleasure.

Recent evidence has also suggested that there may be important information in the dynamics of face action which helps to identify known individuals. Knight and Johnston (1997) showed that famous faces shown in photographic negative were more easily identified if they were shown in motion rather than as individual still images. This observation alone might arise if a moving clip (containing some 25 frames per second) simply provided more static information about different views or expressions than a single still. However, in follow-up experiments in Stirling, Lander *et al.* (submitted) have confirmed and extended this finding by showing that the benefits of motion are found even if the amount of static information is equated between moving and static conditions.

There are a number of different ways that dynamic information might help face recognition. One possibility is that when we become familiar with people we learn their characteristic patterns of speech and expressive gestures, and that access to these dynamic patterns may be helpful in circumstances where the static information is difficult to recognize. Certainly one component of successful impersonation seems to be the reproduction of the movements and mannerisms of a target personality. Another possibility is that motion helps the visual system to build a three-dimensional description of a face which is particularly useful when recognition of the static information is difficult (for example negative images make it difficult to see three-dimensional shape, so movement might be particularly useful in restoring the 3D information in this case). Current research is exploring these and other possible mechanisms by which movement contributes to our processing of facial images.

This chapter has provided a brief overview of some of the complex processes which occur when the human visual system views faces. In the next chapters, we turn to consider how subtle variations in the information encoded in these ways from faces may be used to make important social categorizations, such as male or female, attractive or unattractive, happy or sad.

3 Physical differences between faces: age, sex, and race

3.1 Overview

We can readily categorize individual faces into different types of social group on the basis of their appearance. We are remarkably good at deciding whether faces are male or female, their race, and approximate age. Moreover, such categorizations have consequences for other attributions we make to people. A person with a babyish face, for example, may be judged as immature or dependent. Psychological research is beginning to reveal the visual cues that the human brain uses to perform such categorizations. Powerful image manipulation techniques now give us ways of demonstrating the effects of different types of structural change; quite minor alterations can have dramatic effects on perceived appearance. In this chapter we will describe a number of experiments where the contributions of the overall shape of the face, and of superficial texture and colouration, can be assessed separately. We will show that both of these broad classes of information make important contributions to our judgements of age, sex, and race.

3.2 Age-related differences between faces

In Chapter 1 we considered some of the physical changes which occur as a face grows and ages. We concentrated particularly on the changes in overall head shape which occur during growth in infancy and childhood, describing how a simple geometrical transformation—cardioidal strain—seems to characterize the change in the shape of the skull during that period.

As illustrated in Chapter 1 (Fig. 1.22) and here (Fig. 3.1) the facial features of the young child are relatively lower down the face than those of the adult. Compared with adults, 'baby-faces' have high foreheads, relatively large eyes, and small chins. The bush-baby (Fig. 3.2) gets its name, and even an adult of the species appears peculiarly immature and 'cute', because of its exaggerated baby-face features.

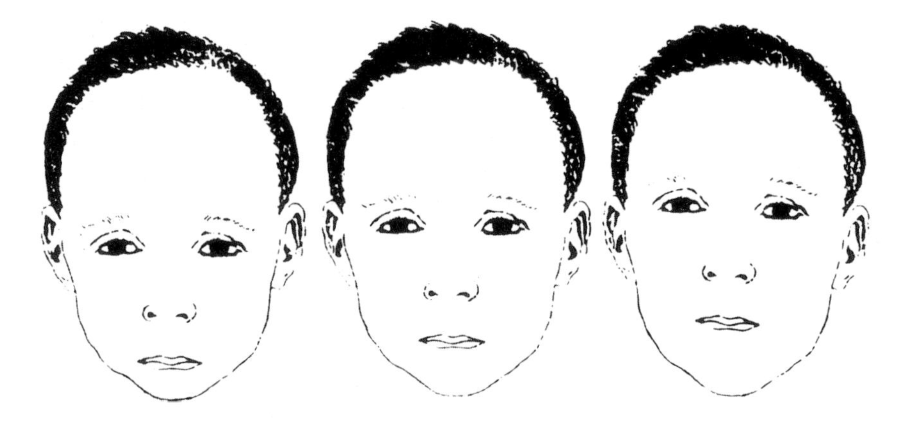

Fig. 3.1 Variations in vertical feature placement create a more babyish face on the left and a more mature one on the right. From Montepare and McArthur (1986).

Fig. 3.2 The baby-faced bush-baby. From Charles-Dominique (1977).

There is evidence that subtle variations in the apparent immaturity of people's faces within a particular age-group can affect the way that other people judge and interact with them.

For example, in a rather alarming and provocative study, McCabe (1984) compared the facial appearances of a sample of children who had suffered physical abuse with the appearance of a control group of children drawn from a similar geographical region who had not suffered harm. The abused children had faces with more adult-like proportions than the non-abused ones—as reflected in relatively short foreheads and long lower faces compared with the control group of children. This finding was replicated with further samples of children and found to apply within each of a number of different age groups spanning toddlers through to teenagers. Thus for 12–15-year-old children as well as for 2–7-year-olds, children with faces which are relatively more adult in proportion were more likely to have suffered from abuse.

There are a number of possible explanations for such a finding, and we should be cautious before drawing simplistic conclusions about the causes of abuse from such a study. None the less, each of the possible explanations merits some discussion since it is possible that any or all of these factors may play some role in some circumstances. One possibility is that children who appear older than their chronological age may tend to be attributed inappropriately grown-up characteristics, and hence may be more likely to be punished for what appear to be immature (but are actually age-appropriate) behaviours. Another possibility, offered by those who favour biologically-based explanations, is that 'baby-faced' attributes act as a trigger that produces care-giving in adults. If key attributes of 'babyness' are lacking, then the infants or children may be more likely to suffer aggression. For example McCabe (1988) draws an analogy between the facial characteristics of the human baby face and the white tail tufts in juvenile chimpanzees. As long as young chimps retain the white tail tufts which signal their youthful status they rarely suffer aggression from adults, but as soon as the tufts turn black—like the adult—then adults will be aggressive towards them. One additional possible explanation for McCabe's results is that the facial characteristics that she measured in her studies co-vary with some other factor or factors which are linked with abuse. For example children with more adult-like faces may be less attractive than those with more baby-like ('cute') faces, and as we will see in Chapter 4, perceived attractiveness can have quite profound effects, at least in experimental situations.

The degree to which a face shows apparently mature or immature proportions has impacts on the impressions of adult as well as children's faces. If adult faces are altered in ways which mimic these 'baby-faced' features, then their apparent age is changed as a result (see Fig. 3.3 for example). Moreover, baby-faced adults are attributed properties which tend to be associated with immaturity. For example McArthur and Apatow (1983/4) manipulated realistic line-drawings of adult faces (see Fig. 3.3) to have more or less babyish characteristics, by varying three separate aspects of appearance; vertical feature placement (baby-faced features are lower down the face); eye size (baby-faces have larger eyes), and feature length (baby-faces have shorter noses). All three separate manipulations affected the impressions made to faces, but their combination had the greatest effect of all. The manipulations affected judgements of physical strength (baby-

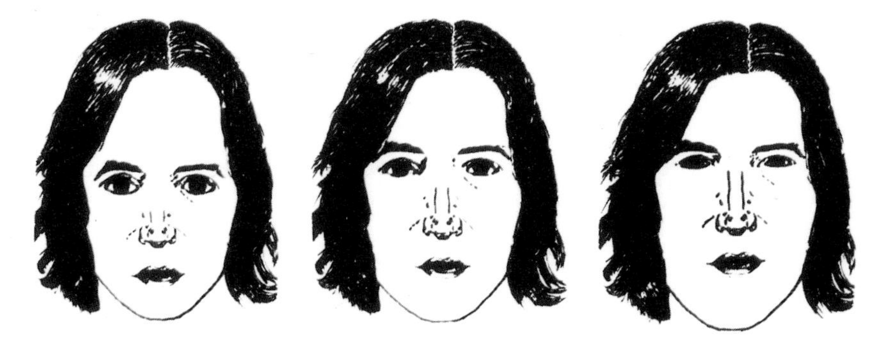

Fig. 3.3 Schematic adult faces. A female face has been altered in a way which has more baby-faced (*left*) or more mature (*right*) features. From McArthur and Apatow (1983/4).

faces were judged as weaker), social submissiveness (baby-faces were judged as submissive), and intellectual naivety (baby-faces were more naive). In a related study using real faces, Berry and McArthur (1985) found that variations in baby-facedness (as measured by variations in chin width and eye size, which were strongly related to rated facial maturity) were positively correlated with the perceived warmth, honesty, kindness, and naivety of the faces.

In a further study (Berry and McArthur 1988) mock jurors were asked to judge the guilt or innocence of a defendant accused of a crime of negligence (forgetting to issue a product warning) or intentional deception (deliberately omitting the warning in order to secure the sale of a product). The other variable in the study was the appearance of the defendant whose photograph was attached to the description of the incident. Photographs of young men were selected in such a way that they were matched on perceived age and attractiveness, but differed in the extent to which the face shape shown had 'baby-faced' or 'mature' proportions. Berry and McArthur predicted that the baby-faced young men would be more likely to be considered guilty of the crime of negligence and the mature-faced young men to the crime of deliberate intent. The results confirmed this prediction.

The studies reviewed above illustrate the impact of a particular variation in facial appearance which usually characterizes changes occurring during the growth of a child. For reasons unrelated to their chronological age, some children, or some adults, have faces which are more baby-like and some which are less baby-like, and these differences can influence other impressions which are made to these faces.

However, these variations in shape may only have a minor influence on the perceived age of adult faces. More important to our judgements of the age of adult faces may be surface features such as hair colour and dispersal, and the skin texture and wrinkles associated with change later in life. Figure 3.4 shows Lady Robert Manners painted when she was a young woman (by Allan Ramsay), and as an elderly lady (by Thomas Lawrence). A comparison of these two pictures highlights the age-related changes which occur during adulthood.

Fig. 3.4 Two portraits of *Lady Robert Manners*, by Allan Ramsay (*left*) and by Thomas Lawrence (*right*). Courtesy of the National Gallery of Scotland.

We are quite good at making relatively subtle differentiations in age based upon cues from surface texture and colour. For example Fig. 3.5 shows two portraits of Sir Walter Scott—the one on the right was painted when Scott was 51 years old and the one on the left is undated. You should have no difficulty in deciding whether the date of the second portrait was likely to be earlier than the first, though you may find it difficult to make explicit the reasons for your decision.

In summary, age-related changes may very crudely be divided into two types—changes in shape, which may occur through growth or weight gain or loss, and changes in the characteristics of the surface texture and colouration of the skin and hair. The changes of both sorts over periods of a decade or two may be rather subtle and difficult to describe. However, there is evidence that human vision is sensitive to both these different kinds of features when judging the age of faces.

To find out which kinds of information are important to human vision, perceptual psychologists often conduct experiments in which different sources of information are systematically concealed or enhanced, or in which different cues are put into conflict with one another. The study of face perception has been greatly enhanced in recent years by the transformations which can be made using computer graphical techniques. We will see a number of examples of this kind of approach in the studies considered in this chapter.

Fig. 3.5 Two portraits of *Sir Walter Scott*, by Sir Henry Raeburn (*right*) and Colvin Smith (*left*). Does he look older or younger on the right? Courtesy of the Scottish National Portrait Gallery.

Burt and Perrett (1995) set out to investigate the relative contribution of gross shape compared with surface texture/colouration in the perception of the age of adult faces. They made use of clever computer graphical techniques to do this. First they collected together a number of different male faces spanning a range of 35 years within seven distinct age groups; 20–24, 25–29, 30–34, 35–39, 40–44, 45–49 and 50–54 years. Participants were reasonably good at judging the age of these original images.

The different faces within an age-band were averaged together to provide a 'composite' face for that age-group, using the morphing techniques described in detail in Chapter 5 (Box 5B). Briefly, by careful alignment of a large set of key landmarks identified on each individual face, faces can be averaged together without blurring due to the mis-alignment of features from different faces (see Fig. 3.6). The apparent ages of these age-group composites were related to the ages of their constituent faces in an orderly way, though each composite was judged as younger in appearance overall than the average age of the faces it contained. This may be because the composite technique softens the impact of the different wrinkles and skin texture changes from each of the individual contributing faces.

By examining how each age-group composite face differed from neighbouring composites, Burt and Perrett were able to describe how the faces of one age group deviated from others in terms of their shape,

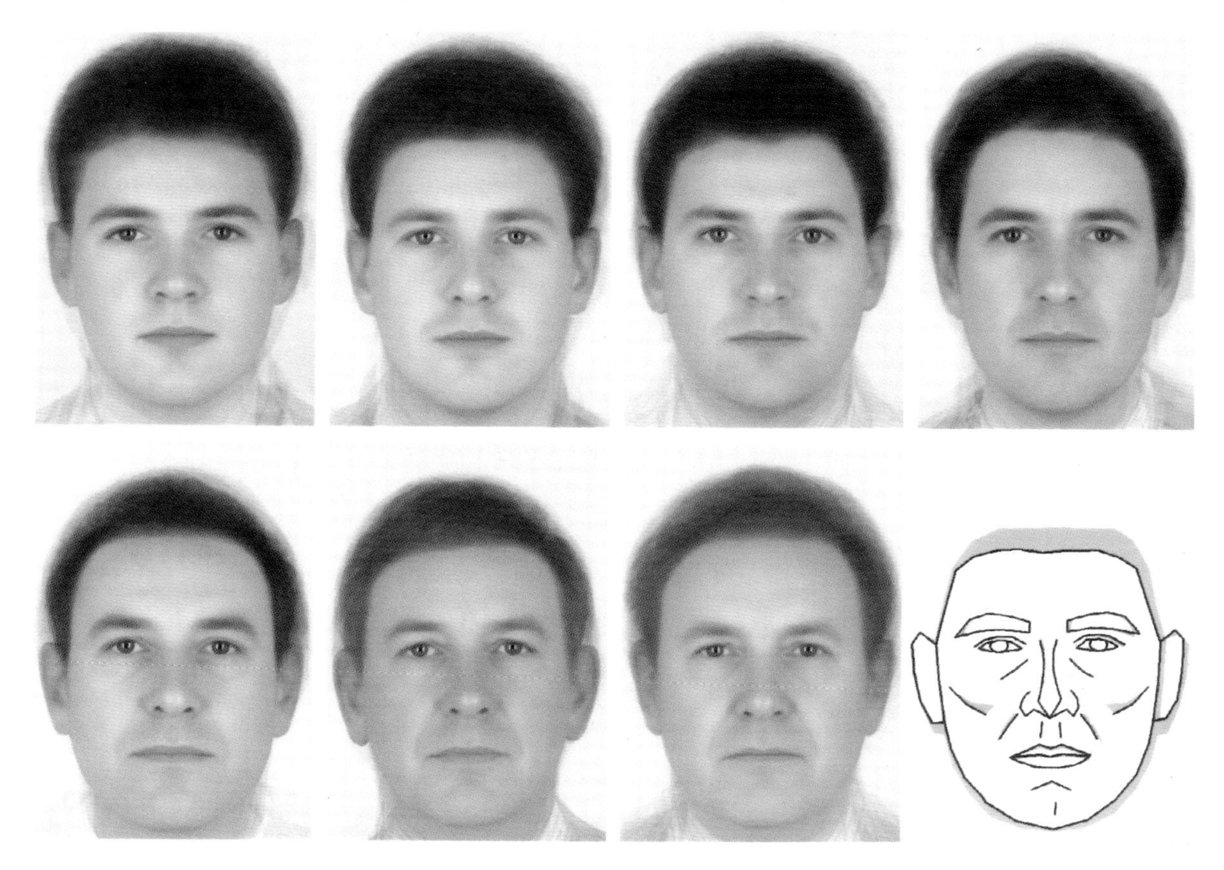

Fig. 3.6 Face blends of seven different age-groups from 20–24 (*top left*) to 50–54 (*third from left in bottom row*). Differences between the 25–29 and 50–54 age groups are shown in the shading in the line-drawn figure at the far right. The positions of the younger age group are shown with the dark lines. The edge of the shaded area shows the movement of these points with age. The older age-group has a higher forehead (receding hair), fatter face, thinner lips, etc. Reproduced from Burt and Perrett (1995).

and in terms of their texture and colour. This made it possible to exaggerate the differences between one age group and the next to produce a 'caricature' of age-related changes (see Box 5B, Chapter 5 for a fuller description of caricature). In Fig. 3.7 the colour and texture differences between the average 50–54-year-old face (right hand panel) and the average male face across all age groups have been exaggerated to give the face shown in the left, where the age-related differences between the older age-group and the mean have been enhanced. Using the same techniques, Burt and Perrett were also able to take any individual face and apply a transformation to move its shape, or its colour characteristics, or both, towards that characteristic of an older or younger age-group. In Fig. 3.8 this technique has been used to age the appearance of an individual face in a most convincing way.

There are a number of practical situations in which it may be important to try to simulate such age-related changes in appearance, and these computer techniques can be used to good effect. For example when children or adults have been missing for several years it is important that images issued of them are somehow brought up to date by incorporating age-related changes. The cardioidal strain trans-

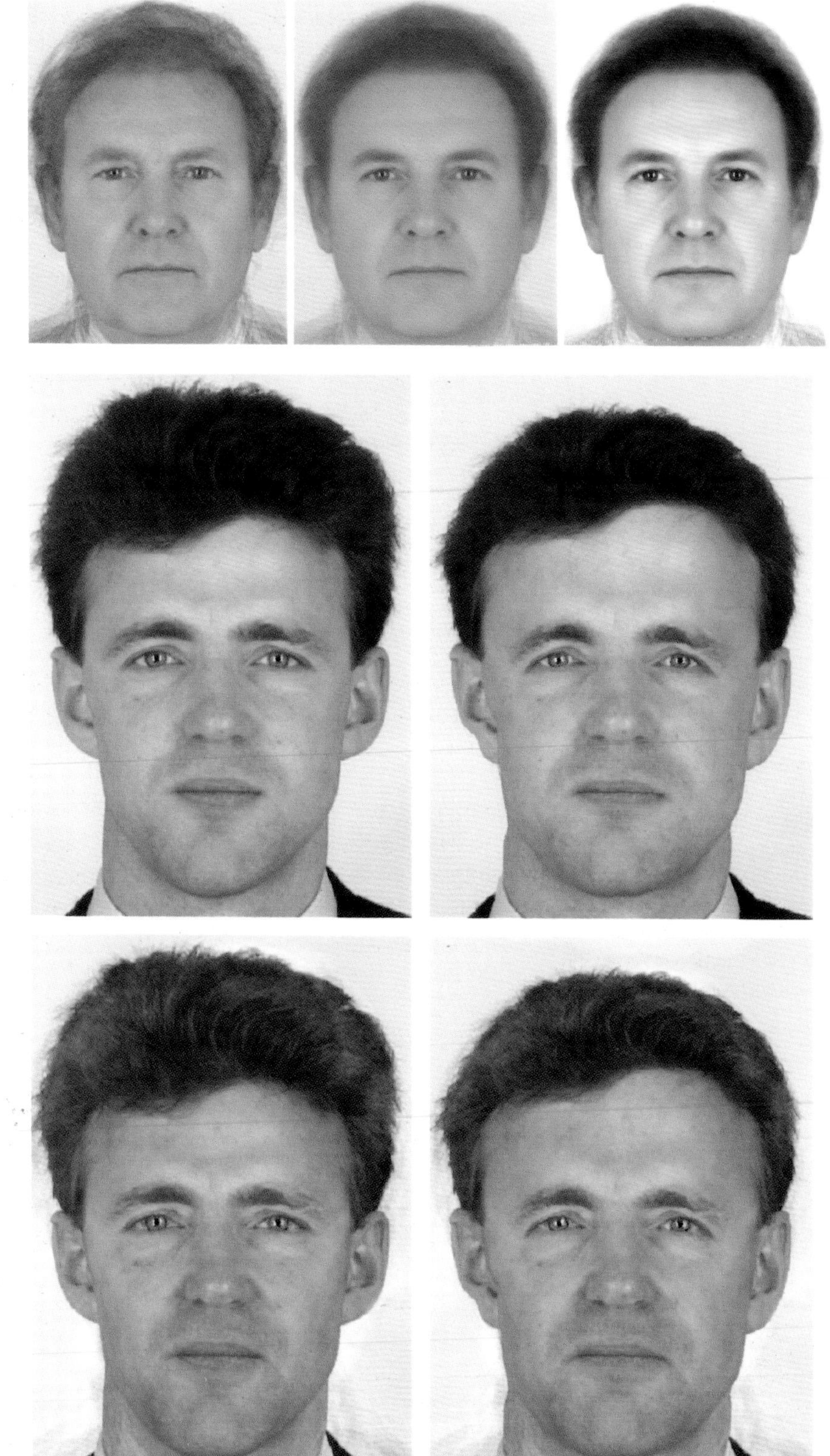

Fig. 3.7 Enhancement of the colour cues to age. The central image shows the blend of the 50–54-year-old age group. The image to the left has exaggerated the differences between this and the average of all age groups. To the right is an image which enhances the colours in the blend by exaggerating differences between the blend and a uniform grey image. Reproduced from Burt and Perrett (1995) with permission.

Fig. 3.8 The original individual face shown in the top left has been aged by transforming its shape (*top right*), its colour/texture (*bottom left*) and both (*bottom right*) in the direction characterizing older faces. Reproduced from Burt and Perrett (1995) with permission.

formation can be applied to photographs of young children to displace their features in the ways which would result from ageing, but the methods described by Burt and Perrett may be more appropriate to ageing photographs of adults.

One interesting possibility, that we are not aware of having been attempted, would be to take the younger picture of a person whose older identity is disputed (see for example Box 3A), and transform the younger picture in the manner which would be expected of average age-related changes to see how this image compares with the actual appearance of the person whose appearance is in dispute.

Fig. 3.9 shows the results of a light-hearted attempt to illustrate this process, and how it might fail, using portraits of Bonnie Prince Charlie. Using the techniques described by Burt and Perrett, David Perrett, Duncan Rowland and their colleagues have taken a youthful portrait and altered its shape and texture/colour characteristics in a way simulating the effects of ageing. Though the result is a convincing picture of an old man, it scarcely resembles the portrait of Charles himself painted at the age of 65 years. This illustrates the importance of making the correct assumptions about other individual factors such as weight gain which would need to be incorporated.

The research we have described so far suggests what sources of information may contribute to our abilities to judge apparent age of faces, and how other judgements made to faces may be influenced in rather subtle ways by the age-related differences in facial appearance. The perception of age is, however, in one respect rather different from the classification of sex or race. When age-related changes to faces are considered we can ask the important and interesting additional question of how well individual identity is preserved across age transformations.

Fig. 3.9 Bonnie Prince Charlie aged approximately 11 years old, has been age-enhanced to produce the image in the centre—a plausible image of an old man. At the far right is a portrait of the Prince aged 65 years old. The central image does not resemble this very closely, largely because of differences in weight and the immature nose shape in the starter image. Resemblance of eyes and mouth is quite reasonable. Original portraits courtesy of the Scottish National Portrait Gallery. Age transformation by Michael Burt, Rachel Edwards, Duncan Rowland and David Perrett, University of St Andrews.

3.3 Recognizing age-transformed faces

We usually have little difficulty in recognizing our friends and families despite major changes in their appearances over the years. The art historian Sir Ernst Gombrich reflects:

> We look at the snapshots of ourselves and of our friends taken a few years ago and we recognize with a shock that we have all changed much more than we tended to notice in the day-to-day business of living. The better we know a person, and the more often we see the face, the less do we notice this transformation except, perhaps, after an illness or another crisis. The feeling of constancy completely predominates over the changing appearance. (Gombrich 1982, pp. 107–8)

When an absence of a period of years breaks the continuity of regular exposure to this slow variation, then recognition can be exceedingly difficult.

Maggie Bruck, Patrick Cavanagh, and Steve Ceci (1991) (see Fig. 3.10) asked people to attempt to match high-school graduation photographs with pictures of the same people taken 25 years later, when they were in their early forties. Participants were shown the yearbook pictures of each person and had to choose which of ten pictures of the forty-something adults they thought matched each picture. Volunteers unfamiliar with any of the people photographed performed with an accuracy of 33 per cent overall, which was higher than the rate of 10 per cent which might be expected by guessing alone, but far from perfect. Participants who had been class-mates at high-school were significantly more accurate (49 per cent correct matching) but still also far from perfect at this task.

Classmates were equally accurate whether the pictures to be matched showed the same or a different viewpoint, while those unfamiliar with any of the faces were better at the matches when the depicted viewpoints coincided. The above-chance performance by the strangers is presumably based in part on the perception of features which are invariant across this range of ages (for example the length and shape of the nose and eyes changes little over early adulthood); which are easier to compare when viewpoints match. In contrast, the performance of the familiar judges may be based more on a viewpoint-independent knowledge built up over multiple exposures to the faces at high school. In addition, Bruck *et al.* (1991) suggest that familiar judges may remember characteristic expressions or poses which may be additional sources of age-invariant information on which to make these matches. Bruck *et al.* excluded from their sample a small number of responses to faces that volunteers reported having seen over the past seventeen years, and so we can assume that the effects of familiarity which were observed did genuinely result from memories of the high-school faces which had been retained over at least this interval.

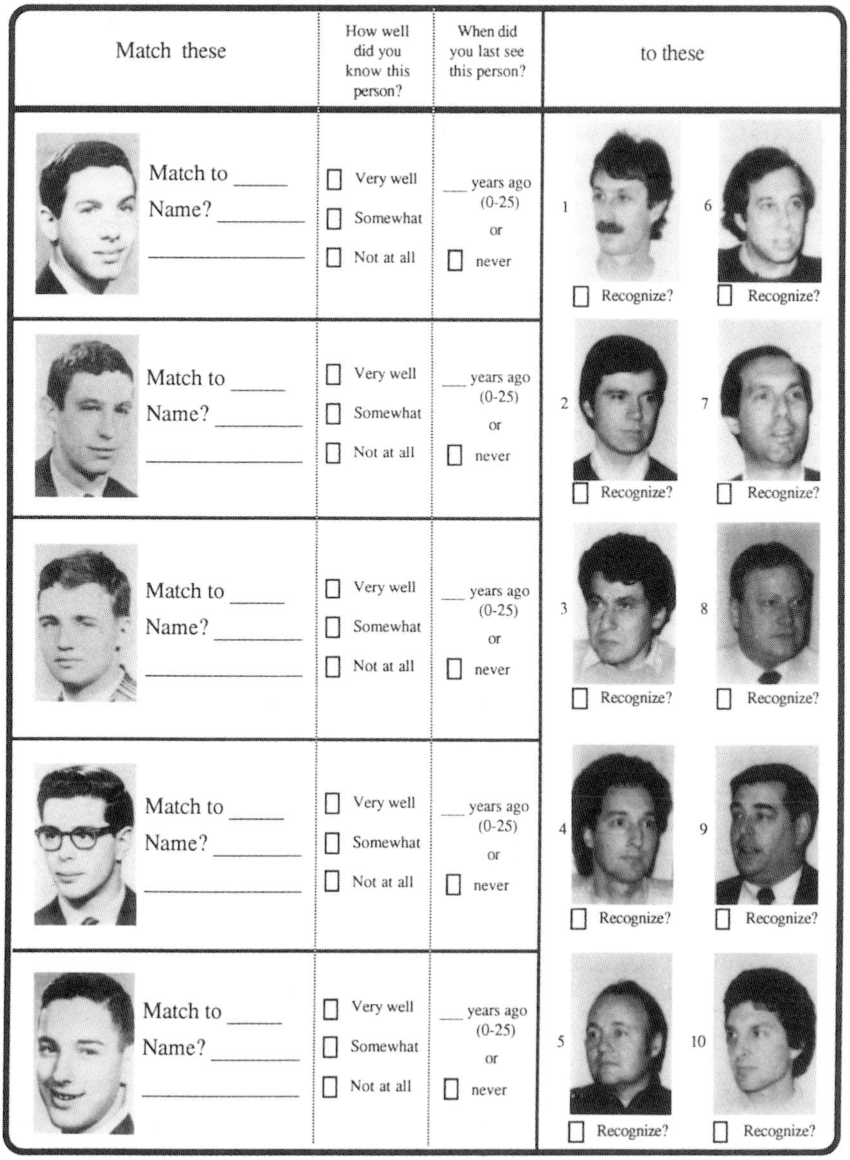

Match these	How well did you know this person?	When did you last see this person?	to these
Match to ____ Name? ________ ________	☐ Very well ☐ Somewhat ☐ Not at all	___ years ago (0-25) or ☐ never	1 ☐ Recognize? 6 ☐ Recognize?
Match to ____ Name? ________ ________	☐ Very well ☐ Somewhat ☐ Not at all	___ years ago (0-25) or ☐ never	2 ☐ Recognize? 7 ☐ Recognize?
Match to ____ Name? ________ ________	☐ Very well ☐ Somewhat ☐ Not at all	___ years ago (0-25) or ☐ never	3 ☐ Recognize? 8 ☐ Recognize?
Match to ____ Name? ________ ________	☐ Very well ☐ Somewhat ☐ Not at all	___ years ago (0-25) or ☐ never	4 ☐ Recognize? 9 ☐ Recognize?
Match to ____ Name? ________ ________	☐ Very well ☐ Somewhat ☐ Not at all	___ years ago (0-25) or ☐ never	5 ☐ Recognize? 10 ☐ Recognize?

Fig. 3.10 Examples of the questionnaire used by Bruck *et al.* (1991). The participants' task was to decide which of the faces numbered 1–10 on the right was the correct match to each of the five faces on the left. After trying this task, you can check the correct matches which are, from top to bottom, faces numbered 7, 1, 5, 9, and 4. Reproduced by permission. Copyright © The Psychonomic Society.

Although the Bruck *et al.* (1991) study was not aimed directly at problems faced in eye-witness testimony, the issue of very long-term memory for faces whose appearance has changed through ageing may be important when identity is disputed many years after an alleged crime has been committed. The most publicized examples have arisen in a small number of notorious war crimes cases.

In the case of Klaus Barbie, victims of 'The butcher of Lyon' spontaneously recognized him when a French television programme broadcast a photograph of a man living in Bolivia under a different name. In this case it seems that viewers from the Lyon region were able to recognize this person without prompting despite the changes in his appearance over more than thirty years.

In the case of 'Ivan the Terrible' (Wagenaar 1988), John Demjanjuk was found guilty by the State of Israel of being responsible for torturing and murdering prisoners at the Treblinka concentration camp some thirty-five years earlier. However, this case highlights a number of worrying aspects of the eyewitness testimony which led to his conviction (as detailed in Box 3A) and points to the difficulties inherent in identification across such large time-scales.

Box 3A: The case of Ivan the Terrible

In 1976, John Demjanjuk, a factory-worker in Ohio, was accused of having been the notorious 'Ivan the Terrible'—responsible for the deaths of hundreds of thousands of Jews in the concentration camp at Treblinka during World War II. Demjanjuk himself had emigrated to the USA in 1951, and denied any connection at all with Treblinka. The case rested entirely on the testimony of a number of eyewitnesses, and raises a number of problems. Willem Wagenaar, whose 1988 book describes the case in detail, is a psychologist who acted as an expert witness to the defence in the case. Survivors of Treblinka were asked if any of an array of photographs was the person they remembered from Treblinka. Among the array, shown here in Fig. 3A.1, was Demjanjuk's 1951 immigration photograph, which at least five survivors (who later testified in court) selected. Some of these same witnesses, plus some new ones, were asked to identify Ivan from a different array, allegedly showing a picture of John (Ivan) in his early 20s when training in Trawniki (Fig. 3A.2). Doubts about this evidence were raised by Wagenaar (1988) since at least eight other survivors (we do not know exactly how many in total) did not pick out Demjanjuk's photograph from either of these arrays.

There were numerous problems with the construction and administration of the arrays used to probe survivors' memories. No adequate attempt was made to match the characteristics of the other faces shown alongside Demjanjuk's against either his visual characteristics, or against his described characteristics. One witness who picked him out from the 1951 photograph said, 'The photo is not quite clear and also the change in age must make a difference. The shape of the face, especially the rounded forehead, strengthens my feeling that it is Ivan. The characteristic short neck on broad shoulders—that's exactly what Ivan looked like.' None of the other faces in the array had round face, short neck, and broad shoulders. Wagenaar gave a set of 25 naive participants the 1951 array with the instructions 'We are looking for a man with a full round face, a short wide neck, a bald pate starting'. All 25 picked out the picture of Ivan as

Fig. 3A.1

the wanted person. This shows that a genuine witness with only a vague recollection of the visual appearance of Ivan the Terrible, but strong motivation to see him apprehended, could have picked out the picture on the basis of its close resemblance to this recollection, and nothing more. This is particularly so given the lapse of time between the alleged war crimes committed by John Demjanjuk and 1951 when his immigration picture was taken. The Trawniki array (Fig. 3A.2), showing the youthful Ivan, was possibly even more unfair, since only two of the eight faces (including Ivan's) faced forwards, and Ivan's face was the only blonde in the whole array.

We know from a number of notorious cases of mistaken identity that it is quite possible for a number of independent witnesses to repeat the same mistaken identity if, for some reason, a person is apprehended with some visual resemblance to the actual culprit. We

Box 3A: Continued

do not know whether or not John Demjanjuk has mistakenly been identified as Ivan. What is of concern is that no adequate attempts seem to have been made to minimize the possibility of mistaken identification in this case, where lapses of time (between the original crime and the attempts to identify Ivan, and between the original crimes and the picture taken of John Demjanjuk in 1951) are already likely to confound the usual processes of person identification.

Fig. 3A.2

3.4 What's the difference between men and women?

We are remarkably accurate at deciding whether faces are male or female. If hairstyle is concealed, men are shown clean-shaven, and cosmetic cues are avoided, people are still about 95 per cent accurate at deciding whether faces are male or female (see Fig. 3.11).

A series of studies has investigated the possible basis of this performance, both by carefully measuring and comparing male and female faces, and by examining how our perceptual judgements are affected when some sources of information are missing.

By measuring large numbers of male and female faces we can

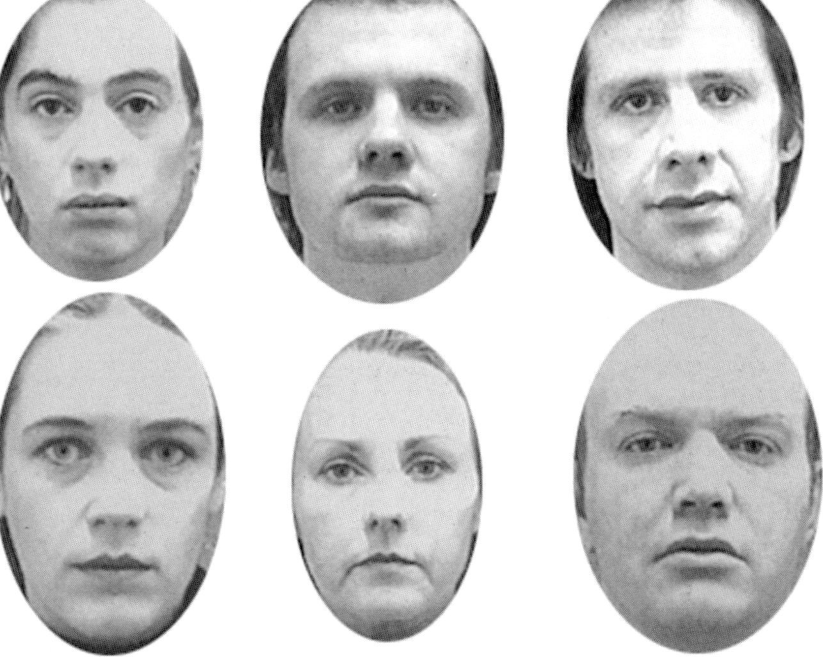

Fig. 3.11 You should find it quite easy to tell whether each of these faces is male or female, but how do you reach these decisions? Photos courtesy of Trish Le Gal, University of Stirling.

identify what information might be used by the human visual system to classify the sex of faces. Identifying what information might be used is only a first step in finding out what is actually useful or necessary for the task. One measurement that differs quite a lot between male and female faces is overall head size, since men are generally taller and broader than women. However, the size of the face is not necessarily a useful cue to determine its sex, and in experiments where faces are all standardized to the same overall size, people are no slower or less accurate at deciding their sex, suggesting that whilst head size is a cue we might conceivably use, in practice it is actually a relatively unimportant cue for human vision. This makes sense in everyday life, where classifying a face (or person) as male or female is something that it is useful to be able to do without other faces there for comparison, and if absolute size were to be used as a cue the visual system would have to compensate for perceived distance in order to use this cue.

To discover what other physical variables might form the basis of human sex judgements, Burton *et al.* (1993) collected pictures of 91 young adult males and 88 young adult females, and made a large number of different measurements of these faces. The sizes of the different features were measured in full-face images (for example the length of the nose or the width of the eyes), as were separations between different features. A number of different ratios were derived from these full-face measurements. In addition, profile photographs of the same people were used to recover some measurements of the 3D shape of the heads, such as the amount by which the nose protruded. On the basis of the measurements made, Burton *et al.* (1993) found that it was possible to classify 94 per cent of these images correctly as male or female (a similar success rate to that achieved by human observers) using a total of just sixteen different measurements. The measurements included some simple, local features such as the thickness of the eyebrows (thicker in men) and the distance between the eyes and brows (greater in women—particularly if eyebrows have been plucked), as well as more complex, three-dimensional measures such as the protruberance of the nose.

A picture of the overall differences in three-dimensional shape which are found between male and female faces can be obtained by comparing 3D surface images of male and female heads obtained using the laser-scanning technique described in Chapter 1 (Box 1C). Alf Linney, Rick Fright and colleagues at UCL were able to average together different surfaces obtained from a number of different male and female faces in order to produce the 'average' male and 'average' female

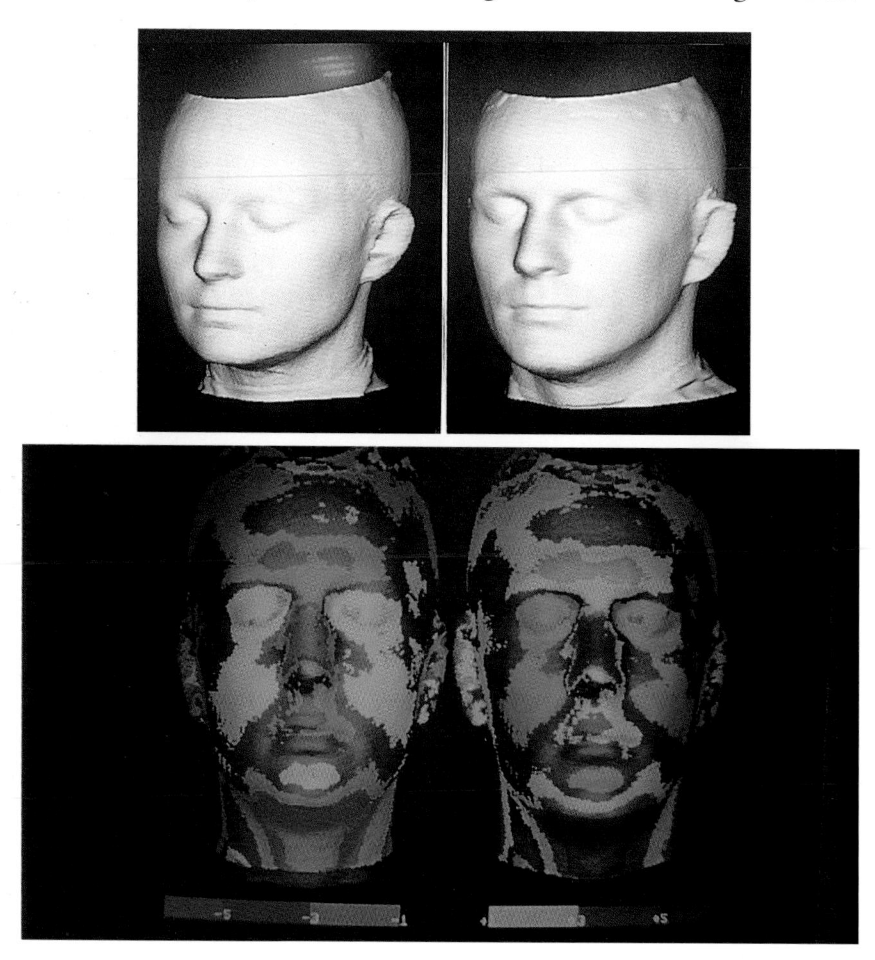

Fig. 3.12 *Upper panel*: average female (*left*) and male (*right*) surface images obtained from laser-scanning. *Lower panel*: a comparison between the average male and female. Lower left shows female-male and lower right shows male-female, with positive and negative differences plotted using the colour scale shown beneath (−5 violet through to +5 red).

surface shown in Fig. 3.12 (Bruce *et al.* 1993). These surfaces were then compared and their differences noted. In Fig. 3.12 these differences are shown by using the colours of the spectrum, with red showing extreme positive differences through to violet showing extreme negative ones. The image on the left shows the differences obtained by subtracting the average male surface image from the average female surface image. The one on the right shows the opposite, with the female surface subtracted from the male one. Thus the left hand image has violet, blue, or green areas (negative differences) where the right hand one has red, orange, or yellow ones (positive differences). Examining these images we see that the male face has a more protruberant nose and brow and broader chin and jaw-line than the female face (red on the right, violet on the left). The female, in contrast, has somewhat more protruberant cheeks and has a fleshier pad on the chin (yellow on the left, green on the right).

Enlow (1982) relates the differences in shape between men and women in the nasal area to their differing oxygen requirements. Because men are larger than women they require a greater air-flow, and hence a differently shaped nasal passage. Examining the differences between the average male and female shape in Fig. 3.12 you should also be able to notice that the average female shape bears a stronger resemblance to the baby-faced shape than does the average male shape. Female faces are more like baby-faces than are male faces—they have smaller chins and noses and their eyes appear larger (a consequence of the lesser brow protruberance).

Figures 3.13 and 3.14 show portraits of two famous couples, where we see good examples of the differences in shape of the male and female

Fig. 3.13 Oskar Kokoschka, *14th Duke and Duchess of Hamilton*. Courtesy of the Scottish National Portrait Gallery. © DACS, 1998.

Fig. 3.14 Avigdor Arikha. *Ludovic Kennedy and Moira Shearer.* Courtesy of the Scottish National Portrait Gallery.

faces. Notice how in each of these pairs of individuals the male has a broader chin and jaw, and heavier brow and nose. The female in each has a more concave nose region and narrower chin.

The discussion so far suggests that, like age, there may be two rather different kinds of features that are useful in the perception of the sex of faces—superficial or 'local' features such as the thickness of eyebrows or the texture of the skin in the beard region (where even clean-shaven men often have visible hair follicles or beard 'shadow'), and overall shape features such as the three-dimensional shape of the nose and brow region. Perceptual experiments support the idea that both these sources of information are important. Bruce *et al.* (1993) compared how accurately people were able to make sex judgements when shown just the three-dimensional surface information derived from laser scanning (see Box 1C and Fig. 3.15) with the accuracy obtained from the same set of faces shown in the manner in which they were scanned—with hair concealed by a bathing cap and with eyes closed. The surface images contain all the same three-dimensional structural information as the photographs of the people being scanned, but lack the local surface information for example about brows and skin texture. Accuracy with the surface images was very much better than chance (84 per cent, for example, when three-quarter pictures were shown), but it fell considerably below that found with the photographs (94 per cent in three-quarter view pictures). This illustrates that the additional texture information (eyebrows, visible hair, stubble, and skin texture)

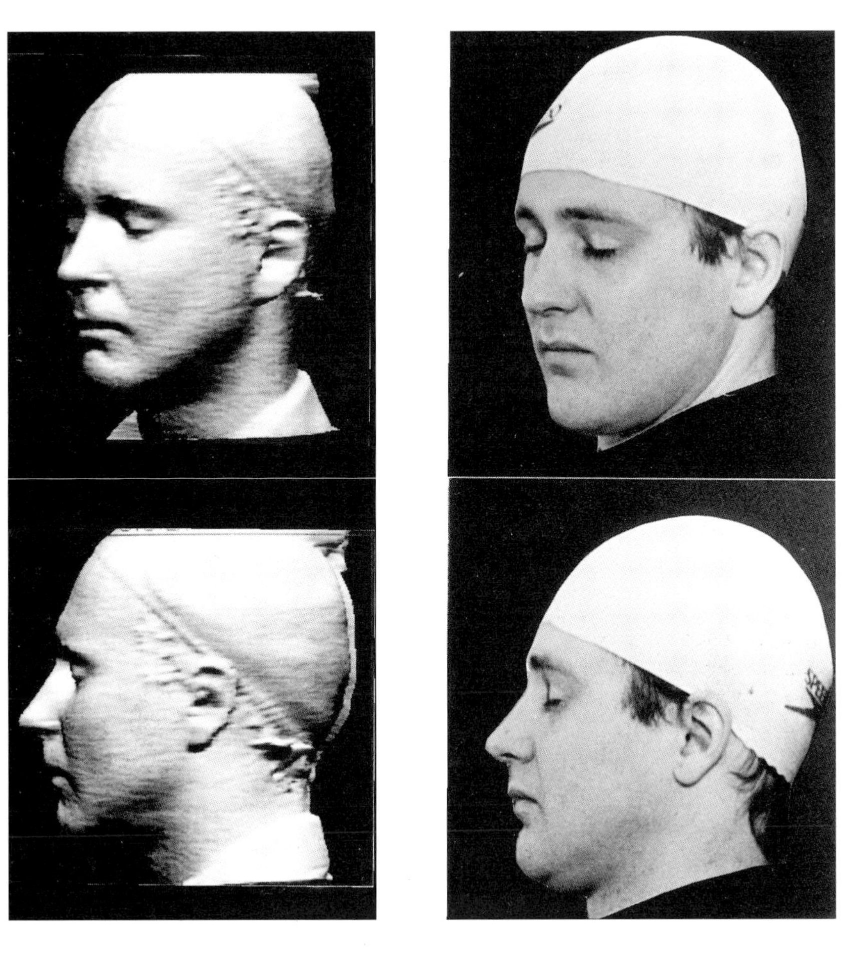

Fig. 3.15 Three-dimensional surfaces obtained from laser scans (*left*) and photographs of the same person as scanned (*right*). The images to the right contain information about surface texture and pigmentation not available in the laser scans. From Bruce *et al*. (1993).

in the photographs adds significant information to that contained within the three-dimensional surface alone. Although the surface images were judged quite accurately in three-quarter view, where their 3D shape could be seen, the accuracy of judging these images dropped a great deal when they were shown in full-face (75 per cent correct), while accuracy with photographs remained at 95 per cent correct in full face images. The local texture cues of eyebrows, stubble etc. are equally visible in full-face and three-quarter photographic images, while cues to three-dimensional shape such as nose and chin protruberance are much less visible in full-face images and must be derived entirely by an analysis of shape-from-shading (see Chapter 2).

The subtle differences in shape and surface texture between male and female faces are often exaggerated by the use of cosmetics to enhance femininity. Fashions in cosmetic application change over the years, but women have often plucked their eyebrows, thus maximizing differences in the hairiness of the brow between men and women. Eye-shadow is used to enhance the apparent size of eyes, exaggerating features related to baby-facedness particularly, and shading may be used on the cheeks to enhance the apparent prominence of the cheek bones, again in line with natural differences in the male and female norm. However, there are enormous variations in the actual shapes of the faces of men and women, and considerable tolerance on the part of the human observer. Some very feminine women have bone structures quite unlike the prototypes we have been discussing.

The portrait of Mary Somerville (Fig. 3.16) is unmistakably feminine in appearance, despite the rather masculine shape and set of the nose and brow, which is highlighted in the profile plaque in Fig. 3.17. Yet if you conceal the hair in 3.17 the profile alone could easily be judged as male. However, such misclassifications are rare when all relevant information is present. The human visual system rapidly weighs up all the evidence from the particular constellation of surface and shape features and comes up with a decision about the person's sex which is remarkably accurate. Usually, of course, there will be additional information from hairstyle, body shape, clothing, and posture which will support the impression gained from the face, and these can help even the most masculine-looking of men to get away with passable female impersonations (e.g. Jack Lemmon in *Some like it hot*; Dustin Hoffman in *Tootsie*) by a combination of cosmetics, clothing, and mannerism.

Indeed, the Scottish artist Sir David Wilkie did not hesitate to use his own face as a model when necessary in his narrative paintings, and in the *Blind fiddler* (Fig. 3.18) was content to do this even when the

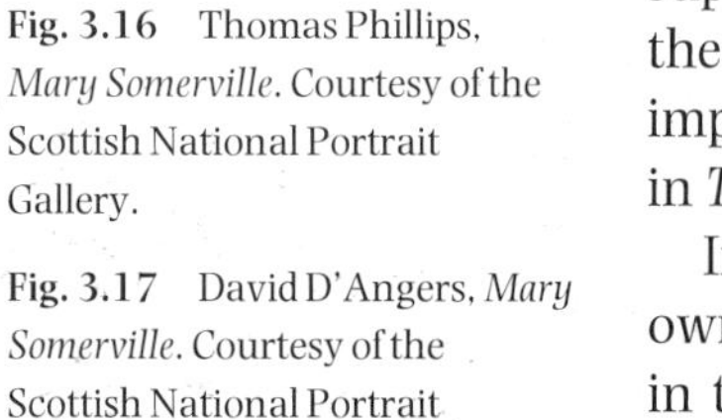

Fig. 3.16 Thomas Phillips, *Mary Somerville*. Courtesy of the Scottish National Portrait Gallery.

Fig. 3.17 David D'Angers, *Mary Somerville*. Courtesy of the Scottish National Portrait Gallery.

Fig. 3.18 Sir David Wilkie's *Blind Fiddler* (courtesy of the Tate Gallery) and detail from the original in which he appears as a servant girl. Wilkie's self-portrait (c. 1804) for comparison. Courtesy of the Scottish National Portrait Gallery.

subject is a female servant (second face from the right; compare with Wilkie's self-portrait below).

One of the best-known cases of female impersonation is the legendary escape of Bonnie Prince Charlie after his defeat by the Hanoverian army at Culloden in 1746. The Prince took refuge in the islands of the Outer Hebrides, but was in increasing danger of being discovered by the large number of troops dispatched to search for him. Flora MacDonald was persuaded to smuggle the Prince across to Skye, disguised as her maid, one Betty Burke 'an excellent spinner of flax and a faithful servant'. Despite his renowned pretty looks, by all accounts, the Prince was unconvincing as a woman. McLynn (1991) suggests he succeeded only because Betty Burke was claimed to be Irish:

> As a female impersonator, the prince left a lot to be desired. The way he moved, his long-striding walk, his very tallness of stature and the general gaucherie of the way he arranged his skirts would have given him away had not the universal prejudice against the Irish worked in his favour. The prevalent notion that Irishwomen were tatterdemalion, hoydenish hobbledehoys predisposed the Scots to be satisfied with the outrageous story that this was an ill-bred female peasant from the bogs. (McLynn 1991, p. 285)

In Fig. 3.19 David Perrett and Duncan Rowland have 'feminized' the face of Bonnie Prince Charlie, to aid his disguise as Betty Burke!

Fig. 3.19 Feminized Bonnie Prince Charlie by Rachel Edwards, David Perrett and Duncan Rowland, University of St Andrews.

3.5 Racial differences

Minor differences in genetics between people from different racial groups result in faces with different characteristic appearances. Skin and hair pigmentation provide the most obvious differences. Celtic Scots have light skin and hair, Japanese have somewhat darker skin and straight black hair, Africans dark skin and black curly hair. The faces of different races differ in average shape as well as colour. Asian faces are flatter than European ones, and African faces have broader noses than Europeans (see Chapter 1 for Enlow's description of the different basic types of face). The portrait of the Arab princess with her black maid (Fig. 3.20) illustrates some of these differences.

Most people report finding it difficult to recognize the faces of people of other races, and probably tend to assume that this is because other race faces are more similar to each other than their own race. However, most psychological experiments have shown that each different race experiences a similar difficulty with faces from other races; for example, Japanese people find European faces hard to recognize just as much as Europeans find Japanese difficult. Therefore it cannot be the case that one race is intrinsically more difficult to recognize than another.

This was shown clearly by Brigham (1986) who combined the data (in a technique known as 'meta-analysis') from fourteen different laboratory studies of cross-race face recognition in which there was a total of 693 black and 752 white participants, each of whom had been asked to identify both black and white faces. They found that the decrement in recognition of other-race faces compared with own-race was highly similar for both the black and the white participants.

To probe further the basis of these differences, we need to analyse the studies more carefully. When participants are asked to recognize faces

Fig. 3.20 Walter Frier, *Arab princess with black maid* after an unknown artist. Courtesy of the Scottish National Portrait Gallery.

in an experiment, or from mug-shots or line-ups, there are two different kinds of errors that can be made. First, a face that was seen in the study phase (in an experiment) or at the scene of the crime (in a real incident) may be missed when memory is tested, or secondly, a face that was not studied/present may wrongly be picked out at test. This latter kind of error is what is termed a 'false positive' and it is of particular concern to the legal profession who need to ensure that innocent people are not wrongly apprehended. (As we will see in Chapter 5, much of the early concern about eyewitness testimony arose because of public concern about these kinds of false identifications.) If an experimental participant, or a witness, sets a very strict criterion for responding that they recognize the face, then they will rarely make false positives but may miss a lot of faces that had earlier been seen. If they set a very lax criterion, on the other hand, they will correctly recognize most faces actually encountered earlier but may make relatively large numbers of false positives.

By examining the relative rates of correct recognitions and false positives it is possible to see if memory is better or worse in some circumstances because the memory trace itself is stronger or weaker (more *discriminable*) or whether it is a person's response *criterion* for saying that they recognize some of the faces that has been shifted to be more lax or more strict. Barkowitz and Brigham (1982) analysed cross-race identification in this way and showed that participants appear to adopt a laxer criterion for faces of another race than for their own, leading to a higher false positive rate. This makes it particularly important that police and the legal system are aware of the difficulties that people have in recognizing faces of other races.

Why then are people so bad at recognizing people of another race? There is little evidence that this arises from simple prejudice or avoidance of people from another race. There is rather more evidence that repeated exposure to faces of a single race has consequences for the perception of other race faces. This theory suggests that people of one race have learned to pay attention to rather subtle characteristics which distinguish different individuals within their own racial group, but have not learned as well the dimensions which are more relevant to other racial groups.

For example, within European faces, hair style and colour are informative cues to which people living in Europe have learned to pay attention as their face recognition skills develop through childhood. This may mean that these same dimensions of variation are used to encode the appearance of all faces. A consequence would be that because all Japanese faces, for example, have more similar values on this dimension, their faces will all look similar to us. Gombrich (1982, p. 114) notes 'It is not only all Chinese who tend to look alike to us but also all men in identical wigs such as members of the eighteenth-century Kit-Cat Club displayed in the National Portrait Gallery in London'. The theory that we attend to the wrong features to discriminate other-race faces suggests that if we had sufficient exposure to such faces we would be able to learn their important dimensions of variation.

There is some weak evidence for this explanation from studies which have examined the face recognition abilities of people who have had varying amounts of contact with the other race. For example, Chiroro and Valentine (1995) found that black African students with considerable exposure to white faces were no worse at recognizing white faces than black ones, while black Africans with little exposure to white faces showed the usual cross-race effect. Interestingly, those with high contact and good recognition of the other race faces were actually somewhat poorer at recognizing faces of their own race, a result which may lend support to certain 'statistical' models of face encoding (see Chapter 5, Box 5A). Unfortunately, however, results obtained with white observers in Chiroro and Valentine's study were not so clear-cut.

It is not just the identification of faces which is difficult in other race faces. O'Toole *et al.* (1996) showed that people find it difficult to tell the sex of other-race faces too. This in turn suggests that there may be subtle variations in the characteristics of male versus female faces between different racial groups and that these differences too may require learning.

What information do we use to categorize a face into a racial group? While our immediate intuition might be that skin colour is the most important cue, potential information is also given by variations in overall shape of faces between different races. In experiments reported by Hill *et al.* (1995) where classification as European or Japanese was required, there was evidence that three-dimensional shape was if anything more important than surface colour for this judgement (see Box 3B).

Box 3B: Effects of 3D shape in judging the sex and race of faces

Harold Hill, Vicki Bruce, and Shigeru Akamatsu (1995) examined the separate effects of three-dimensional shape and surface colour on the perception of sex and race. They made use of a more advanced form of the laser-scanning equipment described in Chapter 1 (Box 1C) for the measurement of 3D surfaces of faces. The scanner used by Hill *et al.* (1995) recovered the colour from each point on the surface as well as recording its 3D position. Each face scanned in this way can therefore be described as an array of 3D surface coordinates (displayed as a smooth surface in Fig. 3B.1), plus an array of colours (displayed

Box 3B: Continued

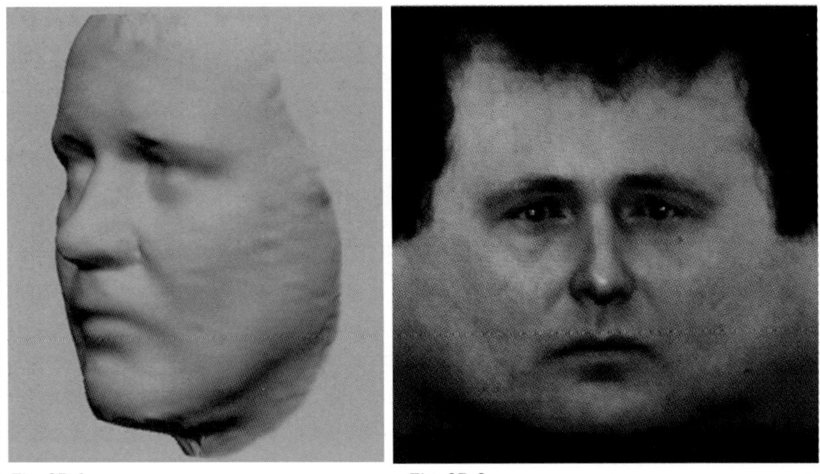

Fig. 3B.1 Fig. 3B.2

as a rectangular panel in Fig. 3B.2). To reconstruct a 3D image of the face, the colour information can be superimposed on the 3D coordinates to give a 3D model on which surface texture and colouration has been reinstated. This allows us to see what happens if the sources of 3D and colour information are *mismatched*, for example a 3D shape from a male face combined with a colour map from a female one; or a 3D shape from a Japanese face combined with a colour map from a European one. By comparing the effects of mismatched versus matched 3D shape and texture information we have another method of finding out which source of information is the more important in determining the sex and race of faces.

Hill *et al.* (1995) obtained scans of four clean-shaven male and four female Japanese and four clean-shaven male and four female European (Caucasian) faces. Using the 3D information alone (e.g. Fig. 3B.1) they found that observers were on average 72 per cent correct at judging the sex of the surfaces. This performance is a little less accurate overall than performance described by Bruce *et al.* 1993, but as half the faces were of another race, and judging faces of another race is more difficult (O'Toole *et al.* 1996) this may explain the discrepancy. However, observers were 88 per cent correct at classifying the 3D surfaces as Japanese or European. When just the colour information was presented in a second experiment, flattened out (as in Fig. 3B.2) to minimize the cues to shape embedded within the colour, observers were highly accurate at judging the sex (97 per cent correct) but less accurate on the race (90 per cent correct). The very high performance on the sex judgement task probably arises because of a number of residual superficial cues in these colour images (for example some visible stubble on several of the male faces, plucked eyebrows on the women, etc). Finally, they examined the effects of combining 3D shape and colour

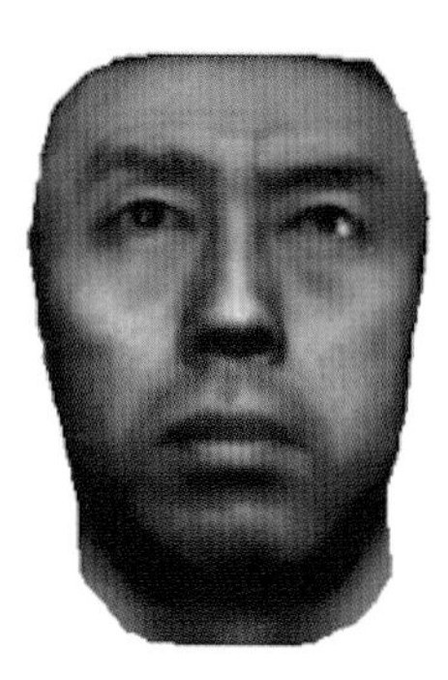
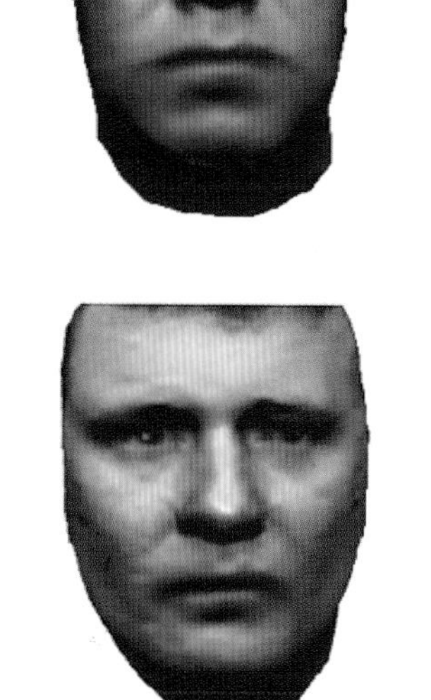
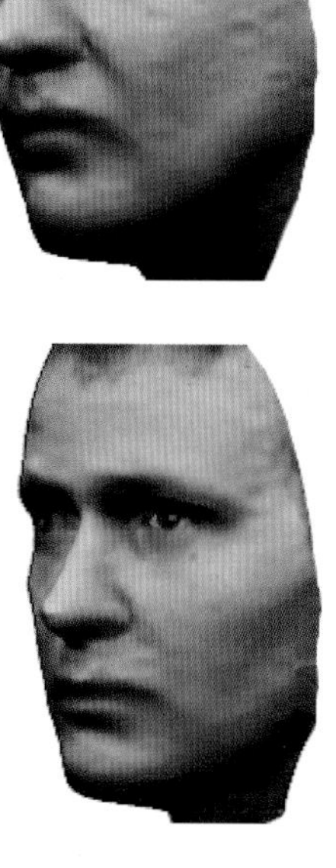
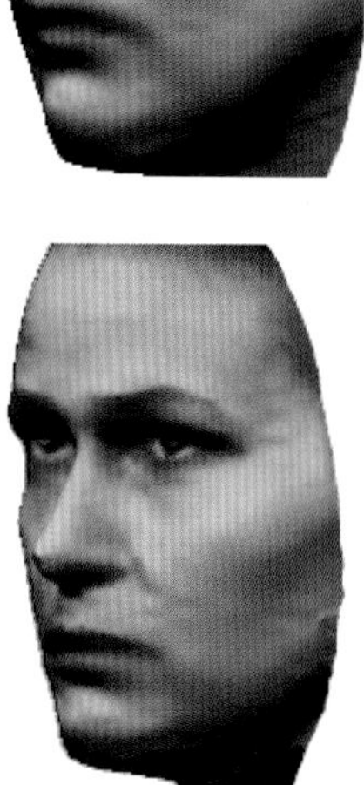

Fig. 3B.3

information so that the sources of information about sex or race matched or mismatched.

Figure 3B.3 shows examples of the matched and mismatched face images used in these experiments. In each block of four, the top left and bottom right images have matched colour and shape, and the top right and bottom left are mismatched. The full-face images show four faces in which race is matched or mismatched:

Top left: shape and colour from male Japanese;
Top right: shape Japanese, colour European;
Bottom left: shape European, colour Japanese;
Bottom right: shape and colour from male European;

The three-quarter view images show faces in which gender is matched or mismatched:

Top left: shape and colour from European male;
Top right: shape European male, colour female;
Bottom left: shape European female, colour male;
Bottom right: shape and colour European female.

When sex judgements were made, the effect of mismatching information from the three-dimensional shape was relatively small, confirming that, with these images, the task was dominated by the superficial cues from the colour cues. However, when made,

Box 3B: Continued

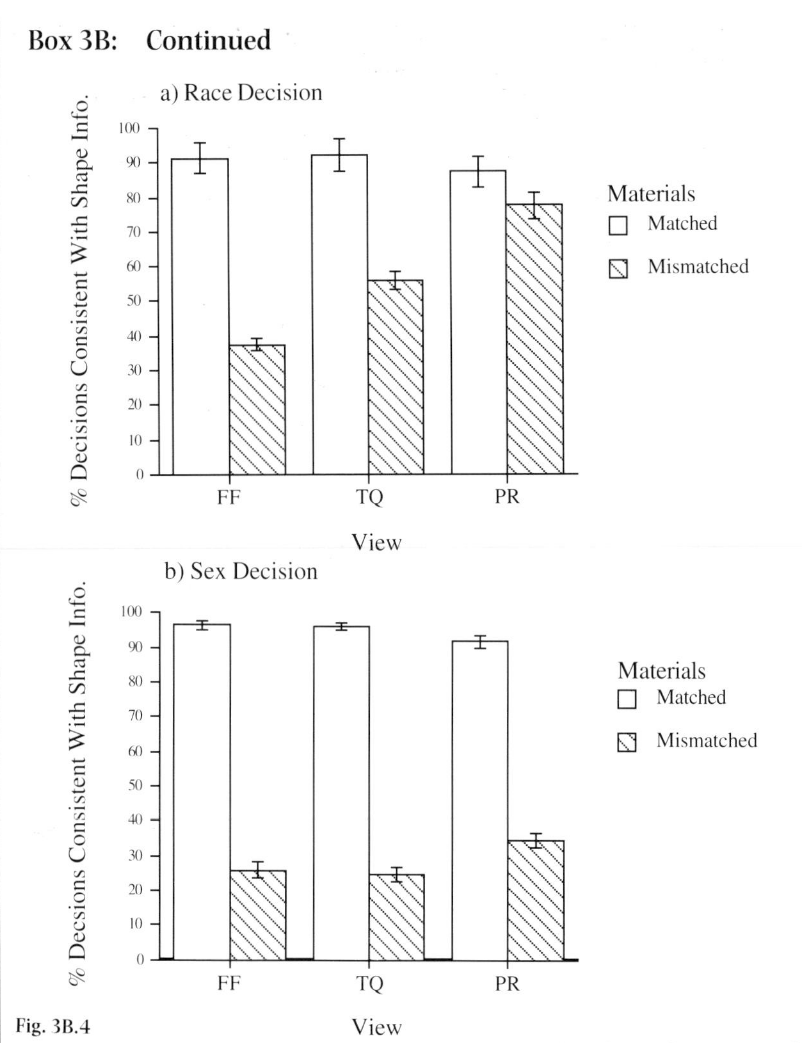

Fig. 3B.4

judgements were influenced substantially by the three-dimensional shape.

The experimental results are shown in Fig. 3B.4. Participants had to decide the race (top panel) or sex (bottom panel) to faces shown in full-face (FF), three-quarter (TQ), or profile views. The graphs plot the percentage of trials where the decision was consistent with the shape information. For matched trials (white bars), where shape and colour are consistent on the dimension judged, then performance is nearly 100 per cent consistent with the shape. When shape and colour mismatch (shaded bars), the lower performance indicates the contribution of the mismatching colour information. The higher the shaded bars, the greater the contribution of the shape to the decision. It can be seen that the shape has a greater effect for race (upper) than sex (lower) decisions, and that its effect is least when faces are shown in full-face (where three-dimensional shape is hardest to see).

(a)

(b)

(c)

Fig. 3.21 *Left*: feminized; *centre*: original; *right*: masculinized Mary Queen of Scots. Original portrait courtesy of the Scottish National Portrait Gallery. Transformed images by Duncan Rowland, David Perrett and Rachel Edwards, University of St Andrews.

3.6 Interactions between age, sex, and race

Although the perception of and consequences for memory of differences in the age, sex, and race of faces have been discussed under separate headings, there are actually many interactions between these different dimensions because of shared characteristics that arise through the ways that different faces are structured and grow.

There is some overlap between the characteristics of a face judged particularly feminine, and the 'baby-faced' characteristics of the juvenile face. This may be one reason why women more than men may feel pressured to use cosmetics, or even cosmetic surgery, in order to look youthful. Figure 3.21 shows a set of images which demonstrate this quite clearly. The central image is a famous portrait of Mary Queen of Scots painted by an unknown artist. Duncan Rowland, David Perrett and Rachel Edwards have used techniques similar to those described in Fig. 3.8 to make the face more feminine (left-hand image) and more masculine (right-hand image) in appearance, by distorting the original image in a manner that moves it closer to the average female or male appearance (see Rowland and Perrett 1995 for details of this technique). What is striking about these images is that the more feminine version looks considerably younger than the original and more masculine versions.

As well as the overlap between the visual features of youthful appearance and femininity, there is some overlap in the characteristics of different racial groups and those of different sexes. Japanese faces are more feminine in appearance, as their nasal areas are less protrusive, while southern European faces are more masculine in appearance. Consistent with this, Hill *et al.* (1995) found that observers were faster and more accurate at judging the sex of three-dimensional surface images of Japanese females and European males. Japanese males and European females were slightly more difficult for observers to judge.

This overlap in physical characteristics may in turn impact upon the kinds of stereotypical attributions that we make to faces. A 'baby-face' may seem feminine, but such faces may also appear dependent or helpless, attributions which may be wholly inappropriate for the individual whose face has these characteristics. In the next chapter we consider further how the appearance of faces may affect the judgements we make about people.

4 The mating game: attractiveness and the sociobiology of faces

4.1 Overview

As well as categorizing and recognizing faces, we also respond to them more or less positively according to how attractive they seem to us. The appearance of an individual face is a complicated result of a number of different kinds of genetic influence, moderated by environmental and cultural factors and individuals' own choices about how to present their faces. Although there are wide differences between individuals and across cultures in what is considered attractive, there is also a surprising core of agreement. A number of hypotheses have been suggested to account for this, but as yet we do not fully understand the interplay of factors that affect attractiveness. We also infer a wide range of other things from faces, sometimes wildly inaccurately, and often without insight into how such impressions are formed.

4.2 One of us: inheritance and family resemblance

The cuckoo uses the unusual reproductive strategy of laying its eggs in other birds' nests—typical hosts are meadow pipits, dunnocks, and reed warblers. The female cuckoo finds the nest of a bird of its target host species, removes an egg from it, and lays one of her own. As many as twelve eggs may be laid, in twelve different nests. When the cuckoo chick hatches, it pushes out the host bird's eggs, and its unwitting foster parents feed its considerable appetite even though they are often small birds and the cuckoo grows to be quite a bit larger. The chick's signals of cheeping and open-mouthed gaping, which are common to several species of birds, suffice to elicit feeding from its overworked hosts.

There is a complex interplay between the behaviour of the cuckoo and its host. Members of potential host species may chase away cuckoos seen in the vicinity of their nests—birds which do this are likely to enjoy better reproductive success, giving a reason for the behaviour to evolve. So the cuckoo has become circumspect in its activities—that way it too

Fig. 4.1 *Stewart Family Tree* by unknown artist. Courtesy of the Scottish National Portrait Gallery.

maximizes its chances. And in some cases it lays eggs which are coloured like those of its host species as well.

Much of the reason behind this gradual escalation of adaptations and counter-adaptations is that the host bird is being tricked into acting against its own interests. From the cuckoo's unusual behaviour came the word cuckold, used to denote the husband of an unfaithful wife. The sense of dishonour which used to be attached even to the thought of marital infidelity is evident in Shakespeare's Othello, and the modern use of paternity tests and alarming statistics on the abuse of step-children show that many people still have deep-seated reactions to nurturing someone else's children. To understand why, we need to know some rudimentary genetics.

The blueprint for an organism's development is encoded in its genes—a set of chemical instructions that control its growth from a fertilized cell. These genes are located on several chromosomes in the cell's nucleus. In organisms with sexual (male–female) reproduction the chromosomes are arranged in pairs, with one member of each pair contributed by the mother, one by the father.

Because the chromosomes come in pairs, both members of most chromosome pairs can carry instructions about the same characteristic. For example, consider the influence of the genes for blue or brown eye colour. The members of the relevant chromosome pair might both have the brown-eyed gene, both have the blue-eyed gene, or have one of each. If there are two brown-eyed genes the person will have brown eyes and with two blue-eyed genes the person will have blue eyes, but it turns out that if there is one blue-eyed and one brown-eyed gene the person will have brown eyes. The brown-eyed gene has a dominant influence, whereas the blue-eyed gene is what is called 'recessive'—it can exert its influence only when there is an identical gene on the other member of the chromosome pair.

This example illustrates one of the key concepts in genetics—the difference between genotype and phenotype. The genotype refers to the genetic specification itself, the phenotype to the characteristics observed in the developed organism. As the eye-colour example shows, because some genes are dominant and others recessive two people with the same phenotype characteristic of brown eyes can actually have different genotypes—one person with two brown-eyed genes, the other person with one brown-eyed and one blue-eyed gene.

The environment can also play a key role in the way in which a given genetic potential is expressed. To use a simple example, we know that children reared in very deprived circumstances do not thrive intellectually or physically—there is always a close interaction between heredity and environment. This applies even to those characteristics such as eye colour which are fixed relatively early in development, but in such cases the relevant environment may be the surrounding intra-uterine cells rather than the physical or social world.

The development of the hard and soft tissues of the skull and face was discussed in Chapter 1.3. The complexity and intricacy of the genetic control of this is evident in the many congenital syndromes which affect the appearance of the face, such as Down's syndrome. Most of the relevant genes have not yet been identified, but clinicians can become remarkably skilled at recognizing their different characteristic effects on a person's appearance (Baraitser and Winter 1983).

Identical twins provide an informative example. Genetically identical (monozygotic) twins are created by the division of a single fertilized egg. They can look very alike, yet because of gene–environment interactions they will always be subtly different, and their relatives and friends do learn to tell them apart. Like any form of perceptual learning, this involves learning to home in on those parts of the faces of a pair of twins which are the most different. Figure 4.2 shows the faces of twin sisters Rosie and Lizzie, together with computer-manipulated images in which the differences between Rosie and Lizzie are reduced or exaggerated. The principles of these image-manipulation techniques will be de-

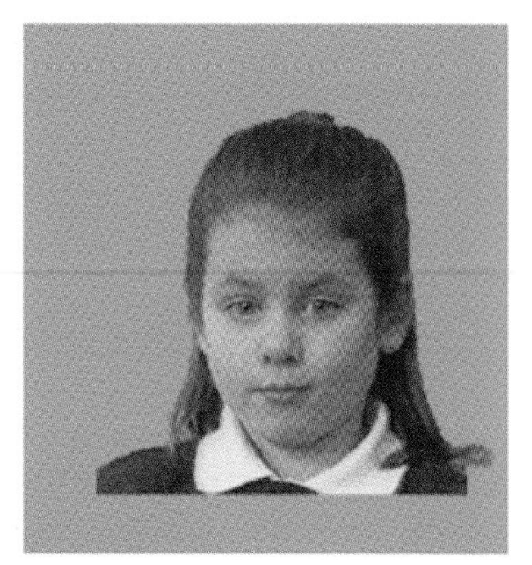

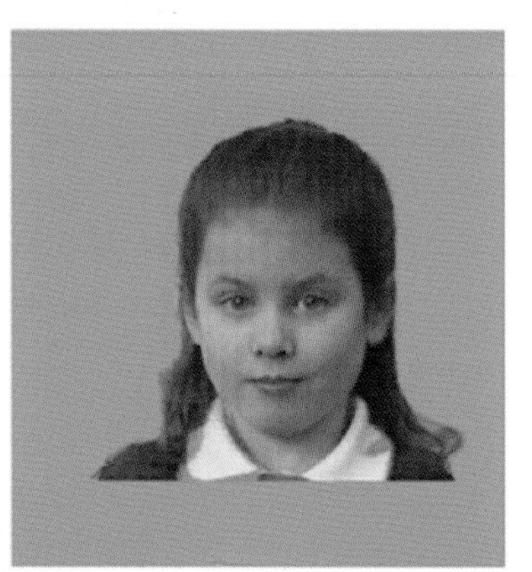

Fig. 4.2 Photographs of twin sisters Rosie (*upper left*) and Lizzie (*upper right*), together with computer-manipulated images in which the differences between Rosie and Lizzie are reduced (*centre image*) or exaggerated (*lower left* and *lower right* images). Thanks to Sarah Stevenage and colleagues at the University of Southampton, the twins and their parents.

scribed in Chapter 5 (Box 5B). Here, they serve to highlight what is common to the twins' faces, and what differences are available for their friends and families to learn. Once the appropriate cues are learnt, of course, the faces start to look quite different, so that relatives will judge Rosie and Lizzie's faces to be less similar than will people seeing them for the first time (Stevenage, in press).

There is no doubt, then, that facial appearance is under a considerable degree of genetic control. Everyone knows this—it is still common to debate which of the baby's features are like its father, and which like its mother.

What gives this an added twist is the fact that since the baby has been carried by its mother, it will automatically have 50 per cent of its genes derived from her genetic make-up. The other 50 per cent come from its natural father. But bringing up a human child is a protracted process, representing a considerable parental investment—this makes it a potentially important issue whether the father who helps nurture the infant is its biological father.

To see why this might be important, we need to understand a little about a discipline which has come to be known as sociobiology. In everyday life, we usually think of reproduction as a way of ensuring the survival of the species. From this (species) perspective, it does not matter whether the baby is looked after by its biological father or not. But several recent findings have suggested that it may be equally near (or nearer) the truth to think of the species as a way of ensuring the continuation and development of particular sets of genes, and from this perspective it is very important whether a person is the baby's natural father (with 50 per cent shared genes) or not (much smaller proportion of shared genes).

This perspective—the 'selfish gene' theory (Dawkins 1976)—has been useful in providing accounts of otherwise puzzling phenomena such as altruism. Many acts of altruism are puzzling because individuals act against their (narrowly defined) self interests, and in ways that do not always seem directly to enhance the prospects of survival of the species. But altruism fits exactly the selfish gene hypothesis, because it turns out that acts of altruism between one individual and another are on average related to the extent to which they have genes in common.

Such ideas are not always popular, because they seem to many people to take away something of the humanity from human beings. In our view, that is a misconception. The claim of selfish gene theory and related sociobiological accounts is not that altruism is nothing but gene survival, or that foster- or step-parents will always be the enemies of their children. On the contrary, it is accepted that through our great capacity for learning, symbolization, and cultural development there will be many counter-examples to these rather basic rules of our animal nature; but these counter-examples should not blind us to the fact that the influences

Fig. 4.3 *James Russell*, Professor of natural philosophy at Edinburgh University, and his son *James*, painted in 1769 by David Martin. Courtesy of the Scottish National Portrait Gallery

of fundamental genetic mechanisms can still be revealed when one looks at what happens across the average of many many instances.

Paternity confidence, then, is a potential issue for males in a monogamous species where parental investment is high. For similar sociobiological reasons, many animal species have developed elaborate mechanisms of kin recognition, and these have been intensively investigated (Fletcher and Michener 1987). It seems likely that kin recognition is just as important to humans, but we have little idea as to exactly how it is achieved (Wells 1987).

You can immediately see the family likeness in the picture of Professor James Russell and his son (Fig. 4.3), or successive generations of Stewart kings (Fig. 4.4). But in scientific terms, very little is known about how we see (or how we think we see) family resemblance. In Fig. 4.4, a cumulative average Stewart face is shown alongside each portrait, to demonstrate that image-manipulation techniques offer considerable promise for investigating such questions.

The eminent Victorian mathematician and scientist Sir Francis Galton explored family likeness by superimposing pictures of different individuals onto the same photographic plate (Galton 1883). This technique is illustrated in Fig. 4.5; each picture was aligned with those superimposed onto it using a common facial feature (in this example, the eyes). In this way, Galton thought he could use photography to emphasize what is common to the different individuals, since such details should remain sharp whereas regions of difference will become

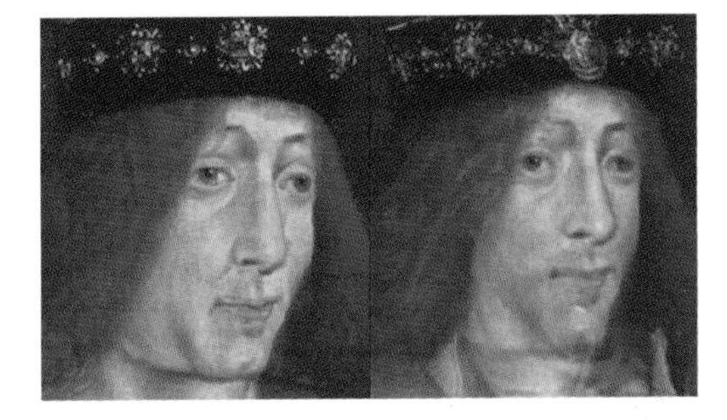

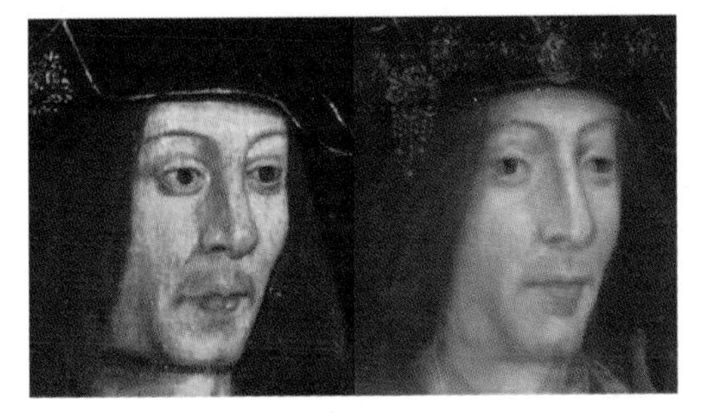

Fig. 4.4 Portraits of James I through to James V, with a cumulative average Stewart face shown alongside each portrait. [The cumulative average is the average of the face in that row and all those above it]. Original portraits, courtesy of the Scottish National Portrait Gallery. Computer manipulations by Peter Hancock, University of Stirling.

Fig. 4.5 Galton's technique, and some examples of the composite portraits he produced.

blurred where the component images do not align. As well as investigating family resemblance, Galton envisaged several other applications of this procedure, one of which was to try to get a more accurate impression of the appearance of historical figures (such as Alexander the Great, who is also shown in Fig. 4.5) by finding the features common to different portraits.

Galton (1883) felt he had some success with this approach. Of the full-face and profile pair of pictures of the two sisters shown in Fig. 4.5, he commented:

> The interest of the pair lies chiefly in their having been made from only two components, and they show how curiously even two faces that have a moderate family likeness will blend into one. That neither of these predominated in the present case will be learned from the following letter by the father of the ladies, who is himself a photographer:-
>
> I am exceedingly obliged for the very curious and interesting composite portraits of my two children. Knowing the faces so well, it caused me quite a surprise when I opened your letter. I put one of the full faces on the table for the mother to pick up casually. She said, 'When did you do this portrait of A? How *like* she is to B! Or *is* it B? I never thought they were so like before.' It has puzzled several people to say whether the profile was

intended for A or B. Then I tried one of them on a friend who has not seen the girls for years. He said, 'Well, it is one of the family for certain, but I don't know which.' (Galton 1883, pp. 8–9)

However, such reactions were not universal. We will give Galton the last word on this for the moment, but we return to his method in Section 4.4:

I have made several other family portraits, which to my eye seem great successes, but must candidly own that the persons whose portraits are blended together seldom seem to care much for the result, except as a curiosity. We are all inclined to assert our individuality, and to stand on our own basis, and to object to being mixed up indiscriminately with others. (Galton 1883, p. 9)

4.3 Changing faces

As the satire of a school of beauty (Fig. 4.6) makes plain, some people have always been willing to go to extraordinary lengths to try to improve their appearance.

Advances in cosmetic surgery offer still more radical possibilities. In 1990, in All Saint's Church, Newcastle, the performance artist Orlan began a project which involved using computer software to create a new self-portrait in which her own features were blended with the chin of Botticelli's Venus, the forehead of the Mona Lisa, the eyes of

Fig. 4.6 Burney: *An Elegant Establishment for Young Ladies*, c. 1800. V&A Picture Library.

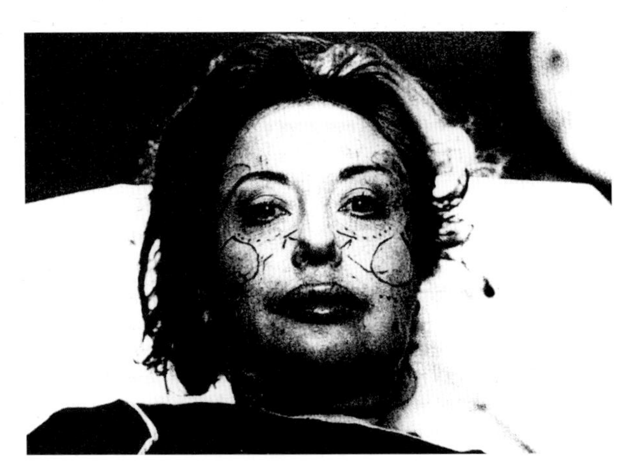

Fig. 4.7 The performance artist Orlan.

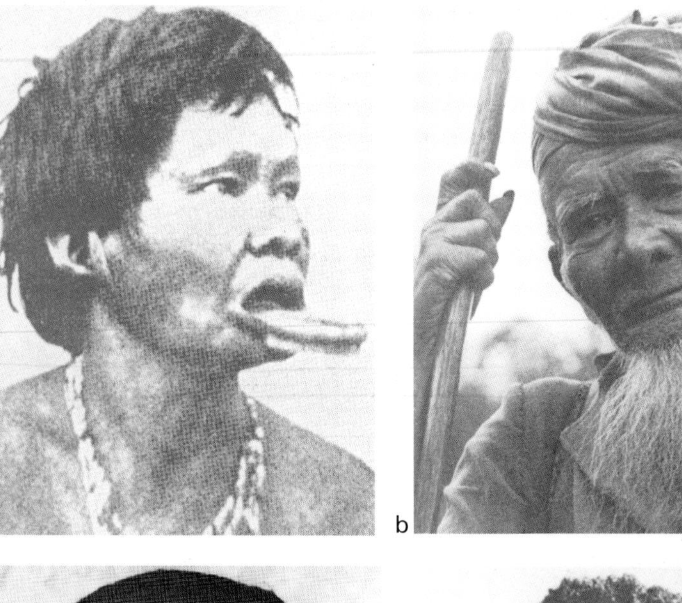

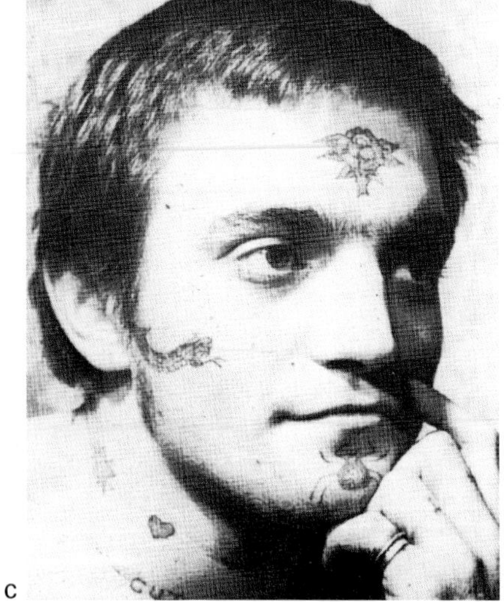

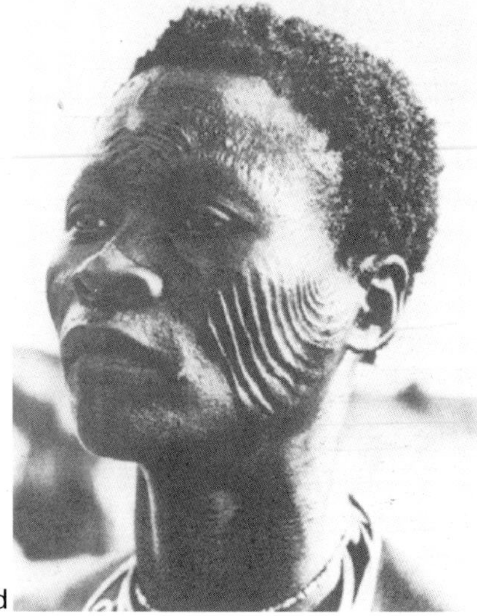

Fig. 4.8 Some ornamental devices which involve exaggeration of existing facial features (*top row*), or the creation of new facial features through scarring or tatooing (*bottom row*). From Liggett (1974). a, b and d courtesy Musée de l'Homme. c courtesy of Popperfoto.

Gérome's Psyche, the mouth of Boucher's Europa, and the nose of a Diana from the School of Fontainebleau. Orlan's 'performance' involves undergoing surgery to reshape her face into this new image—to date she has had nine such operations, some of which were broadcast live to various locations. Figure 4.7 shows Orlan's face prepared for one of these operations—the markings are reminiscent of medieval and physiognomic texts (see Section 4.5).

Even those who are reluctant to suffer this level of privation are generally happy to alter their looks through hairstyling, shaving, and the use of cosmetics or ornaments.

In deciding how to present our faces, we are greatly influenced by culture, fashion, and fad. It seems that almost anything has been in vogue somewhere at some time. A common theme has been to exaggerate particular facial features through the use of ornamentation or cosmetics, but there are also plenty of examples where features are effectively created where they did not exist before, such as through scarring or tattooing. Figure 4.8 gives examples of differing tastes.

Often the features chosen involve the creation or exaggeration of differences between the sexes. So in present-day western societies there are sex differences in typical hairstyles, and women generally use cosmetics whereas men don't. Moreover, the cosmetics are often applied in ways which emphasize sexually dimorphic features such as

Fig. 4.9 The face of Queen Elizabeth I, with 16th century and 20th century cosmetics. Original portrait courtesy of the National Portrait Gallery, London. Computer-manipulated image courtesy of Rachel Edwards, Kieran Lee, David Perrett, and Duncan Rowland, University of St Andrews.

those discussed in Chapter 3—especially differences in lip shape, cheek shape, and the eye region.

To illustrate how fashions in female cosmetics have changed over the centuries, Figure 4.9 shows the face of Queen Elizabeth I with sixteenth-century cosmetics, including the whitened face fashionable at that time, alongside a computer-retouched version in which she is shown with the average skin texture and cosmetics of some twentieth-century women.

The causes of changes in fashion are complex and might be thought inherently unpredictable, but some trends can be discerned. A persistent theme is a link with social status. In the sixteenth century, many people worked outdoors and would have heavily weathered skins, so a pale skin would be a sign of wealth—which may help to explain why people like Elizabeth I used such intensely white cosmetics. In more recent times, tanned skin has acquired status through becoming associated with expensive holidays, though this may be changing again as the perceived risks of sunbathing increase. Similar points can be made concerning changes in preference for fatness or thinness.

A considerable industry has also been built up to try to ameliorate the effects of ageing of the face by surgical alteration or the application of cosmetics to compensate or conceal age-related changes to the skin. The demand for such treatments seems to arise because modern western cultures particularly value youth, and hence a youthful appearance. In other cultures, wisdom and experience which are associated with age are of greater perceived value, thus reducing the desire to appear young.

Many of the chemicals used in cosmetics in past times were highly dangerous. Ceruse, which was applied in liberal quantities to whiten the faces of Queen Elizabeth, Mary Queen of Scots, and other rich ladies in the sixteenth century, was a paste of lead compounds. The urge to beautify oneself has been so great that even when the poisonous nature of such substances was widely recognized, many people continued to use them.

And well they might. For those of less fortunate appearance, the literature on the social psychology of attractiveness makes depressing reading. A remarkable number of our evaluations of people can be shown, on average, to be directly influenced by their physical attractiveness (Bull and Rumsey 1988). These influences range from professed satisfaction with a blind date, through evidence that simply being with an attractive person enhances one's social status, to findings that beautiful people are thought to possess other desirable psychological attributes, and that they are even less likely to be found guilty in criminal trials!

There are many examples in the research literature to show that most of us are susceptible to a stereotype that 'what is beautiful is good'. Some of the classic reports were by Karen Dion and her colleagues—we will use just two examples. In one study (Dion *et al.* 1972) participants were asked to assess people's personal characteristics from photographs of their faces—attractive faces were rated as belonging to individuals with more socially desirable personality characteristics, higher occupational status, greater marital and parental competence, and greater social and professional happiness. In the other study (Dion 1972), a description of an aggressive act was accompanied by a picture of an unattractive or an attractive child. Adults were more likely to attribute the reasons for the aggressive act to character dispositions in the unattractive child and to circumstances for the attractive one—in other words, they tended to assume the unattractive child was nasty but the attractive one was having an off-day.

Although few findings in social psychology are more consistently obtained than the finding that people respond more positively to more attractive than to less attractive individuals, the physical cues which determine attractiveness have proved difficult to specify precisely. We now turn to this issue.

4.4 Attractiveness and beauty

The most long-lived theory of beauty has been the ancient Greek hypothesis that it is a matter of good proportions. Agreement on what these proportions might be was less easy to achieve. In the bust of Apollo (Fig. 4.10) the ratio of the whole face (x) to the height from chin

Fig. 4.10 Bust of Apollo, showing the golden section. Courtesy Mansell Collection.

to eyes (y) is the same as the ratio of the chin to eye section (y) to the forehead (z). This mathematical function, in which the ratio of the whole to the larger part (x:y) is the same as the ratio of the larger part to the smaller part (y:z) was known to Plato—it is called the golden section.

Although the golden section was widely used, lots of other variants of numeric and geometric approaches were tried. The notebooks of Leonardo da Vinci contain many examples of his interest in human proportions, as is indicated by Fig. 4.11. Alternatively, Fig. 4.12 shows how neatly the face of Botticelli's Venus fits an idea that the perfect face is divisible into sevenths, which had been favoured by mediaeval artists. The hair occupies the top seventh, the forehead the next two-sevenths, the nose two-sevenths, then a seventh for the space between nose and mouth, and a seventh for the chin.

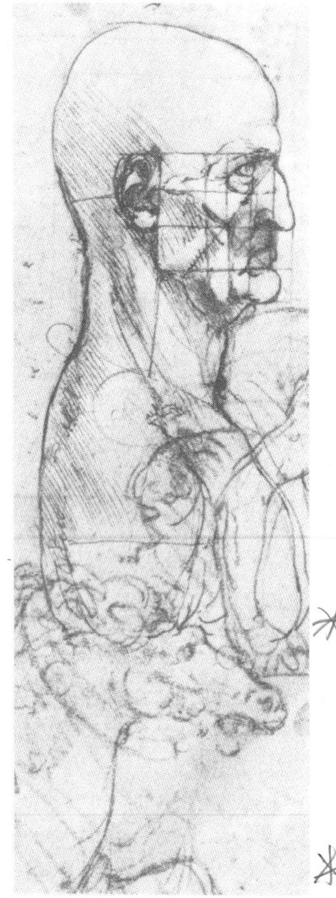

Fig. 4.11 A drawing from one of Leonardo da Vinci's notebooks, showing his interest in facial proportions.

Often, such hypotheses concerning the mathematical ideal were not particularly about people's faces as such. Dividing things into sevenths reflects a long preoccupation with seven as a magic number, and the golden section was considered to apply to the beauty of anything in nature or in art—it was used to evaluate the beauty of a landscape as well as a face.

The search for the mathematical underpinnings of beauty petered out in the eighteenth century, when the philosopher David Hume (1757, pp. 208–9) argued that beauty, 'is no quality in things themselves: it exists merely in the mind which contemplates them; and each mind perceives a different beauty.' The idea that beauty is in the eye of the beholder has since remained the dominant view among the general public, artists, art critics, and philosophers.

The readiness with which we now accept that there are no ideal physical standards of beauty is hardly surprising, given the abundant evidence of cultural differences in what is considered attractive (see Fig. 4.8), changes in the appearance of people across relatively short historical time-spans within western culture (Fig. 4.13), and the substantial differences between the ways in which individuals belonging to different social groups have chosen to present themselves in our own lifetimes (rockers, mods, skinheads, punks, spice girls, and on into the future).

But variations in individual taste, and the fact that the tastes of groups of people may shift markedly if this becomes important in establishing their collective identity, should not be taken to imply that there is no commonality.

A useful start may be to try to distinguish beauty and attractiveness. The psychologist John Liggett (1974) and others have argued that beauty is a quality of the whole person, not just their physical attributes. The success of this has perhaps been undermined by the fact that a similar logic has been used to try to legitimize beauty contests by

Fig. 4.12 The face of Venus from Sandro Botticelli's *The birth of Venus* and *Mars and Venus* can be divided into sevenths. Adapted from Liggett (1974). Courtesy of The Uffizi Gallery, Florence and the National Gallery, London. With the permission of the Ministry of Culture and the Environment.

including judgements of 'personality', and few people have been taken in by that. But we agree with the basic point that it may be useful to distinguish physical (and by implication sexual) attractiveness from less tangible judgements.

Despite the easily demonstrated differences in taste which we have discussed, there is actually a reasonable degree of agreement between people from different cultures as to which faces are attractive if one uses averaged responses (Langlois and Roggman 1990). Hence, although there can certainly be wide variation between the views of any two individuals as to what is attractive, they actually vary around an underlying norm which is surprisingly consistent across cultures. In addition, when babies who are less than a year old are shown faces adults consider attractive or unattractive, they spend longer looking at the attractive faces—as if they also prefer them, even though they could not yet have tapped into cultural aesthetic standards (Langlois *et al.* 1987).

Why should there be an underlying consistency in judgements of attractiveness? It is too early to give a definite answer, but some interesting ideas have been put forward.

In a study which has generated a great deal of interest Judith Langlois and Lori Roggman (1990) developed a computer version of Galton's (Fig. 4.5) composite portrait technique. What they did was to take photographs of faces with standard pose, expression, and lighting. These were then scanned into a computer and adjusted to all have the same distance between the eyes and lips. Each image was then divided into a very large number of tiny squares (pixels), and the brightnesses of corresponding pixels in different faces of the same

Fig. 4.13 Portraits of *Sir David Murray of Gorthy* (by an unknown artist; in the collection of the Royal Museum of Scotland), *Sir Mungo Murray* (detail, by John Michael Wright), *Alexander Murray* (by Allan Ramsay, in the collection of the Royal Museum of Scotland), *James Murray* (by Allan Ramsay), *William Henry Murray* (by Sir William Allan), and *Sir David Murray* (by John Pettie) illustrate the changes in style of self-presentation within the last 400 years. Courtesy of the Scottish National Portrait Gallery and the Royal Museum of Scotland.

sex were averaged to create computer-composite images. Examples of male or female composite faces made from 4, 8, 16, or 32 faces are shown in Fig. 4.14.

When people were asked to judge the attractiveness of these composite faces, they rated them as increasingly attractive the more faces went into each image—in other words, perceived attractiveness increases as one moves downwards in Fig. 4.14. This applies both to the male and to the female faces.

The mathematical consequence of increasing the number of faces used to create a composite image (as in Figure 4.14) is to move the brightness values in that image closer to the average brightness values of the entire set. This happens because the more images are used, the more the idiosyncrasies of particular faces which may be unusual become ironed out.

It seems, then, that moving a facial image closer to the average (in effect, moving it in a downward direction in Figure 4.14) increases its perceived attractiveness. We must now ask, why?

One rather uninteresting possibility is that it may derive from

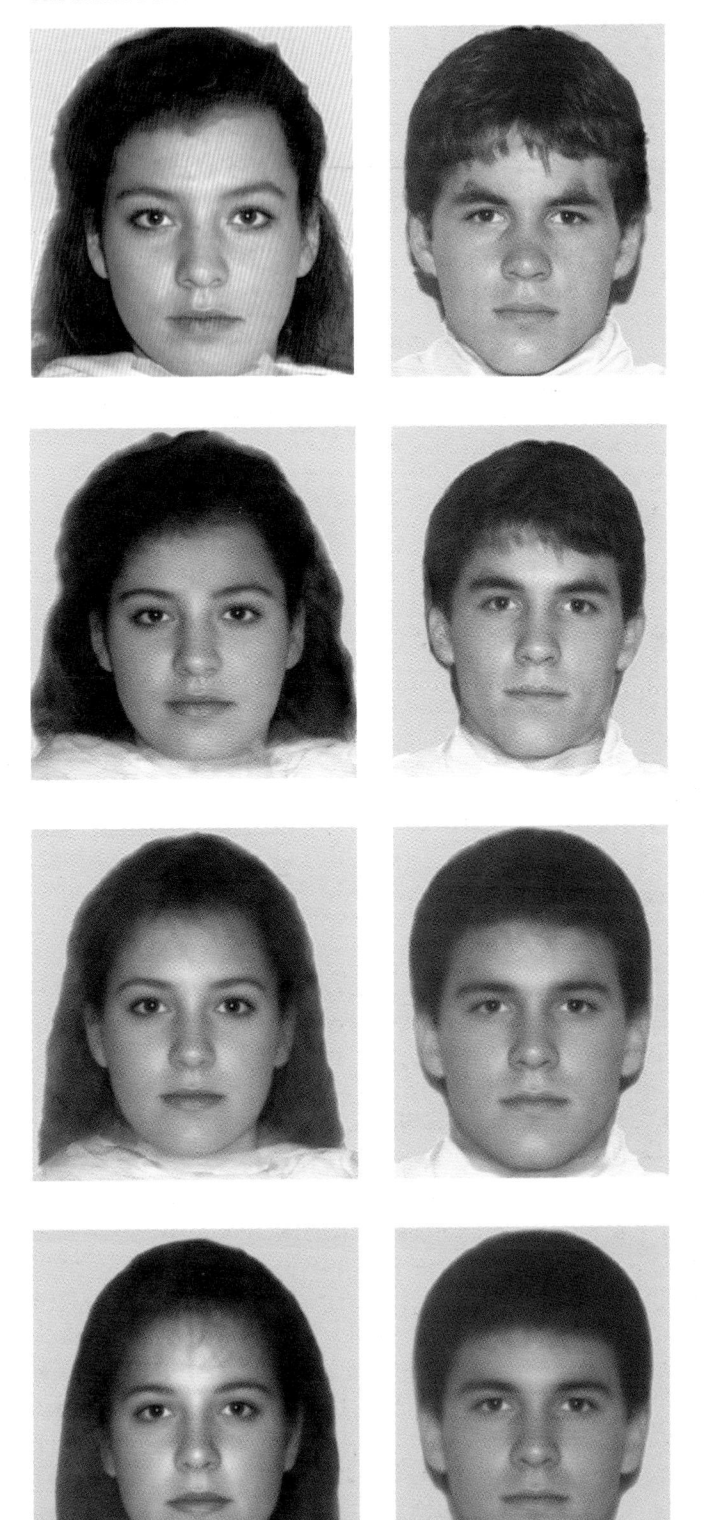

Fig. 4.14 Computer-composite faces. The columns show composite sets created from female faces (*left column*) or from male faces (*right*). From top to bottom, rows show composites created by averaging across four faces, eight faces, sixteen faces, and thirty-two faces. Courtesy Judith Langlois, University of Texas.

this specific method of averaging brightness values. A technique often used in photography and cinema has been to defocus the image a little in close-up shots of the face, to hide surface blemishes and create a slight aura of mystery. Perhaps something similar happens when more faces are put into a computer-composite based on pixel brightness values?

A similar point had been noticed by Galton (1883, p. 224), who remarked that, 'All composites are better looking than their components, because the averaged portrait of many persons is free from the irregularities that variously blemish the looks of each of them.'

We can discount this because the same effect is found when the locations of features in line drawings are moved closer to or further away from their average locations in a set of faces (Rhodes and Tremewan 1996). The technique (known as anticaricaturing) is described later, in Chapter 5 (see Fig. 5.18). These line drawings are free from surface blemishes, yet they still become more attractive as their shapes get closer to the average.

The relation of attractiveness and averageness therefore seems to represent a genuine phenomenon, though it needs to be treated cautiously. So what might be its basis?

One possible contribution is from our preference for things that are familiar to us. The importance of familiarity has been extensively investigated by the social psychologist Robert Zajonc (1980), who considers it as intimately linked to affective reactions which are often the very first reactions of the organism, and which are the dominant reactions for many species.

Importantly, many studies have shown that we will prefer visual stimuli we have seen before even when we have no recollection of having seen them at all (Bornstein 1989)! The mechanisms which lead us to prefer a familiar stimulus are therefore separate from those involved in consciously remembering that we have seen it before—these two kinds of familiarity are quite different.

This form of preference without conscious inference may contribute to perceived attractiveness. Although you may never have seen before the actual composite images from the bottom row of Fig. 4.14, their closeness to the average of the faces you *have* seen may suffice to create a preference based on a comforting sense of ease.

Other potential explanations of the importance of averageness (or near-averageness) to attractiveness come from sociobiological hypotheses. The mechanism used to copy genes during reproduction is very slightly susceptible to error. These gene copying errors, called mutations, are important in introducing new characteristics into the gene pool, but only a minority of the mutant genes turn out to be useful. Langlois and Roggman (1990) had recognized this possibility; they pointed out that individuals with characteristics that are close to

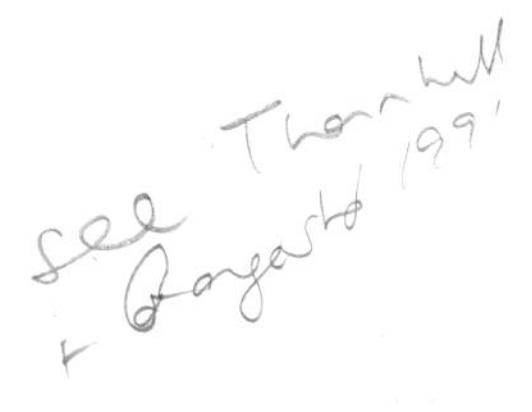

the average of the population might be preferred because they are less likely to carry harmful genetic mutations.

But is this the whole story? Could it really be that attractiveness is little more than averageness?

It seems unlikely. People who become film stars or sex symbols are not noticeably average in appearance, and recent psychological research has also shown the 'attractiveness is averageness' position to be too simple. For example, Japanese and European people agree on the relative attractiveness of faces of both their own and the other race (see Box 4A). Yet Japanese and European people often have much more experience with own-race than other-race faces, and in Section 3.5 we discussed some of the abundant evidence that such differential experience can create difficulties in cross-race recognition. So why does differential experience have less impact on perceived attractiveness?

More direct evidence that there is something other than just averageness involved comes from elegant work reported by David Perrett and his colleagues, described in Box 4A (Perrett *et al.*, 1994). They found that whilst averageness is indeed attractive, it is not optimally attractive. Instead, images created from the average shape of a set of attractive faces were rated as more attractive than images created from the average shape of the set of faces from which the attractive faces had been selected.

The idea of a sociobiological underpinning to certain aspects of attractiveness has undergone considerable development in recent years. Attractive facial features may be attractive because they signal sexual maturity and fertility, and this can in part account for the fact that attractiveness also seems to be linked to features (see Section 3.2) which signal a certain youthfulness (Jones 1995). Another hypothesis is that attractiveness is part of a mechanism which promotes an optimal degree of variety in the gene pool of a species for maximal resistance to potential parasites (Thornhill and Gangestad 1993). Alternatively, a case can be made that attractiveness relates to sociobiological mechanisms involved in family resemblance and paternity confidence, discussed in Section 4.2 (Salter 1996).

It is clear that the exploration of sociobiological hypotheses is at an early stage; at present there seem to be more hypotheses than there are facts to be accounted for. In addition, it should be noted that any evolutionary selection pressures which operate to determine attractiveness may actually work in the opposite direction to selection pressures relating to other facial functions (such as respiration or ingestion, see Chapter 1). This would limit the extent to which attractive faces might deviate from the population average (Perrett *et al.* 1994). But the sociobiological approach clearly holds great promise, and it has already generated some interesting predictions.

An example concerns facial symmetry and attractiveness. Nearly all faces are slightly or not so slightly asymmetric around the vertical

Box 4A: Face shape and attractiveness

David Perrett and his colleagues examined the extent to which attractiveness can be explained by averageness of face shape (Perrett *et al.* 1994).

Figure 4A.1 shows their technique. First, the average shape of the faces of 60 Caucasian females aged 20–30 was calculated by

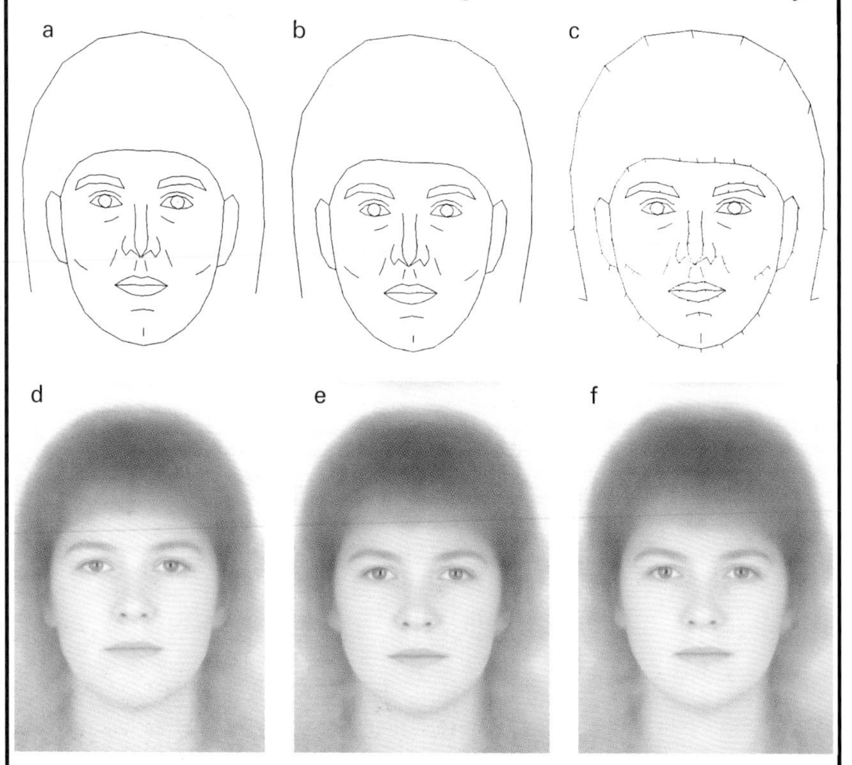

Fig. 4A.1

marking the locations of 224 feature points (such as the end of the nose), and calculating their average positions. This is shape (*a*) in Fig. 4A.1. The same process was then carried out with the fifteen faces rated as having the highest attractiveness. The average shape of these fifteen attractive faces is shape (*b*) in Fig. 4A.1. The differences between the locations of feature points in shapes (*a*) and (*b*) were then calculated, and then each was increased by 50 per cent to create shape (*c*). These changes are indicated by the short lines. Finally, an average pigmentation was mapped onto each face shape, to create face images (*d*)–(*f*).

When British people were asked whether image (*d*) or image (*e*) was the more attractive, they consistently preferred (*e*) to (*d*), and when the choice was between image (*e*) and image (*f*) they preferred (*f*).

These findings show there is something more to attractiveness than just averageness. If attractiveness was entirely due to averageness, there would be no difference between face shapes (*a*) and (*b*), and hence face images (*d*) and (*e*). This is because the highly attractive faces used to create shape (*b*) and image (*e*) would themselves be close to the average of the whole set (shape (*a*) and image (*d*)). Instead, though, the average derived from highly attractive faces is more attractive than the average of the entire set from which they were taken.

The same point was apparent when the difference between the average shape of attractive faces and the average shape of the entire set was increased. The effect of such a change would be to make the resulting faces more different from the average—yet it increased perceived attractiveness.

Perrett *et al.* also found exactly the same pattern of results for Japanese faces (Fig. 4A.2), regardless of whether their attractiveness was judged by Japanese or by British people. There is clearly something pan-cultural about attractiveness.

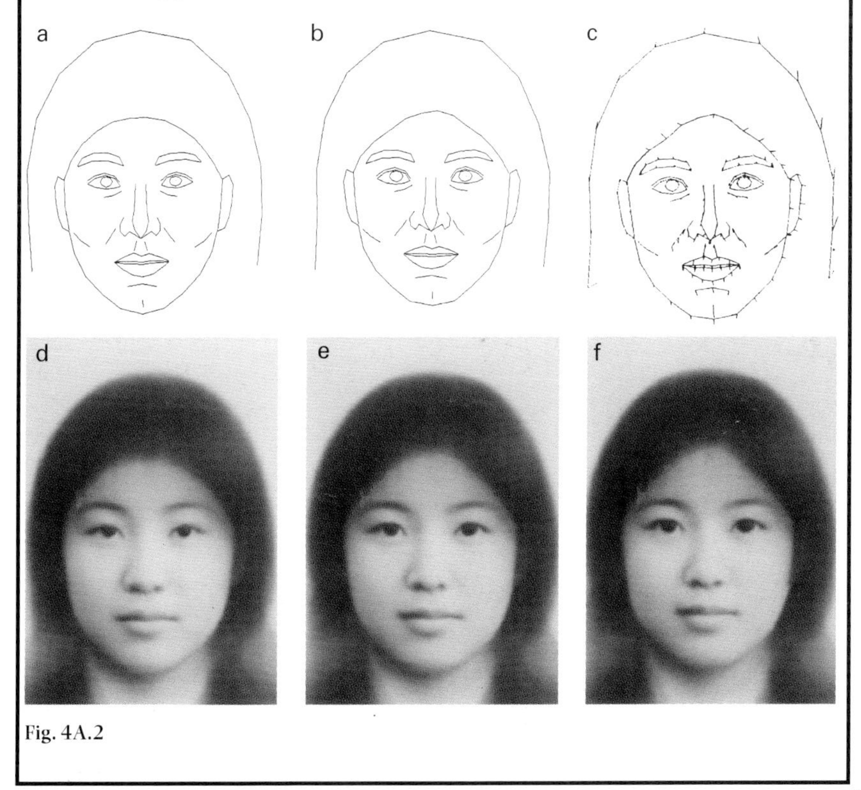

Fig. 4A.2

see p145

midline, but it turns out that even photographs of faces with only slight asymmetry can be made more attractive if they are altered to be perfectly symmetric (see Fig. 4.15).

This preference for symmetric faces is consistent with several variants of sociobiological accounts, but impressively it was predicted

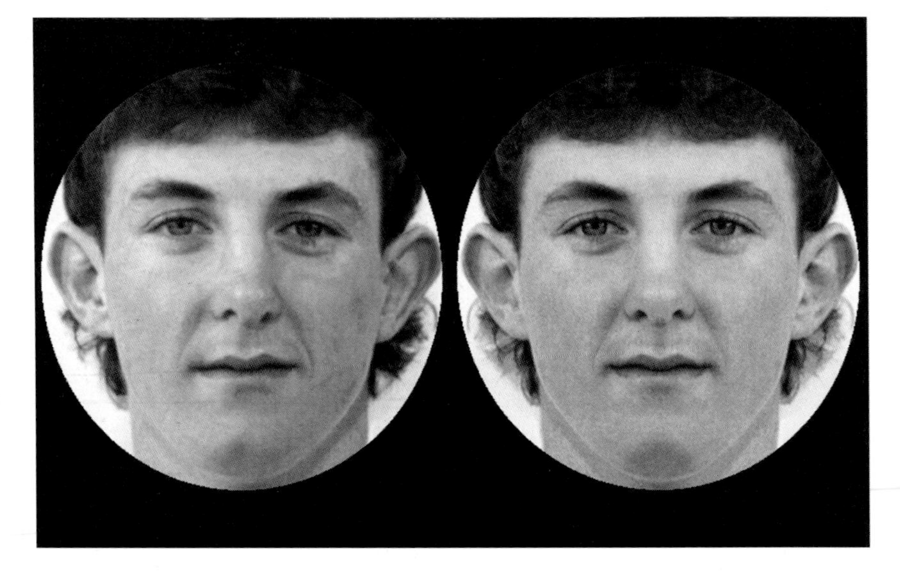

Fig. 4.15 Example demonstrating a contribution of facial symmetry to perceived attractiveness. The face on the left is a normal photograph, whereas that on the right has been altered to make it perfectly symmetric. Images produced by Gill Rhodes, University of Western Australia.

and discovered from the theory that the basis of attractiveness is to be found in the need to produce offspring with optimal parasite resistance (Thornhill and Gangestad 1993). The ability to generate clear, testable predictions is one of the hallmarks of a good scientific theory.

Some of the findings and techniques we have discussed are already finding their way into the repertoires of contemporary artists. Figure 4.16 shows images from an exhibition staged by the artist Rosemarie Trockel. Similar faces were placed on 3000 billboards in Vienna. Trockel used computer software to transform the 'organic beauty' of the model to a more idealized beauty by removing surface blemishes, correcting contours, and making some features more symmetric. Her intention was that the faces 'should be as beautiful as nature and the computer allow'.

4.5 Can we tell other things from the face?

There have been many attempts to tell people's psychological make-up from their faces—the discipline known as 'physiognomy'. The first physiognomic writings are usually attributed to Aristotle. He noted several previous attempts by others, but was cautious about their validity. Aristotle thought that the key to character lay in comparing the features of people with those of animals—a nose might be fat like a pig's and therefore indicate stupidity, or flat like a lion's to indicate generosity, and so on.

In the Middle Ages, opinion veered more toward the astrological,

Fig. 4.16 Computer-manipulated images from an exhibition staged by the artist Rosemarie Trockel. Courtesy of Museum in Progress, Vienna.

and elaborate systems were devised to deduce an individual's fate from planetary influences on the features (Fig. 4.17).

Aristotle's views were resurrected in the Renaissance by Giambattista della Porta, whose *De humana physiognomonia* (1586) synthesized Hippocrates' typology of four temperaments (sanguine, phlegmatic, melancholic, choleric) with the idea that these could be revealed in the facial features, and again suggested that the correct interpretative technique involved comparison to animal species (see Fig. 1.2).

The most famous physiognomist was Johann Kaspar Lavater, who worked in Zurich in the eighteenth century and attained great fame, being consulted and befriended by the poet Goethe, and having an influence which extended into the depiction of character in nineteenth-

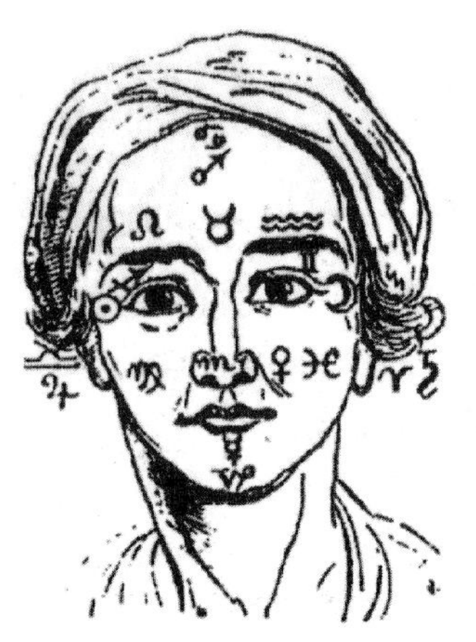

Fig. 4.17 Face-reading in the Middle Ages; the idea was that a person's fate could be derived from planetary influences. From Landau (1989).

century fiction (Tytler 1982). Lavater (1793) argued that character could be read from facial features. He considered the nose to be the indicator of taste, sensibility, and feeling; the lips of mildness and anger, love and hatred; the chin the degree and species of sensuality; the neck the flexibility or sincerity of the personality, and so on.

Although Lavater claimed that his views were based on observation, he also insisted that physiognomy required a special talent to interpret these observations, and even maintained that this could only be developed by beautiful people:

> No one whose person is not well formed, can become a good physiognomist. . . . As the most virtuous can best determine on virtue, and the just on justice; so can the most handsome countenances on the goodness, beauty, and noble traits of the human countenance, and consequently on its defects and ignoble properties. The scarcity of human beauty is the reason why physiognomy is so much decried, and finds so many opponents. (Lavater 1793, p. 85)

The claim that special qualities were needed in a good physiognomist no doubt contributed to Lavater's influence at the time, but ultimately proved the undoing of his system after his death. Without its inventor, and with no basis in science or systematic record, who was to decide which were the skilled and which the unskilled practitioners?

There were attempts to establish a more scientific basis for physiognomy in the nineteenth century. An example is Camper's attempt to measure intellect from the angle of the nose (Fig. 4.18). This was initially grounded in the fact that the changes in facial shape across age gradually alter this angle as a child grows up (see Chapter 1). From this came the incorrect deduction that more intelligent individuals would have more 'developed' faces, and that this could be measured from the angle of the nose. The same technique was then used to make entirely specious comparisons of different races.

Fig. 4.18 Camper's attempt to measure intellect from the angle of the nose. From Cooper and Cooper (1983).

A different (but to modern opinion almost as absurd) tack was taken by Galton (1883). He decided to use his technique of composite portraiture (Fig. 4.5) to investigate the 'criminal face'.

Many of us hold the opinion that we can weigh up whether someone is a shifty character or a bad lot. Figure 4.19 shows life masks of William Burke and William Hare, and a drawing made from them. Burke and Hare were Irish navvies who had come to Edinburgh seeking work during the industrial revolution. They became the most notorious criminals in Scottish history (Edwards 1980). In the early nineteenth century, there was a lucrative trade in robbing the graves of the newly deceased to sell the bodies to anatomists for dissection—a practice known as resurrectionism. In 1827, when one of the lodgers in Hare's house died leaving some rent unpaid, Burke and Hare developed a new variant of this business: they filled the deceased's coffin with bark from a nearby tannery and sold the body to the

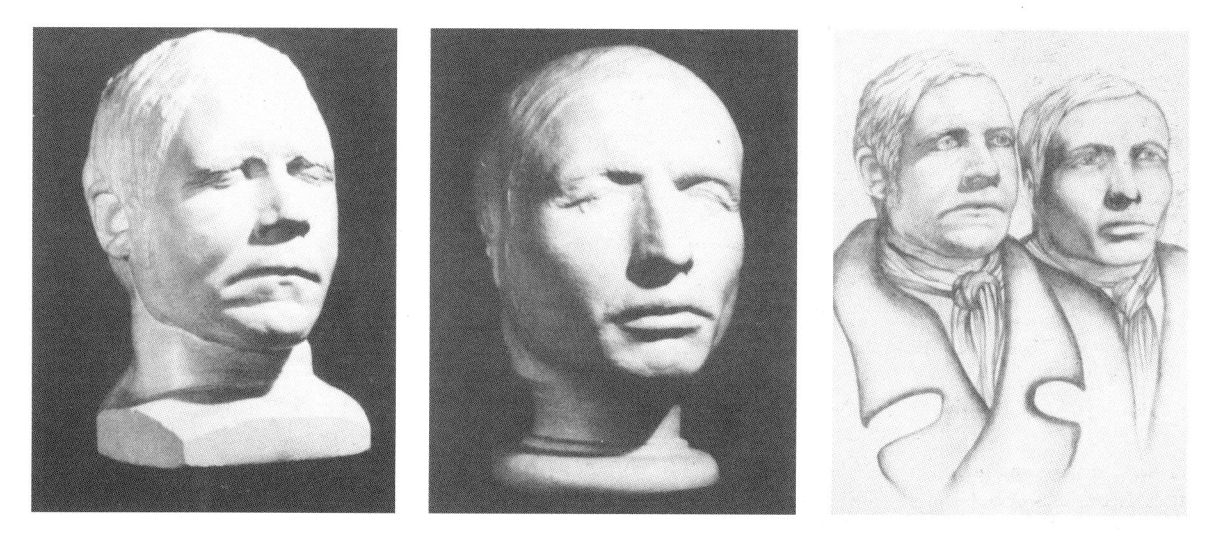

Fig. 4.19 Life masks and drawing of Burke and Hare. From Kaufman (1988).

anatomist Dr Robert Knox (Fig. 4.20). From this coincidental beginning, Burke and Hare graduated to murder. They would get their victims drunk, smother them to make it seem they had died of natural causes, then take the body to Dr Knox's house. To assist the delivery side of their business, they invested in a horse and cart.

Burke and Hare's career ended on Halloween 1828, when the frantic cries of their final victim were overheard. The trial began on Christmas Eve 1828. It attracted enormous public interest, including the close attention of Sir Walter Scott. Burke and Hare, along with Hare's wife Margaret and Burke's mistress Helen McDougall were accused of murdering sixteen people. Only Burke was convicted. He was hanged on 28 January 1829 and his body was given to Dr Knox's

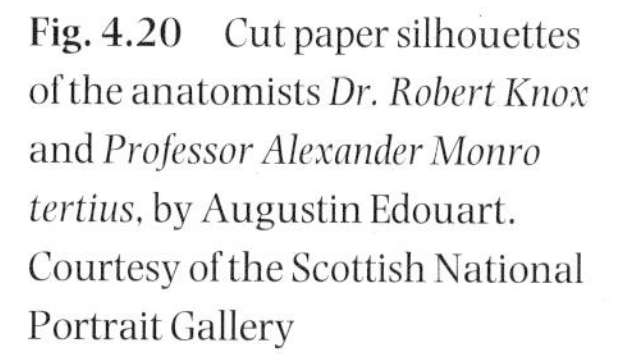

Fig. 4.20 Cut paper silhouettes of the anatomists *Dr. Robert Knox* and *Professor Alexander Monro tertius*, by Augustin Edouart. Courtesy of the Scottish National Portrait Gallery

rival, the then Professor of Anatomy at Edinburgh University, Alexander Monro *tertius* (Fig. 4.20) who dissected and lectured on the cadaver the next day, after which it was put on public display and reputedly viewed by 30 000 people. It was ironic that Monro should be given this task, since it was probably him that Burke and Hare had been seeking when they had their first corpse to sell, but on that fateful day they had been redirected by a protégé of Knox's.

Hare and his wife walked free after he turned 'king's evidence', and the case against McDougall was deemed 'not proven'. Hare was escorted to the English border under a false identity, but this proved to be a difficult trip when he was recognized, and he was last seen walking along the road from Carlisle to Newcastle. Grisly rumours circulated as to what happened to him after that, but nothing is known.

There was a political undertone in that the lower-class immigrants Burke and Hare were tried, whereas the influential Dr Knox was not charged with any offence and continued his career, even though the final victim was found on his dissecting table and he should surely have been suspicious. Burke loyally refused to incriminate Knox, saying only that when he purchased one of the bodies 'Dr Knox approved of it being so fresh, but did not ask any questions'. Asking no questions may have been enough for the law, but public opinion was less forgiving—there were demonstrations outside Knox's house, and his effigy was later burnt at Portobello.

Knowing all this, it becomes difficult not to see stereotyped criminal characteristics in the faces of Burke, Hare, and even indications of a certain lack of moral fibre in the silhouette of Dr Knox. But is there any evidence of the validity of such impressions?

Fig. 4.21 Galton's composites of criminal faces.

For his own researches into the criminal face, Galton enlisted the help of Sir Edmund du Cane, HM Director of Prisons, who allowed access to photographs of inmates taken by the prison authorities. Figure 4.21 shows two composite portraits. One is of the faces of men convicted of murder and violent crime, the other of thieves.

Galton (1883, pp. 10–11) held great hopes for this technique, since he considered that 'It is unhappily a fact that fairly distinct types of criminals breeding true to their kind have become established, and are one of the saddest disfigurements of modern civilization.'

This type of approach reached its zenith in the work of Cesare Lombroso (1911), who claimed that the 'born criminal' was characterized by facial asymmetry, a low sloping forehead, prominent brows and anomalous teeth.

However, results have not favoured such theories. Instead, it seems more likely that to the extent to which such characteristics were present in institutionalized criminals, they were also shared by many underprivileged but law-abiding people of the time.

Galton's own results were probably a great disappointment to him;

the composite portraits (Fig. 4.21) look only like somewhat undernourished men, as would be expected if there was no validity to Galton's premise. Rather than give up the idea, though, Galton reasoned somewhat desperately that the composites, 'are interesting negatively rather than positively. They produce faces of a mean description, with no villainy written on them. The individual faces are villainous enough, but they are villainous in different ways, and when they are combined, the individual peculiarities disappear, and the common humanity of a low type is all that is left.'

The results of studies of other characteristics such as personality or intelligence have been equally negative. There are no grounds for thinking that we can infer any of these accurately from people's faces (Shepherd 1989).

But there is more to it than that. Consider an instructive study carried out by Stuart Cook (1939). He took photographs under standard lighting conditions and gave an intelligence test to 150 male students who had just started at University. When he asked people to estimate the intelligence of these students from their photographs, their estimates were completely unrelated to the intelligence test scores or the students' performance on their courses. However, Cook also demonstrated that even though they therefore did not seem to be valid, people's estimates tended to agree with each other! In other words, there was something about the faces they were fairly reliably picking out, but this did not seem to be valid for the task of estimating intelligence. As far as Cook could tell, the factors which seemed to lead to judgements of high intelligence (in 1939) were symmetry of facial features, seriousness of expression, and tidiness of hair and appearance.

The literature on psychological judgements to faces is full of similar examples. It seems that while we may not be accurate in our judgements, there is none the less a good deal of agreement about the judgements we make. Thus, for example, the rather seamy history of the 'criminal face' has shown that there is little validity in the ability to predict criminal behaviour from facial appearance, but remarkable agreement between different observers. Such agreement is undoubtedly perpetuated by stereotypes derived from several sources, including the media portrayal of different types of character. Actors with villains' faces will tend to play villains' roles, thus reinforcing the impressions that all of us hold of the face of villainy.

More subtle judgements about faces can also be made with a high degree of agreement. Ray Bull and his colleagues (Bull and Hawkes 1982; Bull *et al.* 1983) demonstrated that people tended to agree on their judgements of the political views of faces shown. In each study, photographs of clean-shaven male British Conservative and Labour politicians were collected, with careful matching of the groups in terms of their approximate age, the presence or absence of spectacles, facial expression, and so forth. Volunteers, who were unaware that the faces were those of politicians, were asked to rate each face for its intelligence, sincerity, social class, political inclination, and attractiveness. There was a fair degree of agreement between observers in the judgements made of the political inclination of the faces. The rub was again that not all these judgements were accurate. In Bull and Hawkes (1982) study of six faces which were judged to be most extremely 'Conservative', two were of Labour politicians, and of five judged clearly 'Labour', two were actually Conservative. However, while actual political allegiance did not discriminate between the two extremes, other associated factors did. The 'Conservative' faces were judged as more intelligent, of higher social class, and as more attractive than the 'Labour' faces, and these judgements did not differ according to the political inclination of those making the judgements. Bull *et al.* (1983) found a similar pattern of effects.

These studies indicate that judgements about which political party a person supported were closely associated with their rated attractiveness and apparent intelligence and social class, but these factors were not themselves affected by the political persuasion of the people doing the judging. This latter aspect of these findings contrasts with an earlier study by Jahoda (1954) who found strong differences in the stereotypes of Conservative and Labour politicians depending on the political views of the observer. Jahoda quoted the following as typical of a Conservative observer 'The ones with breeding in their features are Conservatives. Socialists are rough-looking types', compared with 'The fat and stupid ones are Conservatives. Labour people have a frank and open appearance' from a Labour supporter.

It would be interesting to repeat these studies now. Political parties are themselves much more aware of the impact of personal appearance on voters' impressions and (perhaps) their votes, and much more effort goes into manipulating these impressions. As a result, non-stereotypical images may be projected (for example Tony Blair as leader of New Labour in the United Kingdom, with his attractive, intelligent, and high social class demeanour) which may in turn influence the stereotypes we hold.

It is somewhat worrying that all of us probably use stereotypes so freely and unthinkingly. Common denominators for many of these impressions seem to be attractiveness, sex, and age (Shepherd 1989), but research also shows that other cues can also be picked up quite easily and non-consciously (see Box 4B). The best we can do is probably to try to be more aware that this happens, in the hope of counteracting at least the worst consequences.

Box 4B: Non-conscious learning of social cues

Whereas we can judge physical characteristics such as age and sex very accurately from seen faces (see Chapter 3), research on social perception shows that we are very inaccurate at judging personality, intelligence and a whole host of other psychological characteristics from appearance. Interestingly, however, even though these psychological judgements are inaccurate, they are often very reliable from one person to another. People agree on what makes a face *look* intelligent, honest, reliable, and so on. Even though they may not be able to put this knowledge into words, it seems to involve widely-shared stereotypes.

How do we acquire such stereotypes? The answer may be 'surprisingly easily'. Pawel Lewicki (1986) carried out experiments where people were exposed to remarkably simple co-occurrences of certain types of social information with certain physical features, to see how readily these would be picked up.

Figure 4B.1 shows the faces Lewicki used. The top row faces were carefully matched to the bottom row faces for perceived attractiveness, and to have comparable ranges of pose, expression, gaze direction, clothing, etc. What differs is hair length; the people in the top row have longer hair than those in the bottom row.

In one of Lewicki's experiments, participants were shown three of the faces from the top row, one at a time, while a brief vignette of each person was read out. These vignettes were actually invented, but the experiment was carried out in a manner which would make people suppose that the vignettes were genuine, and that the photographs were of people who were remarkably *kind*. For example, 'No one could ever call her self-centred. She does a lot for other people; she is sensitive and helpful. She knows how to treat each individual so as to make him or her feel really good.' (Lewicki 1986, p. 146)

Fig. 4B.1

Box 4B: Continued

A further three faces from the bottom row were also presented, together with vignettes which emphasized how *capable* each person was. For example; 'She is very intelligent and effective. She knows very well how to make the best use of her particular talents, so she usually wins. She likes to be on a tight time schedule and she hates to waste her time.' (Lewicki 1986, p. 146)

So, during this procedure, three faces with long hair were shown accompanied by descriptions emphasizing their kindness, and three faces with short hair accompanied by descriptions emphasizing their capability.

Later on, participants were shown the other two faces from the top row, and the remaining two faces from the bottom row, and they were asked to evaluate whether each person was 'kind', and whether each was 'capable'. The times they took to make such decisions to these entirely novel faces were measured, and are shown as Condition I in Fig. 4B.2, where they are plotted with solid lines.

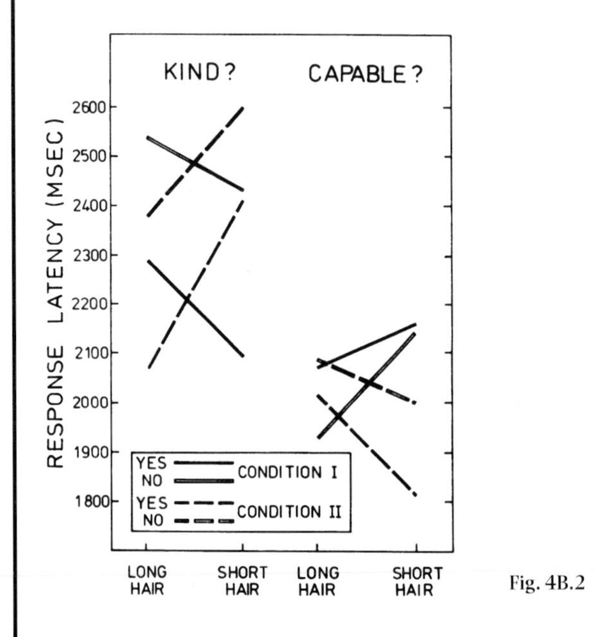

Fig. 4B.2

The result of the experiment is clear. When people were judging the kindness of new faces, they spent longer evaluating those with long hair, and when judging capability they spent longer evaluating those with short hair, regardless of the eventual decision they reached. It therefore seems that from as few as three training trials with kind long-haired faces and three trials with capable short-haired faces the participants had become sensitive to the potential link between hair-length and these traits. Yet, when quizzed about this, no one explicitly stated that they used such a rule. It was acquired in a non-conscious manner—influencing people's judgements without them being aware of the source of the influence.

As a check that the results did not reflect pre-existing stereotypes participants brought to the experiment, Lewicki also included Condition II, in which a different group of participants were exposed to the opposite set of contingencies, with long-haired faces being characterized as 'capable', and short-haired as 'kind'. This exactly reversed the pattern of response times to the new faces (see Fig. 4B.2) showing that it was indeed what subjects had learnt during the experiment (albeit non-consciously) which influenced their behaviour.

5 Whose face is it? How individual faces are recognized

5.1 Overview

Recognizing the identities of people we know is a basic and important social act, and recognition from the face is an ability at which we become very skilled as we grow up. None the less it is puzzling how we achieve this. If all faces are essentially similar (see Chapter 1) how do they also convey our individual identities? Galton (1883) expressed the problem as follows:

> The difference in human features must be reckoned great, inasmuch as they enable us to distinguish a single known face among those of thousands of strangers, though they are mostly too minute for measurement. At the same time, they are exceedingly numerous. The general expression of a face is the sum of a multitude of small details, which are viewed in such rapid succession that we seem to perceive them all at a single glance. If any one of them disagrees with the recollected traits of a known face, the eye is quick at observing it, and it dwells upon the difference. One small discordance overweighs a multitude of similarities and suggests a general unlikeness. (Galton 1883, p. 3)

Since Galton's speculations about the process, the sources of information used in face recognition have been carefully explored, giving useful insights into how we achieve this feat. Interestingly, some of the work has direct applicability to the interpretation of portraits and caricatures. Because faces are the most important key to identity, reconstructions and drawings of the face also play an important role in detective work.

5.2 Face features and configuration

When we are asked to describe a face, or to speculate on how individual faces may be represented in memory, there is a temptation to think of the face in terms of a list of separate features such as 'large green eyes' or 'hooked nose'. This tendency is undoubtedly created in part because our language has discrete vocabulary items for the different functional parts of the face. However, these linguistic terms may have arisen because these different features serve different sensory functions—the eyes see, the jaws chew—as we discussed in Chapter 1.

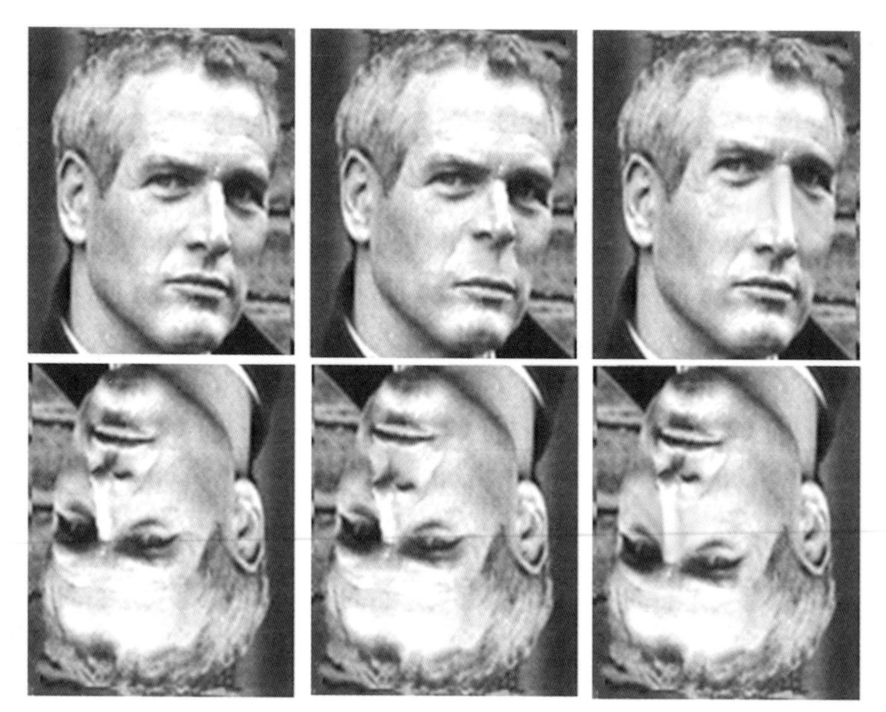

Fig. 5.1(a) You should find it easy to detect which image shows the real Paul Newman, though the distortions are much easier to see in the upright than in the inverted images. Figure created by Helmut Leder, University of Fribourg.

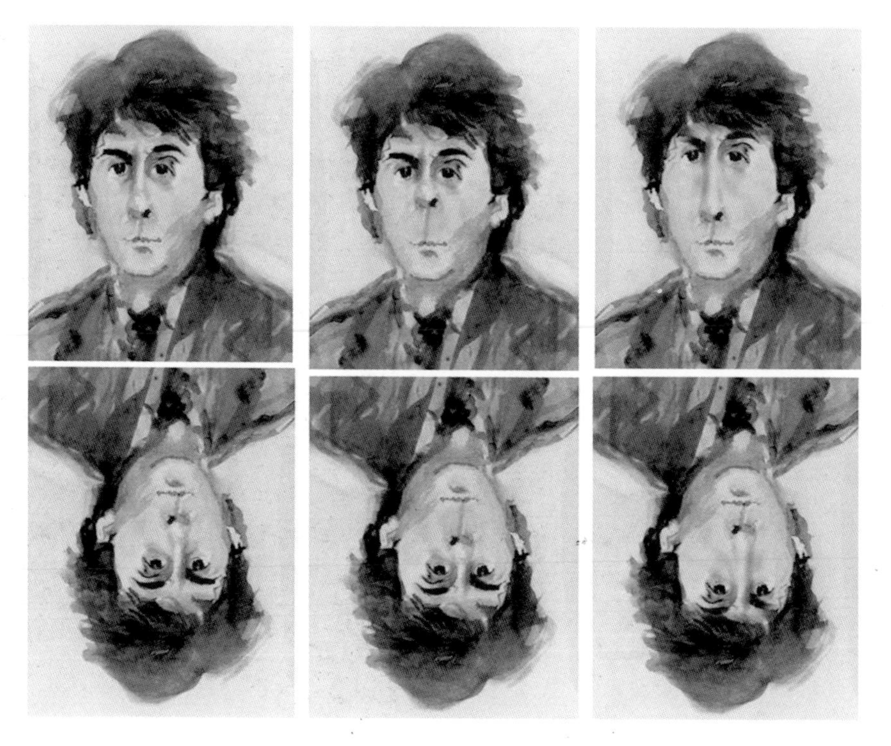

Fig. 5.1 (b) Similar manipulations have been made to a portrait of actor Tom Conti (original shown at top left, by Ishbel McWhirter, courtesy of the Scottish National Portrait Gallery). The distortions made in the centre and right hand panels are again more obvious in the upright than in the inverted images. Figure created by Helmut Leder, University of Fribourg.

It is not necessarily the case, though, that the visual system describes faces in the same way. Indeed, there is a good deal of evidence that face patterns are treated more as wholes or as interrelationships between different features, than simply as a list of their features. In Fig. 5.1 the face 'features' are kept the same, but distances between features have been altered to produce striking differences in appearance. As we discuss later, it has proved remarkably difficult to produce good likenesses of faces using 'kits' of face features such as the Photofit kit. This may be because such kits do not naturally tap the processes that human brains use to describe and retrieve faces.

One strong piece of experimental evidence that we do not process features independently from each other comes from the composite technique used in experiments by Andy Young and his colleagues (Young *et al.* 1987). They divided faces horizontally into upper and lower halves. Although people were quite accurate at identifying the isolated top half of a face when it was seen on its own, when it was combined with the wrong lower half it became extremely difficult to recognize to whom the upper features belonged (see Fig. 5.2). It seems that the impression created by, for example, the eyes and forehead is modified by the features seen elsewhere in the face. This 'composite' effect was found only when the two half-faces were closely aligned. If the two halves were both present, but presented so that they did not

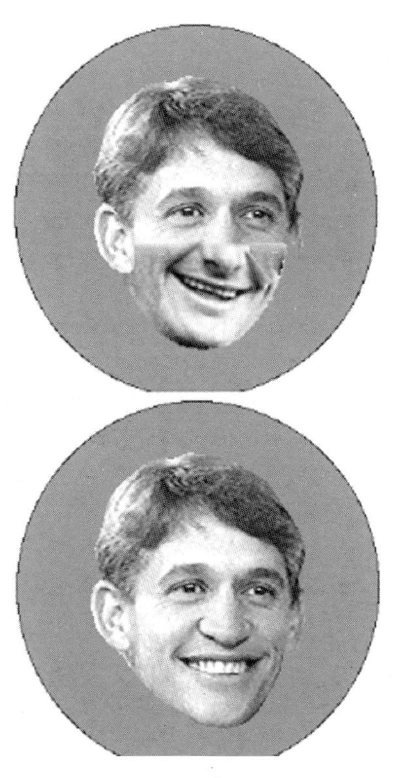

Fig. 5.2 Gazzaker? The composite face (top left) has been created from the top half of Gary Linneker's face and the bottom half of Paul Gascoigne's. Even though there are differences in the shading patterns, a plausible new face identity emerges. When the two images are offset (*top right*) the component identities are easier to see. Figure created by Derek Carson, University of Stirling.

line up to form a whole face, then identification of each half was unaffected by the presence of the other one.

There is a good deal of other evidence that the whole face is more than the sum of its parts. For example, Tanaka and Farah (1993) asked volunteers to learn the names of a set of faces constructed from a 'kit' of face features so that each face had different features (see Fig. 5.3). Later, the volunteers had to try to identify which face feature belonged to a particular 'target' character. For example they would be asked to answer the question 'Which is Larry?' and shown two alternatives to choose from, where the only difference lay in the specific nose which was depicted. People were much better at doing this when the nose was shown in the context of the whole face than when the nose was shown in isolation (so the question became 'Which is Larry's nose?') or within a face context where the features were jumbled up.

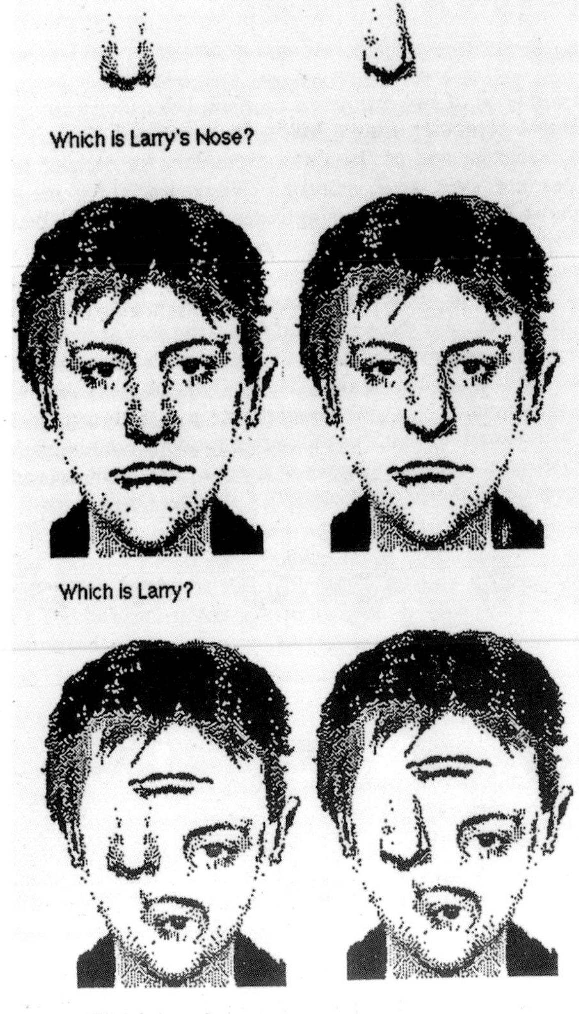

Fig. 5.3 Isolated features, whole face and jumbled face images used by Tanaka and Farah (1993).

Recent theories of face recognition have taken such observations alongside other evidence (see Chapter 2) that patterns of light and dark are crucial for face recognition, to suggest that memory for faces is based upon storage of variations in the image intensities of whole facial patterns, rather than on a more abstract description of face features and their individual measurements. According to this view, the brain stores faces as a summary based upon the statistical variation of whole face images in the population. Box 5A describes this theory in more detail.

Box 5A: Principal components analysis of face images

As we described in Chapter 2, a computer stores a (monochrome) image of a face as the intensity or 'grey' level in each of the pixels used to display the image (for colour images the intensity of three different colours is separately coded). With a typical face image spanning many thousands of pixels, face images take up a great deal of memory. However, by analysing the statistical variation across different pixels in a large number of different face images, a more economical way of coding individual faces can be derived.

Across a series of faces, there will be variation in the intensity shown within each pixel. For example, some men have receding hairlines, and so the pixels at the upper forehead will be light (skin), while others have a full head of dark hair and the corresponding pixels may be dark. By analysing the patterns of correlation between the grey levels in all the different pixels across a series of faces, the principal components of this variation can be extracted.

This may be made clearer with an example from a different domain. Suppose we test a large number of children aged 10 to 16 years on the time it takes them to swim one length of a pool using different swimming strokes, and the time it takes them to walk or run different distances. If we correlate all these different measures of walking, running, and swimming along with the height and weight of each child, we shall extract some underlying dimensions that account for the variation more economically. The first component in our hypothetical example would probably correspond to the 'age' of the child—as age rises so children will tend to be taller, heavier, and be able to move faster, however tested. However, there would be other components too. Once age is accounted for, a second component might reflect overall athletic ability; children may tend to be relatively good or relatively poor at all the field and pool tests. Further components might reflect swimming or running abilities, and so forth.

Returning to faces, individual faces are like the different children in our example above, and the grey level of each pixel is like the score on

Box 5A: Continued

each different test. From this large amount of data, a more economical set of components can be extracted which accounts for the majority of the variance in the image pixels. The extent to which each pixel within the set of faces contributes to a particular component can be represented graphically by depicting extreme contributions by white and black. Such graphic ways of depicting the different components of variation are called 'eigenfaces'. The top row of Fig. 5A.1 shows the first four eigenfaces extracted from a set of 174 male faces, aligned with each other by bringing their eyes into correspondence.

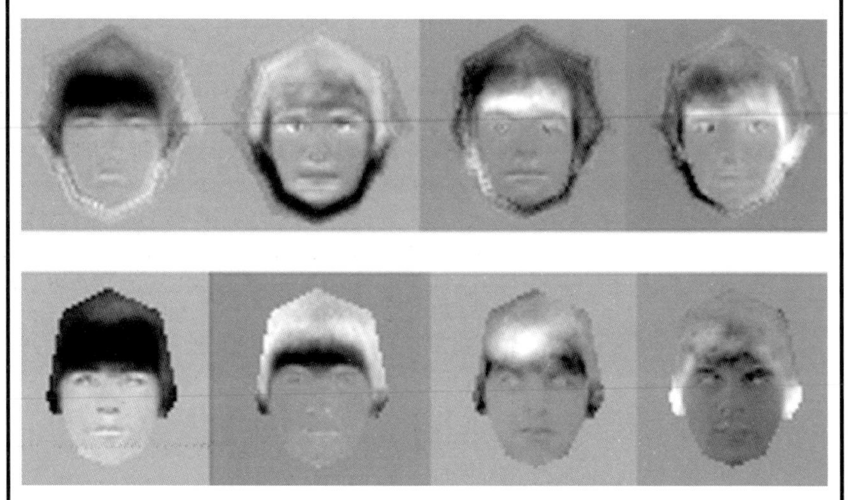

Fig. 5A.1

The lower-order eigenfaces, which capture the most variance between the set of faces, seem to extract interesting overall variations between sub-groups of faces. For example in Fig. 5A.1, the first eigenface captures gross variations in hairstyle which was one of the most variant properties in the set of young male faces analysed here. The second captures variations in face size especially round the chin. If the set of images analysed contains female as well as male faces, the early components seem to code overall variation in male-female properties (O'Toole *et al.* 1993), and so forth. Higher-order eigenfaces seem to capture idiosyncratic details of individual, or small numbers, of faces within the set.

If a set of eigenfaces is derived from a set of face images, then any face can be described as an appropriately weighted sum of this set of eigenfaces. The eigenface representation can thus provide an economical method of coding large numbers of faces. In addition to storing the eigenface images, only the weights for each individual face are needed.

Of more interest to neuroscientists is evidence that the human brain may itself do something rather like an eigenface analysis when storing faces. Alice O'Toole and her colleagues (1994) have shown that there is a strong correlation between human memory for individual faces and how well they can be reconstructed using principal components analysis (PCA). Moreover, O'Toole *et al.* (1994) showed that a PCA-based model produced a good simulation of the other-race effect (see Chapter 3). A PCA-based model exposed mainly to faces of one race will code faces of another race less successfully.

For eigenface methods to work well, faces must be brought into alignment, otherwise, for example, the grey level in the pixel for the tip of the nose in one face will be compared with that for the upper lip of another. Clearly, if there is to be a sensible interpretation, like must be compared with like. To achieve a good alignment of face images, faces can be morphed to a common shape before PCA is conducted (see Boxes 5B and 6A for more details of morphing), and analysis conducted both of the grey levels in the 'shape-free' (morphed) images, and of the shape vectors (the transformations needed to restore the original shape to the face). The lower row in Fig. 5A.1 shows the first four eigenfaces extracted after all the 174 male faces were morphed to a common shape. Note the absence in the lower row of any variations around the bottom of the face. All four eigenfaces here seem to be coding aspects of hairstyle. Peter Hancock and colleagues at the University of Stirling (Hancock *et al.* 1996, 1998) have shown that separating shape information in this way provides a PCA model which improves the correlation between the PCA-based memory for individual faces, and human memory for, and perceived similarity between, the same face images.

Principal components analysis is one of a number of techniques for coding faces in terms of their *image-features* rather than more abstract descriptions of face features (such as nose length) or three-dimensional shape. Other techniques are based more closely on the known image-filtering properties of the human visual system (see for example Lades *et al.* 1993), and current research is investigating which of these models most closely resembles the way that the human brain codes and stores faces in memory (Hancock *et al.* 1998).

Not all face features are remembered equally well. When unfamiliar faces must be recognized, the external features of hairstyle and head shape dominate memory, perhaps because these occupy such a large part of the image (see Box 5A). As faces become familiar, there is a shift in memory so that the internal face features become relatively more salient. This may be because hairstyle varies across encounters with familiar faces while internal features do not, and internal features must be attended to in face-to-face communication but not external ones.

5.3 Orientation: why are upside-down faces so hard to recognize?

Our skill at recognizing faces seems to depend upon an increasing sensitivity to the configuration of the face as a whole—the spacing and relationship between different features. This ability to perceive subtle aspects of face configuration depends upon faces being seen in their normal, upright orientation. For example in Fig. 5.1, you will find it more difficult to notice the changes in the placement of the face features of Paul Newman and Tom Conti in the inverted panels. Upside-down faces look strange to us, and we find them very difficult to recognize. In typical experiments recognition of familiar faces may be 95 per cent accurate when they are upright but drop to only 50–60 per cent accuracy when the same faces are shown upside-down. Try looking at the 'Sergeant Pepper' album cover shown in Fig. 5.4 and see how many of the faces you can identify. Now turn the book the other way up and try again.

Our difficulties in recognizing upside-down faces seem to arise because we are relatively insensitive to the spatial relationships between the features of upside-down faces, and it is these spatial relationships which carry much of the information about personal identity. One striking demonstration of our insensitivity to spatial relationships in upside-down faces comes from the 'Margaret Thatcher' illusion, discovered by Peter Thompson (Thompson 1980). Here, the eyes and mouth in a smiling face are cut out and turned upside down (see Fig. 5.5: of course it does not *need* to be Margaret Thatcher's face for the illusion to work, but it does need to be an expressive face). When the result is viewed with the entire face in its usual orientation, the face appears to have a grotesque expression. However, if it is viewed upside-down it is difficult to see that there is anything at all abnormal about the face. It appears that the relationship between the different face features which is so apparent in the upright head is almost invisible in the inverted one.

James Bartlett and Jean Searcy (1993; Searcy and Bartlett 1996) have explored this effect systematically by asking volunteers to rate how grotesque different faces appear to be when presented upright and upside-down. Grotesqueness due to spatial alterations such as those in the Thatcher illusion, or simpler changes such as those illustrated in Fig. 5.1, is virtually abolished by inversion. However, if local changes are made to faces, by blacking out teeth to produce grotesque 'vampire' faces, these remain grotesque when seen upside-down.

Consistent with the suggestion that it is spatial relationships between different features which cannot be seen when faces are turned upside-down, the composite effect described earlier (see Fig. 5.2) disappears when the face is inverted. Young *et al.* (1987) showed that

Fig. 5.4 Upside-down faces. Peter Blake's 1966 cover of The Beatles' *Sergeant Pepper's Lonely Hearts Club Band*. © Peter Blake. All rights reserved, DACS, 1998.

people actually become better at identifying one of the half-faces in an inverted composite than in an upright one! While this may seem paradoxical, given that upside-down faces are harder to identify, the effect can be explained as follows. The reason that half a face within a composite is difficult to identify when the composite is shown upright is because a new face identity arises from the combination of the upper and lower face features. If the relationship between these different halves of the face is made difficult to see, by inversion, then it is relatively easier to access the identity belonging to the top or bottom features alone.

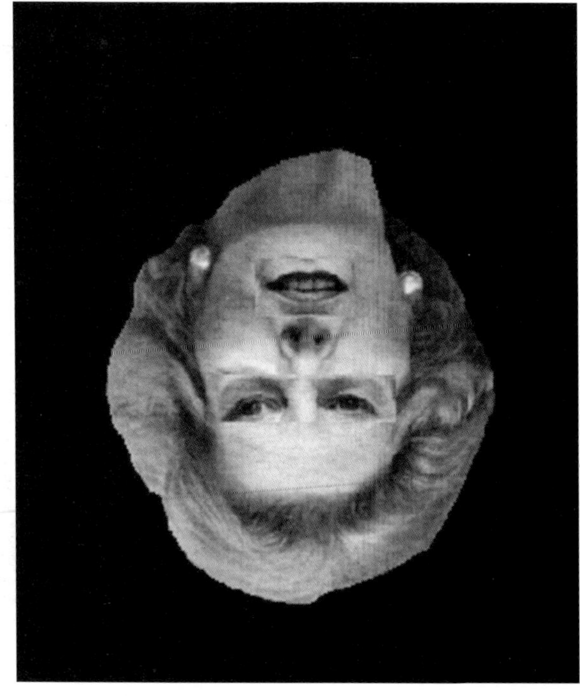

Fig. 5.5 The Margaret Thatcher illusion. The upside-down face shown here looks reasonably normal. It is possible to tell that something has been done to the image on the right, but not easy to know exactly what is unusual about it. Turn the page upside-down and a very different impression is created!

Our stored visual memories of faces therefore seem to be based upon several distinct types of information. Information about isolated features allows us to identify people from individual distinctive pieces of local information, such as the former Soviet President Gorbachev's birthmark. Accurate information about the spatial relationships between different features is also represented in memory, which is why the faces in Fig. 5.1 appear so distorted when their features are displaced. Finally, experiments such as Tanaka and Farah's (1993), and models of storage such as that of Principal Components Analysis (Box 5A), suggest that the face may be stored as a non-decomposed or abstract 'holistic' image, rather than (or perhaps as well as) being stored as a list of features or their relationships. Figure 5.6 summarizes these different forms of information upon which our visual memories for faces appear to be based.

Our stored memories of faces are remarkably resilient. In Fig. 5.7, Dali conveys a perfect likeness of the actress Mae West despite depicting each of her features as an item of interior decor. This image shows how powerfully the face schema overcomes other distorting influences, and shows that even individual identity can be preserved in such circumstances. The artist Giuseppe Arcimboldo (Fig. 5.8) made use of a similar device in constructing faces from arrangements of fruit and vegetables. The right panel of Fig. 5.8 is particularly striking, since the face is almost completely invisible until the page is inverted.

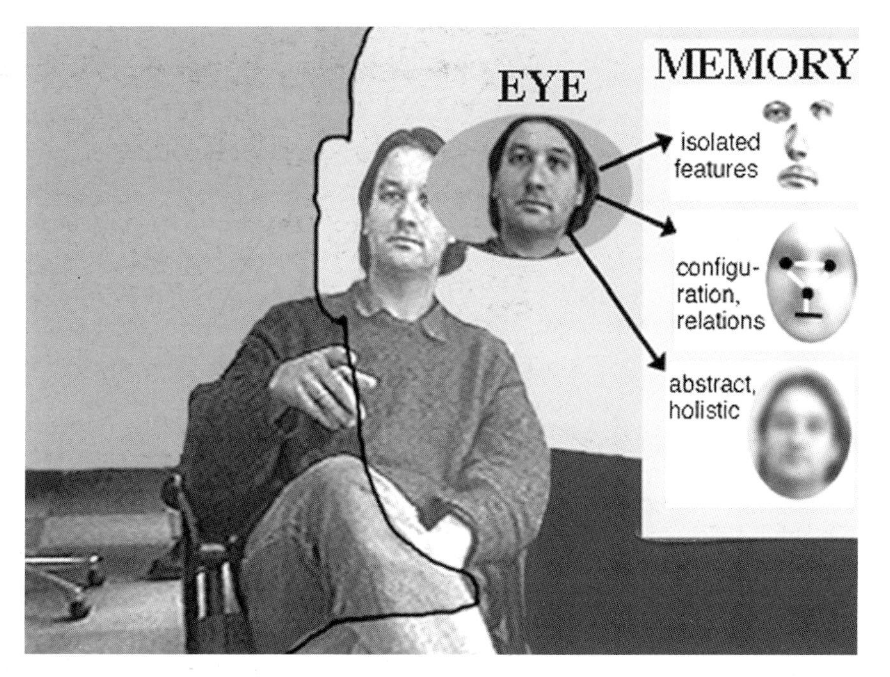

Fig. 5.6 A schematic illustration of the different kinds of information that may form the basis of our visual memory for faces. The figure shows Helmut Leder, University of Fribourg, who created this image.

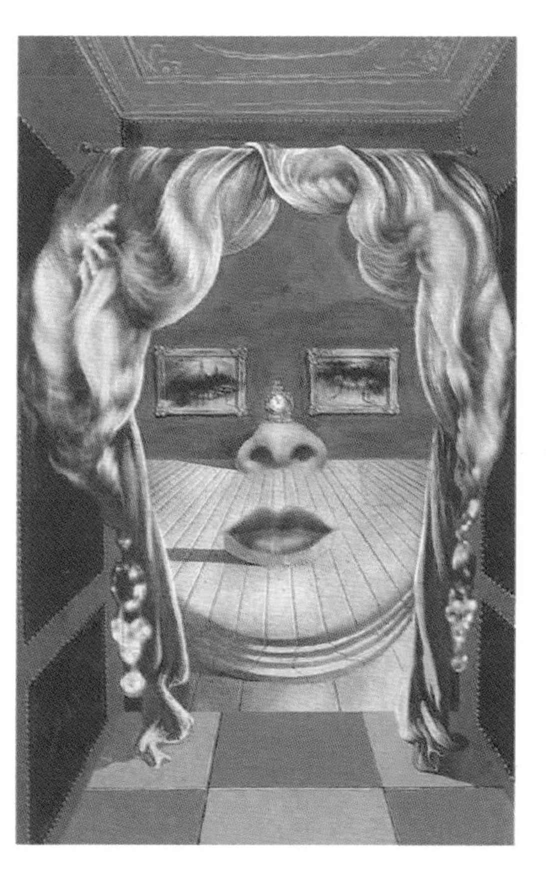

Fig. 5.7 Salvador Dali, *Mae West*. Art Institute of Chicago, Gift of Mrs Gilbert W. Chapman in memory of Charles B. Godspeed. © ADAGP, Paris and DACS, London, 1998.

Fig. 5.8 Two paintings by Giuseppe Arcimboldo. In the left panel (*Rudolph II as Vertumnus*, c.1570. Skoklosters Slott, Sweden) the face is clearly visible in the arrangement of fruit and vegetables. The right panel (*The vegetable gardener*, Museo Civico Ala Ponzone, Cremona) appears to be a simple arrangement of garden produce until the page is inverted to reveal the hidden face.

5.4 Shape and structure

In Chapter 1 we described how the appearance of an individual face results from a subtle interaction between underlying bone structure, the thickness of the fat layer between bone and skin, and the texture and pigmentation of skin and hair. These different sources of information, from the three-dimensional shape of the face to the superficial features of skin colouration, all contribute to the process of identification.

The three-dimensional shape of the face is revealed from a combination of the surface features which are visible and the pattern of light and shade which can be used to derive surface information (see Chapter 2). Some evidence that three-dimensional shape may be important in face recognition comes from the advantage found for three-quarter views rather than full-face images in face recognition (Fig. 5.9). A three-quarter view reveals more about the way that a face is structured in depth. The shape of the nose in particular is very difficult to see in a full-face image and much clearer from an angle. Interestingly, full-face portraits are rare. Artists seem naturally to prefer a viewpoint which reveals more about the shape of the face.

Fig. 5.9 *Sir Daniel Wilson* by Sir George Reid. Courtesy of the Scottish National Portrait Gallery.

However, where experiments have shown advantages for the three-quarter view in face recognition, it tends to be in memory for previously unfamiliar faces. Familiar face recognition seems to be equally easy from full-face and three-quarter views (Bruce *et al.* 1987). This suggests that the three-quarter view may be useful because it allows generalization to a broader range of views than does a full-face image. When a face is already familiar our more frequent exposure to the full-face image when people are interacting, or shown in conversation on television, may help offset any natural advantages given by the angled view.

Although the effects of viewpoint suggest some role for three-dimensional shapes in face recognition, it is actually remarkably difficult to identify faces when *only* the three-dimensional shape is given. Bruce *et al.* (1991) asked a number of their university colleagues to have their faces measured using a laser scanner (see Box 1C, Chapter 1). Surface images of these faces were then depicted using the techniques described earlier, and the resulting images shown to friends and students. Identification rates were remarkably low, and much lower for female faces than for male ones. This demonstrates the importance of superficial features and colouration for our normal recognition processes.

Of course classical sculptures lack these features, though hairstyle is sculpted along with the face. Figure 5.10 shows a picture of Robert Cunninghame Graham (by Raeburn) which includes a bust of someone else (Charles James Fox), which could easily be mistaken for the same person because of the lack of pigmentation. Some classical artists used to paint their busts, presumably to enhance the resulting likeness.

Interestingly, one of the main difficulties with classical busts (and

Fig. 5.10 *Robert Cunninghame Graham*. Attributed to Sir Henry Raeburn. Courtesy of the Scottish National Portrait Gallery.

modern laser scans) of this kind may be the lightness of the pupil area in the eye. In Fig. 5.11, a photograph of a bust of George Combe (left) has had dark pupils added back (centre). The result is immediately more life-like. Classical sculptors sometimes tried to rectify the lifeless quality of their busts by etching circles onto the eye, as has been shown in the right hand side of Fig. 5.11. However, the result is much less impressive than the addition of dark pupils.

Figure 5.12 shows a delightful three-dimensional portrait of the fashion designer Jean Muir, in which a combination of pigmentation and three-dimensional shape is used effectively to convey her likeness. Figure 5.13 shows an unusual Japanese portrait sculpture of the Muromachi period. Not only were the eyes and lips painted onto the wooden sculpture, but the actual hair of the dead Zen monk was originally implanted in the scalp and around the mouth of the face!

Just as upside-down faces are difficult to recognize, so are photographic negatives. Bruce and Langton (1994) compared the effects of inverting and negating faces and found negation had an even more detrimental effect than inverting the image. In Chapter 2 we considered different reasons why photographic negatives may be difficult to identify, and concluded that at least part of the difficulty may arise because negatives make it difficult to derive a representation of three-

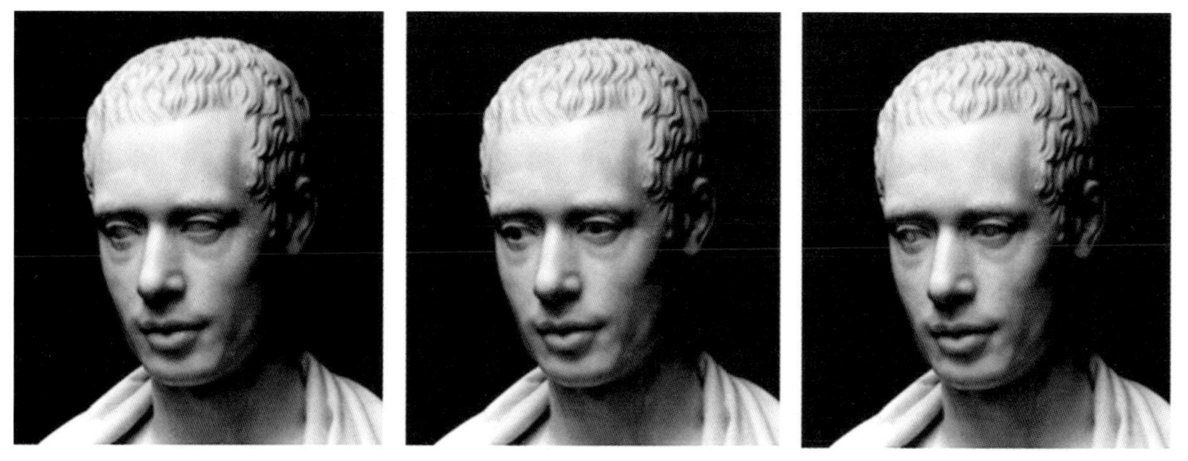

Fig. 5.11 Original (Top) and computer-manipulated versions of the bust of *George Combe* by Lawrence Macdonald, courtesy of the Scottish National Portrait Gallery. *Bottom*: original bust of Combe (*left*) with pupils darkened (*centre*) or added by etching (*right*). Images produced by Helmut Leder, University of Fribourg.

dimensional shape from shading. However, other factors are also likely to contribute. The reversal of brightness in the eyes may make it difficult to encode the face, in the same way that the classical sculptures seem to suffer so much from their 'white' eyes (see Fig. 5.11). Moreover, negative images reverse the brightness of important pigmented areas, so that light skin becomes dark, and dark hair becomes light, potentially 'disguising' the face.

5.5 Distinctiveness and caricature

Although all faces are built to the same basic template, and individual differences between them can be very subtle, some faces deviate more from the average or prototype face, while other faces have a more average or 'typical' appearance. Experiments on human face recognition have shown that faces which are more deviant or

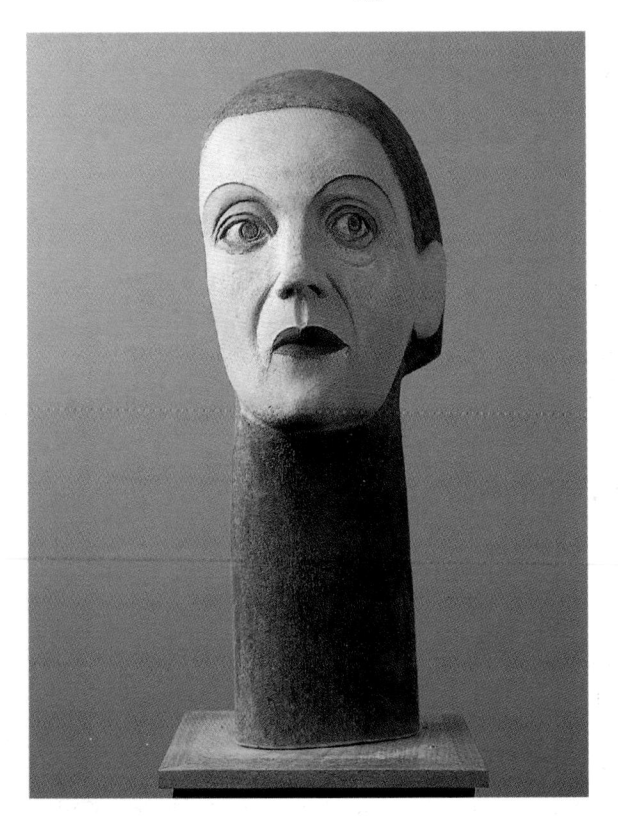

Fig. 5.12 *Jean Muir* by Glenys Barton. Courtesy of the Scottish National Portrait Gallery.

Fig. 5.13 *Ikkyu Osho* (c. 1481). Detail. From Mori (1977).

'distinctive' in appearance are recognized more quickly and accurately than those which are more typical.

Effects of distinctiveness can be revealed in a number of different tasks using faces. With famous faces, it has been found that those which are rated as distinctive in appearance can be recognized as familiar more quickly than those which are rated as more typical in appearance, even though performance is extremely accurate on all the faces. So, a face like that of Prince Charles will be recognized as familiar faster than that of Tony Blair. Using unfamiliar faces, distinctiveness gives advantages on tests of memory. Studied faces which were rated as distinctive are more likely to be recognized correctly when the task is to decide which of a set of faces were studied earlier. If distinctive faces are placed among the non-studied items in the test series, they are less likely to be remembered falsely than more typical faces.

However, when the task is changed from one of recognition to that of classifying the face as a face (where normally arranged faces are interspersed with jumbled faces or other non-face objects) then for both famous and unfamiliar faces, typical faces have the advantage. So, people are slower to decide that, for example, Prince Charles' face is a face, than to decide that Tony Blair's face is a face (Valentine and Bruce 1986).

One simple way of understanding the effects of distinctiveness is the 'face space' framework described by Tim Valentine (e.g. Valentine

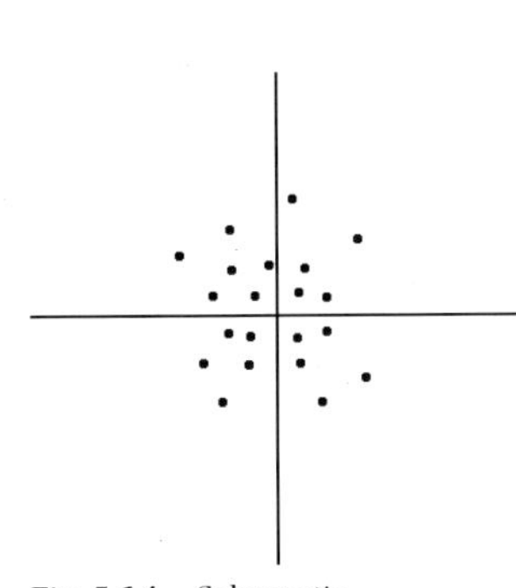

Fig. 5.14 Schematic representation of 'face space', after Valentine (1991).

1991, see Fig. 5.14). This theory proposes that any face can be described by its value along each of a number of dimensions of facial variation. Figure 5.14 simplifies this by showing just two dimensions, but the expectation is that there are a large number of dimensions needed to fully characterize facial appearance. Dimensions could be simple features such as nose length, or global characteristics such as age or face elongation. Faces which are rated as more typical will tend to have values on the dimensions which are true of many faces (e.g. a nose of average length), while those which are rated more distinctive will have more extreme values (e.g. very long or very short noses). In the diagram in Fig. 5.14, then, typical faces will tend to cluster more closely together within the space, while distinctive ones are scattered around the periphery.

The task of identifying a face requires a comparison of the dimensions of a to-be-recognized face with dimensional descriptions of faces which have already been stored, to see whether a stored face can be found which shares this same set of physical dimensions. Where there are many stored faces with similar physical characteristics it will be more difficult to distinguish true from false matches. More precision will be needed on each of the physical dimensions and so the process may take longer, or be more prone to error. There will be few competing descriptions in the area of space occupied by distinctive faces and so matching the description can be achieved more readily. However, if the task is to classify the pattern as a face (rather than a nonface) then the task is to see whether a test pattern conforms to the characteristic of the basic face template. A test pattern which is a typical face will resemble a large number of similar faces within the face space, allowing a relatively fast positive response to be made. A distinctive face, in contrast, will resemble few other patterns in the space, and therefore will get less immediate support for the positive decision, which will therefore take longer to reach.

The face-space metaphor is a simple and useful way to think about typicality and distinctiveness, but in a multi-dimensional rather than two-dimensional space the predicted distribution of faces gets rather more complicated than might be supposed. In particular, it can be shown that *extremely* typical faces (i.e. those which are near-average on all their dimensions) will actually be rather rare (Burton and Vokey 1998). This seems to mesh with Sir Francis Galton's (1883) anecdotal observation about the rarity of what he describes as a typical 'John Bull' (Englishman):

> One fine Sunday afternoon I sat with a friend by the walk in Kensington Gardens that leads to the bridge, and which on such occasions is thronged by promenaders. It was agreed between us that whichever first caught sight of a typical John Bull should call the attention of the other. We sat and watched keenly for many minutes, but neither of us found occasion to utter a word. (Galton 1883, p. 4)

Fig. 5.15 Veridical line-drawing of ex-President Nixon (*left*), and caricature of Nixon (*right*). Reproduced from Perkins (1975).

Distinctiveness may help us to understand how caricatures work; by exaggerating an individual's idiosyncratic features, caricatures exploit distinctiveness by making that face less like others. Caricatures of ex-President Richard Nixon (Fig. 5.15) for example, exaggerated the extended bulb of his nose, and the bags under his eyes. In so doing, they were making the representation non-veridical. Nixon's nose was not *that* big, nor his eyes *so* baggy, yet they rendered the representation better able to characterize Nixon than other faces. Goldman and Hagen (1978) studied caricatures produced of Richard Nixon by seventeen different artists during 1972–1973, and found that there was a great deal of consistency across different artists in terms of which features of the face were distorted in their drawings, though considerable variation in the extent of the distortions.

In terms of the face space framework, the caricature produces a description which is more extreme (further out in space) than the actual Nixon, but this representation is less likely to be confused with any face other than Nixon's. This insight can be used to create line-drawn and photograph-quality caricatures by computer.

Susan Brennan (1985) first reported a technique for generating caricatures automatically by computer, and the technique rests on the suggestion that a caricature 'is a symbol that exaggerates measurements relative to individuating norms' (Perkins 1975). Brennan digitized photographs of faces and located a set of 186 key points which described the major face features. For example points within the set of 186 included the inner and outer corner of each eye and a small set of additional points along the upper and lower eyelids. These points could be linked together to form a line-drawing of the original face, as shown in Fig. 5.16. If the same set of points is measured for a large number of faces, and the coordinates of all these faces are scaled to coincide at the pupils, then the average locations of

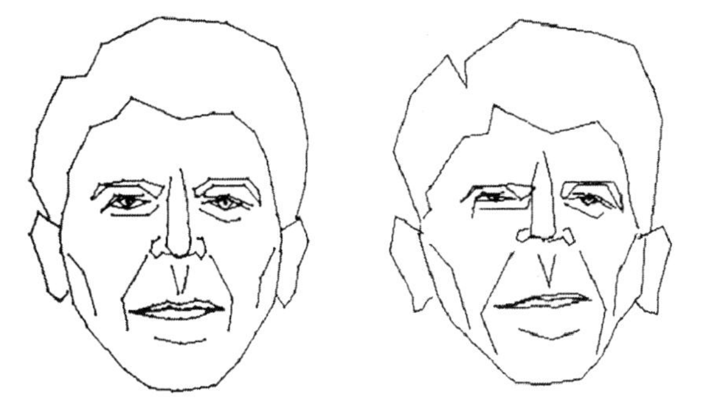

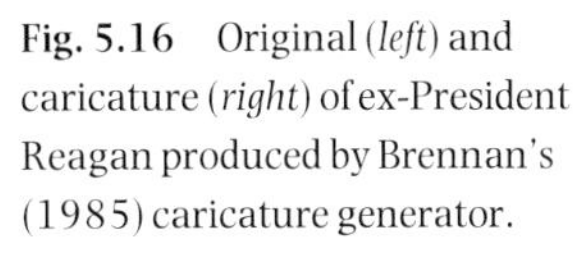

Fig. 5.16 Original (*left*) and caricature (*right*) of ex-President Reagan produced by Brennan's (1985) caricature generator.

each point can be computed, to produce a line-drawing of the 'average' or 'norm' face. To caricature a specific face, the locations of the individual points for that face are compared with the average and exaggerated by multiplication. The effect of this multiplication is that large deviations from the norm are increased more than small deviations, so producing distortions which are greatest for the most deviant aspects of the face. Figure 5.16 shows a resulting caricature of ex-President Ronald Reagan's face, which is much more recognizable than the line drawing to the left which is based on Reagan's uncaricatured point locations. Figure 5.17 shows the results of applying this process to a number of different famous faces and then fitting a line drawing to the points using smooth curves rather than straight lines.

Rhodes *et al.* (1987) produced line-drawn caricatures like this of faces which were personally familiar to the participants in their experiments and were able to confirm that positive caricaturing of this kind made line drawings of faces more recognizable. It is also possible to make faces less recognizable by making 'anticaricatures', by shifting their features towards the average face—in effect making a face more typical in appearance. Faces can be caricatured and anticaricatured to greater or lesser extent by exaggerating or reducing the difference between the face and the norm by smaller or larger percentage changes. Figure 5.18 shows a sequence of caricatures and anti-caricatures of the British comedian, Rowan Atkinson, which clearly demonstrates the enhanced likeness achieved with modest degrees of positive caricaturing. These images have been additionally enhanced, and become better representations as a result, by filling in the dark areas of the hair (cf. Chapter 2 where we discussed the importance of preserving areas of light and dark in face images).

The technique of describing a face by a fixed set of key coordinates and smoothly moving the face towards or away from some other set of coordinates (in this case, the coordinates corresponding to the average face) is the principle which underlies the now common technique of computer 'morphing' which allows a smooth transformation from one image (for example a dog face) through to another (for example a

PRINCE CHARLES

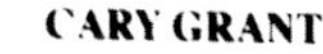

LEONARD BREZHNEV

CLINT EASTWOOD

RONALD REAGAN

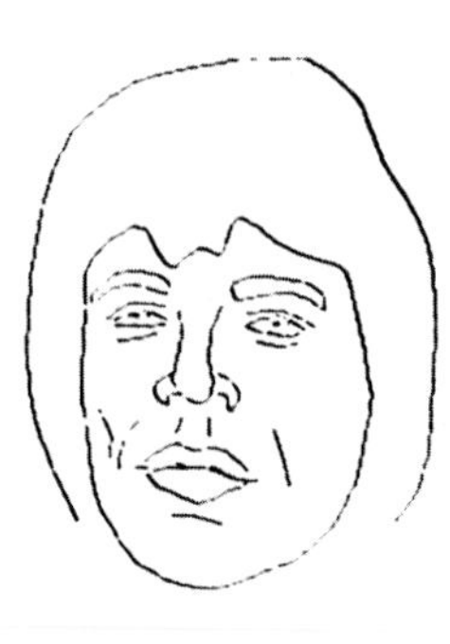

MICK JAGGER

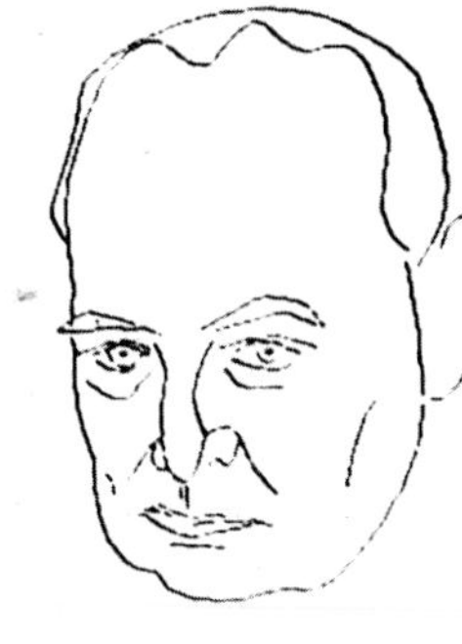

WINSTON CHURCHILL

RICHARD NIXON

Fig. 5.17 Caricatures of famous faces produced with Brennan's (1985) caricature generator. From Rhodes and Tremewan (1996). Reproduced by permission of Psychology Press.

man's face; see 'Scotty Dog' Chapter 1, Fig. 1.4(b)). This technique was anticipated by early caricature artists as illustrated by the political cartoon *Les Poires* (Fig. 5.19).

This famous cartoon was produced by Charles Philipon in 1834, in protest at his prosecution for adopting '*La Poire*' as an emblem for King Louis-Philippe, an emblem which played upon the pear-like shape of the monarch, using the slang word for 'Fathead'. In *Les Poire* the

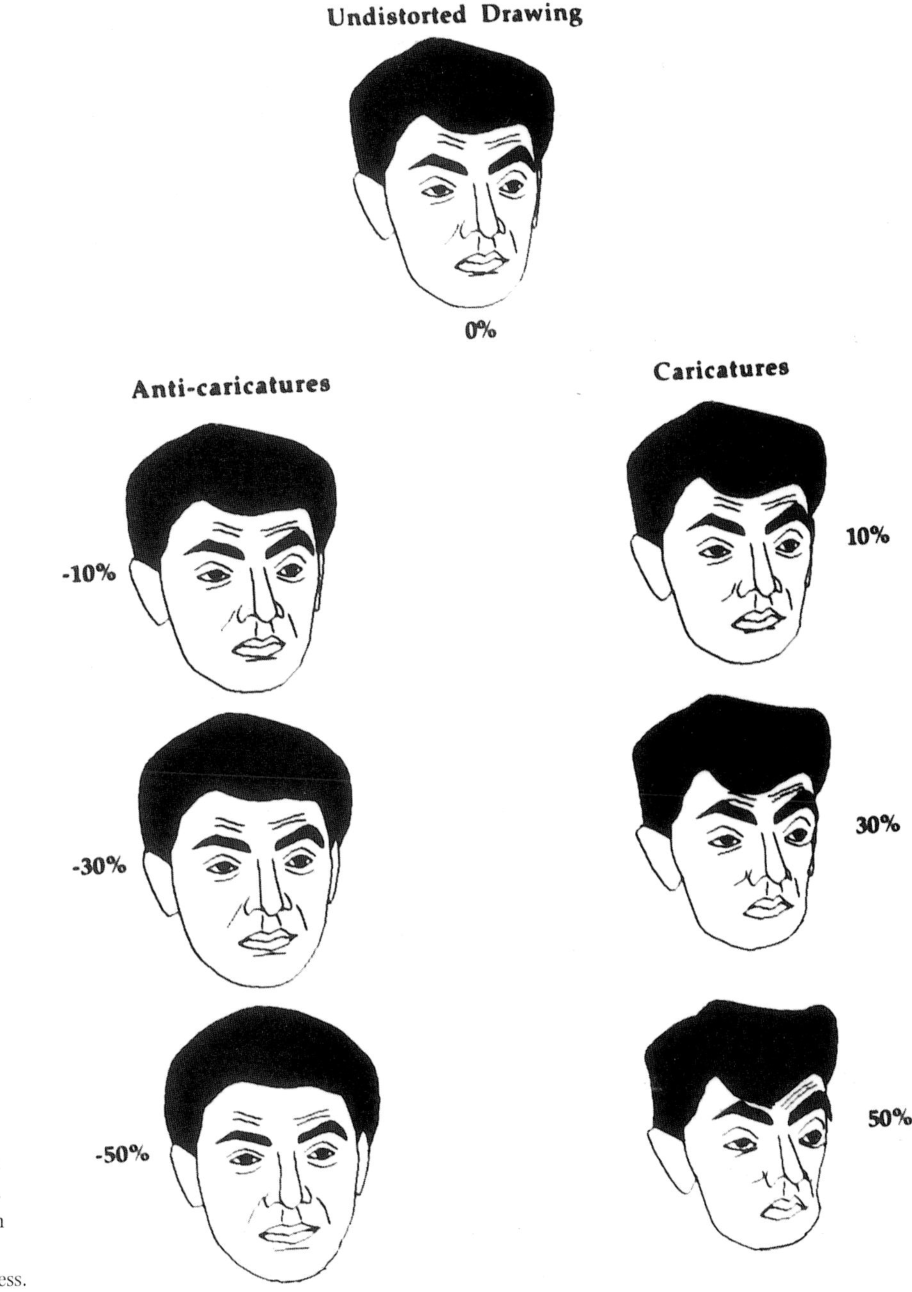

Fig. 5.18 Caricatures and anticaricatures of the British comedian, Rowan Atkinson. From Rhodes and Tremewan (1996). Reproduced by permission of Psychology Press.

gradual 'morph' between the image of the King and of a pear was created by Philipon to protest that all pear-like images would have to be banned in order to avoid offense. The censors were apparently persuaded by this argument and proceeded to ban pears! (Rhodes 1996).

Using high-speed modern graphics computers it is now possible to

LES POIRES,
Vendues pour payer les 6,000 fr. d'amende du journal le *Charivari.*
(CHEZ ALBERT, GALERIE VERO-DODAT)
Si, pour reconnaître le monarque dans une caricature, vous n'attendez pas qu'il soit désigné autrement que par la ressemblance, vous
tomberez dans l'absurde. Voyez ces croquis informes, auxquels j'aurais peut être dû borner ma défense :

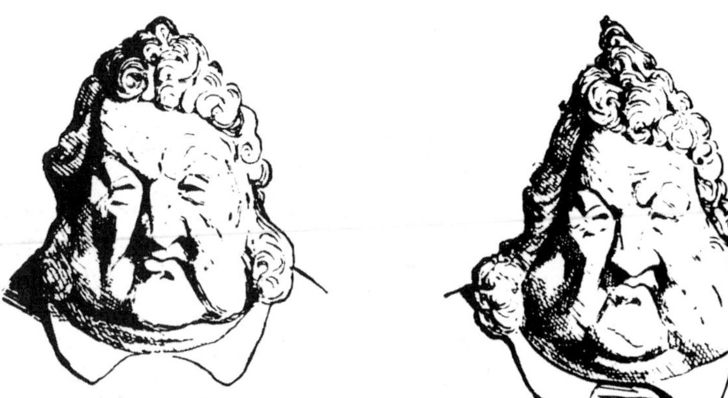

Ce croquis ressemble à Louis-Philippe, vous condamnerez donc ?

Alors il faudra condamner celui-ci, qui ressemble au premier.

Et enfin, si vous êtes conséquens, vous ne sauriez absoudre cette
poire, qui ressemble aux croquis précédens

Fig. 5.19 Charle Philipon's *Les Poires*.

produce computer-caricatures (exaggerations away from the norm) and computer morphs (blends between two different images) in full photographic-quality images rather than just in line-drawings. A number of the novel illustrations in this book were produced at the Universities of St Andrews and Stirling using such techniques, and the techniques themselves have become popular in film and video work, often used dynamically to create clever effects.

Intriguingly, a computer-caricature of a photograph of a person's face produces an image which in some circumstances is judged as more like the person than their actual photo! Box 5B describes the

Box 5B: Photographic caricatures

To produce photographic caricatures by computer, Philip Benson and David Perrett (Benson and Perrett 1991*a,b*; Benson *et al.* 1992) initially proceeded by the same method as Susan Brennan—marking key points out on the computer image of the original face and comparing the locations of these points with those of a comparison face. Figure 5B.1 shows the marking out of the points on a photograph of Nicholas Parsons' face. The inset boxes provide a colour-coded 'key' for the human operator of which points should be marked on the face.

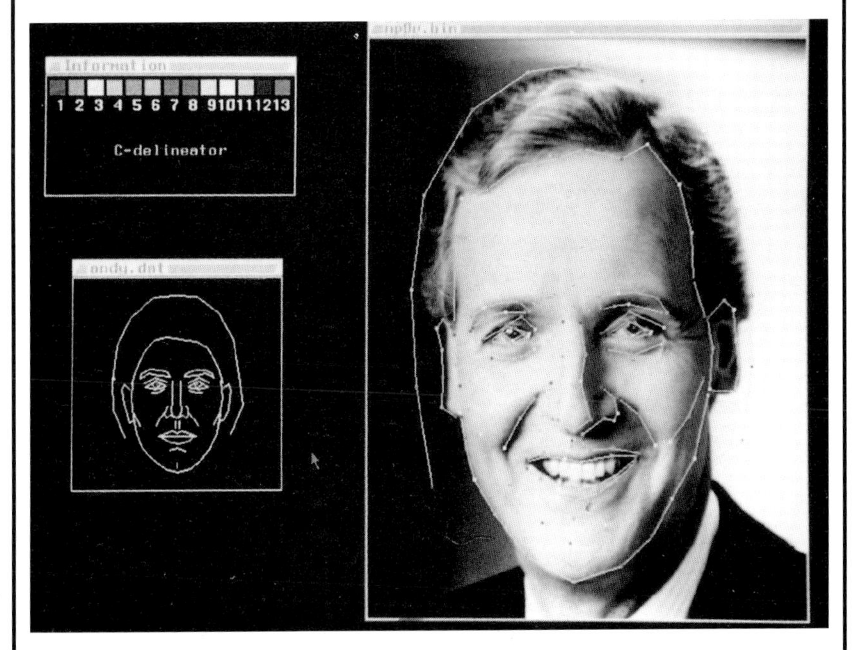

Fig. 5B.1

To produce caricatures, the comparison face would be formed from the average coordinate locations of a larger set of faces. The locations of the points on the target face can then be made to deviate more from the average face (to produce a caricature) or can be moved towards the average face (to produce an anticaricature).

To produce such caricatures in full photographic-quality images there must be a smooth adjustment to the surface information which is aligned with the relocation of the key points. The same set of key points is used to mark out a set of triangular regions within which the grey or colour values of the pixels are transformed smoothly in proportion to the stretching of their triangular boundaries. This is not completely automatic as some interventions must be made to avoid overlap of triangles after extreme caricature distortions.

Box 5B: Continued

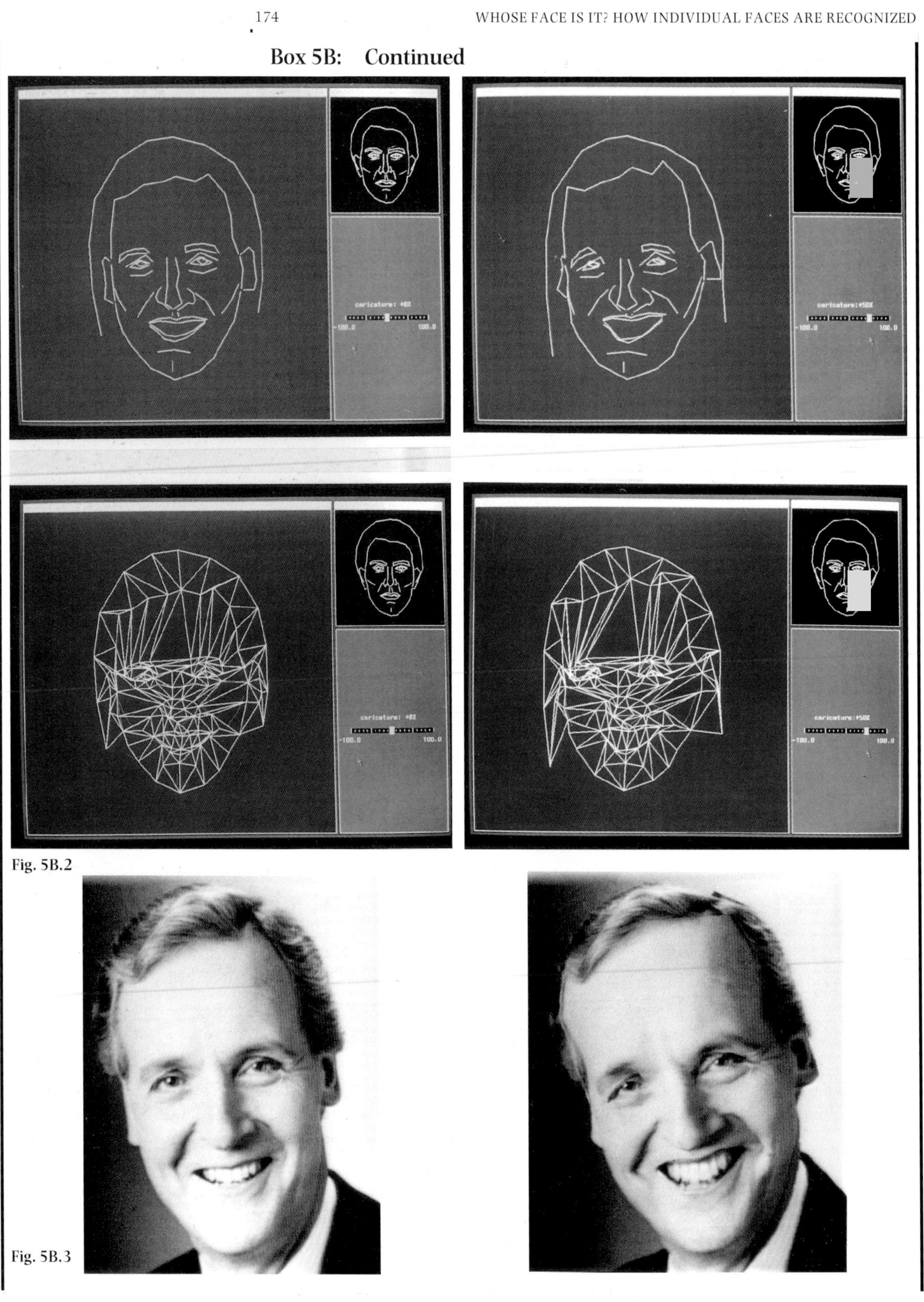

Fig. 5B.2

Fig. 5B.3

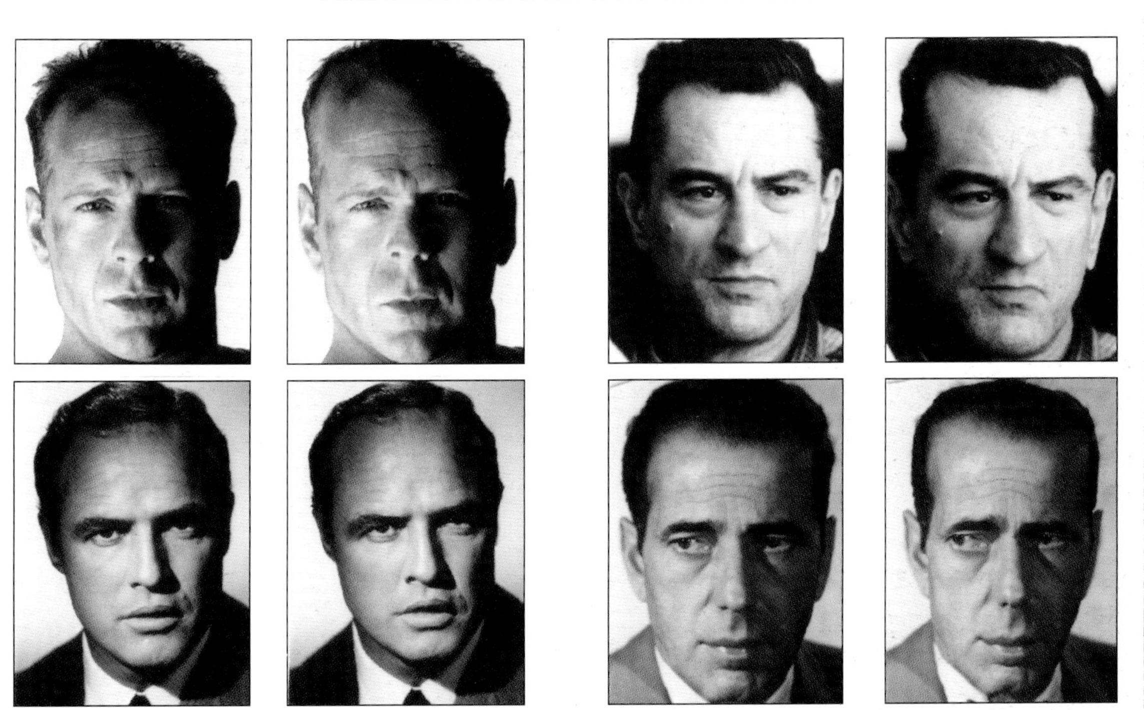

Fig. 5B.4

Figure 5B.2 shows the key coordinates and triangulation of the original face (left) and the changes to these after applying a 50 per cent positive caricature (obtained by increasing the difference between each point and its corresponding location in the average by 50 per cent). The results of this process are shown in Fig. 5B.3. The original face is on the left, and the 50 per cent caricature on the right.

Benson and Perrett (1991*b*) evaluated the results of this process in experiments where participants were asked to select the best likeness of famous faces from a set of alternatives. The average of their choices showed a small degree of positive caricaturing. Furthermore, when verifying face identities, responses to positive caricatures were also somewhat faster than responses to the original images. Figure 5B.4 shows the original (left) and caricatured versions of four American filmstars produced by Kieran Lee at St. Andrews. You can assess for yourself whether the caricatures provide a more recognizable image of each individual.

To produce *morphs* between different images involves the same process as caricaturing. The same set of key points is identified on each of the two images, and their average computed. The average brightness within each pixel in each triangle is adjusted to reflect the contribution of each of the images. Intermediate stages in the morph can be produced by taking weighted means of the feature locations and the pixel values. Box 6A provides further detail of this process as applied to the morphing of different facial expressions. Details of more elaborate image manipulations can be found in Rowland and Perrett (1995).

techniques and results obtained in experiments using computer-generated caricatures.

5.6 Eyewitness identification and forensic reconstruction

The face is not the only route to person identification. People can be recognized by their voices, clothing, characteristic posture or gait, but the face is probably the most reliable means of identification. Laboratory experiments have shown that people are remarkably accurate at remembering briefly viewed, previously unfamiliar faces, typically scoring over 90 per cent correct when asked to decide which of a large set of faces were previously studied.

This high accuracy at recognizing pictures of unfamiliar faces seems to be at odds with our sometimes embarrassing failures to recognize or identify faces in everyday life. Young *et al.* (1985) asked a small number of volunteers to keep diary records of all those occasions in their daily lives when they experienced difficulties or made errors in person identification. Over an eight week period, nearly 1000 such incidents were recorded (see Box 5C for details). In a large number of these incidents, errors were made in recognizing the face. Such confusions are embarrassing in our daily lives, but can be much more damaging still when the errors are made by witnesses to a crime.

One important factor in everyday face memory which was evident in the performance observed in Young *et al.*'s (1985) study is the important role played by context in our recognition of faces. We may fail to recognize even highly familiar faces encountered unexpectedly or in an inappropriate context. The Australian psychologist Don Thomson contrived to engineer the following encounter

> The parents of one of my students had flown from Australia to London. Soon after, and unbeknown to the parents, their daughter and a companion also travelled to London. The daughter and I arranged that she stand at a bus stop near her parents' lodgings; the companion was to stand some distance away and observe. The daughter and the companion later reported that when the parents emerged from their lodgings and saw their daughter they stopped abruptly. The father then approached the daughter and said hello to her. She turned to face him, and as instructed, looked straight through him. His greeting choked in his throat and he lamely concluded, 'I am terribly sorry, I thought you were someone else'. Thomson (1986, p. 121)

Thomson himself fell foul of a dreadful trick of context when he was falsely accused of raping a woman. The unfortunate woman had been raped while a television programme was being broadcast in which Thomson appeared discussing eyewitness testimony. His face became associated in her mind with the incident, and this led to the false accusation. Happily for Thomson, the live TV broadcast provided him a perfect alibi.

Box 5C: Studying naturalistic errors and difficulties in person identification

Andy Young, Dennis Hay, and Andy Ellis (1985) persuaded 22 people to keep diaries for an eight-week period, in which they kept systematic records of errors or difficulties that they had in recognizing people by any means, including face and voice recognition. During the two-month period, the majority of the diarists' recorded incidents were of the following types:

1. *Where a familiar person went unrecognized.*
One hundred and fourteen separate incidents of this kind were reported. For example 'I was going through the doors to B floor of the library when a friend said "hello". I at first ignored him, thinking that he must have been talking to the person behind me.'

2. *Where a person was misidentified.*
There were 314 separate incidents here. Sometimes an unfamiliar person was misidentified as a familiar one: 'I was waiting for the phone. A lot of people were walking past. I thought one of them was my boyfriend'; or one familiar person was identified as another 'I was watching television. I knew that both Ralph Richardson and John Gielgud were present at the function being shown; but I mistook John Gielgud for Ralph Richardson. I realized my mistake when the real Ralph Richardson stood up'.

3. *Where a person seemed familiar, but the diarist did not know why.*
Two hundred and thirty-three incidents of this type were recorded. In some the person was eventually identified successfully 'I was in the back, waiting to be served. I saw a person and I knew there was something familiar immediately. After a few seconds I realized she was from a shop on campus or a secretary in one of the departments. I eventually remembered by a process of elimination.' At other times there was no resolution 'I stopped a passer-by to ask directions. She looked familiar and then she spoke as though she knew me. When she looked as if she knew me I pretended to recognize her too, but didn't ask her name. After 10 minutes I stopped trying to think who she was.'

4. *Where only partial details about a familiar person could be retrieved.*
There were 190 incidents of this type, most often involving failure to remember a person's name 'I saw another student walking past, but I couldn't remember his name, even though I'd been talking about him only a few days ago. Someone had to tell me it.'

Box 5C: Continued

These were the major categories of difficulty reported. Importantly, some patterns of difficulty were never observed. For example, there were no cases where someone remembered somebody's name, but their face did not seem familiar. Follow-up studies (Hay *et al.* 1991) have provoked errors in person recognition by showing large numbers of faces to be identified in the laboratory. In such circumstances, naming errors are also very common, but people never produce names without knowing that the person is familiar to them, and knowing something about why they are familiar (for example the occupation of the person).

Fig. 5.20 Bill Clinton and Al Gore? From Sinha and Poggio (1996).

Figure 5.20 illustrates the power of context in our recognition processes. At first glance, the reader will see President Bill Clinton with his deputy Al Gore. Only on closer inspection does it become clear that both faces are in fact Clinton's. The President's internal features have been transposed into Al Gore's head outline. The power of expectation, plus the tendency for part faces to recombine to form new identities (see Section 5.2) mean that this is likely to have gone unnoticed without some prompting to the reader to inspect the image more closely.

Much of the recent research interest in the processes of human face recognition was stimulated by some high-profile cases of mistaken identity in criminal investigations in the UK which prompted a public enquiry chaired by Lord Devlin (1976). In one such case, Laszlo Virag was convicted of an armed offence on the testimony of eight separate witnesses who picked Mr Virag from identification parades and/or from photographs. One police witness testified that 'his face is imprinted in my brain'. Subsequently, Georges Payen, a man with a passing but not striking resemblance to Mr Virag (see Fig. 5.21), was convicted of this and a number of other related offences. While it is particularly disturbing that so many different witnesses should have made the mistaken identification of Mr Virag, the fact that identification of once-glimpsed faces is error-prone should not be surprising given the regularity with which there are breakdowns in identification in much easier circumstances.

Why is there this apparent inconsistency between accurate memory for faces in laboratory experiments and inaccurate memory in everyday life? One important factor is that laboratory experiments tend to test memory for identical pictures of faces, and thus confound picture memory with true face memory. If recognition memory for faces is tested with different pictures of the same people shown in the study and test phases, recognition accuracy drops dramatically. For example, in one study by Vicki Bruce (1982), recognition was 90 per cent correct when faces were shown in identical views at study and test, but dropped to only 60 per cent correct when the viewpoint and expression were changed from study to test, even though the particular faces used in that study had distinctive hairstyles and clothing was not obscured. Other factors which have been shown to influence face recognition, and which are likely to affect the eyewitness, are a change in the context between the place where the face was originally

Fig. 5.21 Laszlo Virag and Georges Payen.

encountered and where memory for the face is subsequently probed, with context broadly defined to include such things as the clothing of the person to be identified. Moreover, most criminals attempt to disguise their faces somewhat when committing a crime, even if only by wearing a hat or hood over their hair. Hairstyle is such an important determinant of the representation of an unfamiliar face that even crude disguises of this sort will make the later memory task much more difficult. One striking aspect of the comparison between Mr Virag and Mr Payen was that their resemblance to each other was much closer for the lower half of their faces, and in the original incident the 'Gunman of Liverpool' wore a hat, thus concealing those aspects of appearance which would most readily have distinguished Mr Virag.

While all the above factors are likely to contribute to poorer memory for faces in everyday life than in the laboratory, another important set of factors which affects the reliability of investigatory and forensic evidence involving faces arises through the methods that are typically used to probe witness memories. These methods fall broadly into two groups: reconstruction and identification, and there are problems within each group.

Reconstruction is used when criminal investigators attempt to produce, from a witness description, some image of the suspect for circulation to other police forces or to the general public. The traditional method for producing such likenesses used a police artist, who somehow converted the witnesses' description into a sketched portrait, which could then be refined in discussion with the witness. However, the disadvantage of this method is that trained police artists are relatively few and far between. It is therefore not surprising that there has been such enthusiasm for alternative methods that can be used by a witness working relatively independently.

A number of methods have been devised in which a witness, working alone or with a trained operator, tries to construct the target face using a 'kit' of isolated facial features. This method is reminiscent of Leonardo da Vinci's attempt to construct an inventory of all possible face features in order to instruct people how to draw portraits from a single glance at a face (the letters refer to illustrations which accompanied the text):

> To start with the nose: there are three shapes - (A) straight, (B) concave and (C) convex. Among the straight there are only four varieties, that is (A1) long, or (A2) short, and (A3) at a sharp angle or (A4) at a wide angle. The (B) concave noses are of three kinds of which some (B1) have a dip in the upper part, some (B2) in the middle, and others (B3) below. The (C) convex noses again show three varieties: some (C1) have a hump high up, others (C2) in the middle and others (C3) at the lower end. (Leonardo's *Trattato*, quoted by Gombrich 1976.)

However, as noted by Ernst Gombrich (1976) even da Vinci risked becoming exhausted with the complexity of this undertaking:

> The middle parts of the nose, which form the bridge, can differ in eight ways: they can be (1) equally straight, equally concave or equally convex, or (2) unequally straight, concave or convex, (3) straight on the upper and concave on the lower part; (4) straight above and convex below; (5) concave above and straight below; (6) concave above and convex below; (7) convex above and straight below; (8) convex above and concave below. (*Trattato, op cit.*)

And this was just for noses seen in profile! A further eleven types were listed for front views.

Many centuries later, the photographer Jacques Penry invented the 'Photofit', which comprises sets of many hundreds of variants of face features. A trained operator helps the witness to select appropriate features and to refine the emerging representation. However, although performance can be improved with trained and sensitive operators, overall evaluation of the likenesses produced by kits of this kind has shown them to be very poor. For example, in one study by Hadyn Ellis and others (1978), participants rated the likenesses between the constructions produced by volunteer 'witnesses' and the people they were attempting to construct. The constructions were made by the witnesses themselves drawing the faces, or using Photofit, and they were made either with the target face *in view* or from memory. When the target face was in view (that is, the task of the witness was simply to copy the face), then witnesses' own drawings were rated as much better likenesses than were those produced by Photofit. When reconstruction was from memory, the rated likeness obtained with both techniques dropped, and there was a slight but not significant advantage for Photofit.

Why is performance with Photofit so poor? One problem is that the kits assume that a face representation can be deconstructed into its component parts, but we saw earlier in this chapter that representations for face recognition may be more holistic than this. Moreover, relationships between different features are at least as important as the features themselves, but Photofit has only limited opportunities for the manipulation of such factors. Finally, there are a small number of what Ellis (1986) termed 'cardinal' features, which are regularly and reliably used by people to describe and categorize faces. These are face shape, hair, and age, all global dimensions which are difficult to manipulate directly with a Photofit kit.

The rapid development of powerful graphics computers at relatively low cost has made it possible to develop more interactive systems which appear to remedy some of these deficiencies. E-fit was developed jointly by researchers at the University of Aberdeen, the Home Office, and in the private sector. While still based around a basic 'kit' of photographic parts of faces, there is much more opportunity for global manipulations, blending, and elaboration than with Photo-fit, and the resulting images are much more realistic (see Fig. 5.22).

As well as reconstruction, the other main way in which witnesses

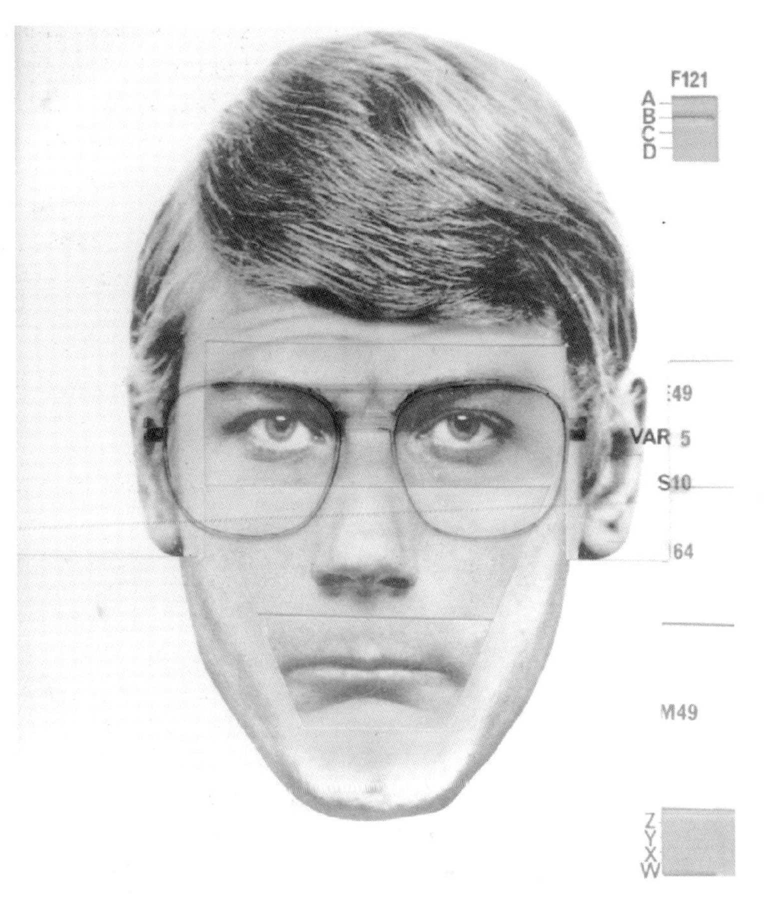

Fig. 5.22 E-fit and Photo-fit constructions of the face of John Major. Courtesy of John Shepherd, University of Aberdeen.

may have their memory for faces probed is via a recognition test of some sort, and this may involve looking at photographs in an album of 'mug-shots' or looking at a live 'line-up' of potential offenders staged at the police station. Both techniques have problems.

The problem with searching through mug-shot files is that they may be very large, and psychological experiments have shown that the chances of correctly recognizing a target face fall off dramatically at later positions within a long list. In effect, the process of looking at a list of potential suspects' faces interferes with your memory for the perpetrator. One way round this problem is to pre-select a relatively small number of faces for the witness to inspect. A Home-Office-funded project at the University of Aberdeen (e.g. Shepherd 1986) developed a system where an initial witness description could be coded by computer, and used to select from the mug-shot file a smaller number of photographs for inspection by the witness. Evaluation of such a system showed that it could produce more accurate performance from witnesses than was possible using the full mug-shot album.

Line-ups pose a range of other problems. One problem is that the volunteers who form the rest of the line may not be well-matched against the suspect on some key dimension. For example, if the witness remembers that the villain was 'very tall', and the suspect is the tallest in the line-up, then the witness may be inclined to pick out this person on this basis alone. One recommendation is that photographs of line-ups should be given to people with no connection to the original incident, along with the original witness description. A fair line-up is one where a 'dummy' witness armed with the description is not able to guess who the suspect might be with better than chance success (see Box 3A, Chapter 3, for an account of a similar test of a photographic line-up used in the case of Ivan the Terrible). Such a procedure would also help ensure that there were no other subtle clues to the likely suspect (e.g. if he were unshaven having spent a night in custody and others in the line-up were clean-shaven this could act as a clue to the likely suspect). Eyewitnesses are very motivated to pick someone from a line-up. They know that the police do not go to the trouble of arranging such a thing unless they have a suspect and thus it is extremely important that the line-ups be constructed and administered wisely.

Finally, it is important that the same witnesses are not asked to examine photographs *and* a line-up, since there may be unintended transference of familiarity gleaned from the photographs. People are much better at knowing that a face is familiar than at knowing why it is familiar. Several notorious cases of mistaken identity involved witnesses who identified from line-ups faces that they had already been shown in photographs.

Recent years have seen a huge rise in the number of crimes where video evidence of the identity of the suspect has been gathered on a security video camera at the time of the criminal act. It may seem that this reduces the need to worry about human factors in face identification, but in fact such images simply raise a new set of problems. Images captured by security cameras are often of extremely poor quality, and thus much interpretation is involved in determining to whom a captured face belongs. Cameras may be set to scan a large public area, and so the images of any individual may be of very poor resolution, or the camera may be at the wrong angle from which to perceive the face. Because video images are costly to store, only a sparse sample of the frames may be captured, and these may miss the most recognizable view of a person's face. For these reasons, video evidence has not provided a simple solution to identification problems. In a number of court cases in recent years, the identity of people caught on video has been hotly debated by expert witnesses for prosecution and defence, each using different methods to attempt to prove that the suspect is, or is not, the person shown on the tape. Advances in the methods that can be used to prove identity in these cases must await further research into the best ways to measure and compare faces in a way which does not result in false matches.

5.7 Stages in person identification: from face to name

Young *et al.*'s (1985) diary study of everyday errors in recognizing people supported the idea that there were a small number of stages that led to successful identification of a familiar person. First, the face had to be recognized as familiar; that is, matched against a record of that person's facial appearance stored in memory. Second, information about why that face was familiar was retrieved; where the person was known from or what they did for a living. The final stage, and one which was surprisingly prone to error, was retrieving the name of the person.

Lots of other evidence supports this simple three-stage model. For example, if people are asked to make rapid decisions about a series of faces, they are quickest to decide whether or not each face is familiar, a little slower to decide whether or not each face belongs to some occupational group (e.g. politicians), and slowest of all to decide whether each face belongs to a person with a particular name (e.g. John). Moreover, the kinds of deficit in person identification which arise from brain damage are also consistent with this sequence. Some patients are unable to recognize people by their faces at all, and all faces appear unfamiliar to them—a failure at the first stage. Others, usually with memory problems, find appropriate faces familiar but cannot recall why, and a third class of patients have particular problems recalling people's names.

So, one of the most frequently reported problems in person identification is forgetting a person's name. You may be able to experience these difficulties for yourself by looking again at Fig. 5.4, this time the right way up, and trying to name each of the famous individuals depicted. Some names will be easily retrieved, but you may struggle for others. In particular, you may find that you experience the tantalizing state of a name being on the 'tip of your tongue'. You are sure that you know it, but it feels as though there is some mental block preventing its retrieval. Why are names so much harder to remember than other pieces of information?

Psychological research has shown that the difficulty of remembering names is not because those words which are people's names are particularly problematic. McWeeny *et al.* (1986) compared how easily people could learn that new faces had particular names or particular occupations, using exactly the same words for each, for example 'This is *Mr* Baker' compared with 'This is *a* baker', and still found much better memory for the occupations than the names.

The difficulty with naming seems to require an explanation in terms of the kinds of memory locations or structures used to store names compared with other kinds of information. One experiment illustrating this was conducted by Bob Johnston and Vicki Bruce in 1990. The experiment used only eight faces, repeated throughout the experiment. Four of the faces were of people called John and four called James. Half the faces were of British and half of American celebrities, and half were of dead people and half were of people still alive (two examples were John Lennon, a dead British celebrity, and John Wayne, a dead American one). The participants were shown pairs of faces and asked to verify as quickly as possible whether the two faces matched or mismatched on a dimension specified separately for each block of trials. In one block of trials the faces were to be matched by name; on another they were to be matched on nationality; and on a further block they were to be matched in terms of whether both faces were dead or both alive. Johnston and Bruce found that all the tasks involving matching for names were conducted more slowly than any of those assessing nationality or dead–alive, even though intuitively it seems harder to retrieve information about whether faces are dead than whether they are called John.

These results can be explained within the three-stage model of person identification. On this model, any task involving name retrieval must take longer than any task involving retrieval of information about identity, simply because name retrieval involves an additional, and time-consuming stage. However, the reasons why names—so central to our identities—may be delegated to this subordinate stage in person identification, remain mysterious, and the correct account of how information about personal identities is stored and retrieved is currently an active area of research and theoretical debate.

6 Messages from the face: lip-reading, gaze, and expression

6.1 Overview

As well as being used to recognize people, the face forms a source of many useful social signals. We use momentary configurations of the lips and tongue to help us to understand speech. We use patterns of gaze to regulate conversational turns, and to indicate where another person is directing their attention. Head gestures such as nodding or frowning add significantly to the information transmitted through speech, and it can be shown that telephone conversations are structured rather differently from face-to-face interactions in order to compensate for the lack of non-verbal cues. Emotional expressions provide information about another person's feelings, and the other social messages from the face may inform about other aspects of a person's mental state. A portrait or caricature artist must convey more than just a physical likeness, including mood and gaze appropriate to the subject.

6.2 Facial expressions of basic emotions: the universality thesis

The facial muscles were described in Section 1.5. Our highly mobile facial features allow us to send and receive many social signals, the most widely investigated of which have been facial expressions of emotion.

The anatomy of the muscles was discussed by Sir Charles Bell; the first edition of his book was published in 1806, but he later prepared an expanded version (Bell 1844). Bell arrived at several plausible hypotheses about how the facial muscles function to signal emotion, but the only source of evidence he had was careful observation and deduction. Observation should never be belittled, but it has limitations with such a complex system.

The earliest systematic experimentation on muscle movements was by the French physician Duchenne (1862) who studied a man who had lost feeling in his face through damage to the nerves. By applying

Fig. 6.1 Dr Duchenne and his patient. The picture shows the man's cheeks being stimulated electrically.

small electric currents to the sites of the different facial muscles, Duchenne could make them move independently, as shown in Fig. 6.1. (This unfortunate man could not feel the electric current.)

The effects of moving some of the facial muscles are indicated in Fig. 6.2, but most readily recognizable facial expressions of emotion involve the simultaneous activation of more than one muscle group (Eibl-Eibesfeldt 1989; Ekman 1972).

The landmark book on facial expressions was by Darwin (1872). His view was that certain facial expressions are biologically created, and that these will be universally recognizable. Certainly, perusal of Darwin's illustrations (see examples in Fig. 6.3) shows that the meanings of these facial expressions have not changed in the last 125 years. Darwin was not the first to propose the universality thesis, but he gave it an entirely new twist by linking it to evolution. For previous authorities, such as Bell (1844), universal aspects of facial expressions were taken to reflect the fingerprints of the creator, rather than the products of evolutionary history.

Whereas nowadays it would be unusual to invoke the creator in a scientific hypothesis, a sceptic might want to argue that one of the

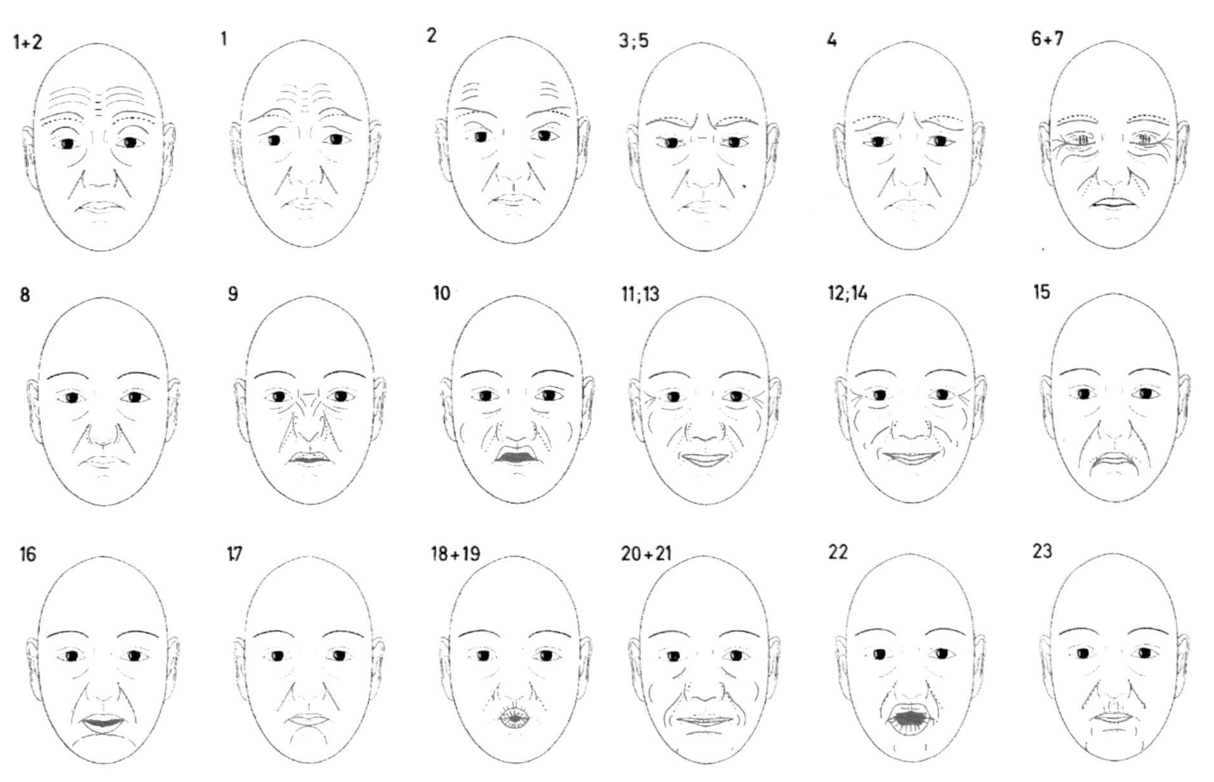

Fig. 6.2 Effects of contractions of various facial muscles (adapted from Hjortsjö 1969). The changes produced by the muscle are marked in red, with its resting position shown dotted. 1+2: frontal muscle (*m. frontalis*); 1: medial part of frontal muscle (*m. frontalis pars medialis*); 2: lateral part of frontal muscle (*m. frontalis pars lateralis*); 3: glabella depressor (*m. procerus* or *m. depressor glabellae*); 4: eyebrow wrinkler (*m. corrugator supercilii*); 5: eyebrow depressor (*m. depressor supercilii*); 6+7: sphincter muscle of the eye (*m. orbicularis oculi*); 6: orbital part of eye sphincter muscle (*m. orbicularis oculi pars orbitalis*); 7: eyelid part of eye sphincter muscle (*m. orbicularis oculi pars palpebralis*); 8: nasal muscle (*m. nasalis*); 9: upper lip and nasal wing elevator (*m. levator labii superioris alaeque nasi*); 10: upper lip elevator (*m. levator labii superioris*); 11: lesser zygomatic muscle (*m. zygomaticus minor*); 12: greater zygomatic muscle (*m. zygomaticus major*); 13: elevator of the angle of the mouth (*m. caninus*); 14: smiling muscle (*m. risorius*); 15: depressor of the angle of the mouth (*m. triangularis*); 16: lower lip depressor (*m. depressor labii inferioris*); 17: chin muscle (*m. mentalis*); 18+19: incisive muscles of the upper and lower lips (*mm. incisivi labii superioris et inferioris*); 20+21: cheek muscle (*m. buccinator*); 22: lip part of sphincter muscle of the mouth (*m. orbicularis oris pars labialis*); 23: marginal part of sphincter muscle of the mouth (*m. orbicularis oris pars marginalis*).

reasons we can still recognize the emotions from the pictures in Darwin's book is not our biological endowment but the fact that printing, photography, cinema, and television are so dominant in western culture that they have forced us all to learn a fixed set of conventions. To make a proper test of the idea that the expressions of emotion are universal, Paul Ekman therefore visited a preliterate culture in New Guinea, where people had never seen photographs, magazines, cinema, or television, and had been visited by few outsiders. They would therefore have had no significant opportunity to learn about the facial expressions of people from other cultures. Figure 6.3 shows people from this culture demonstrating what their faces would look like in response to the stories 'your friend has come and you

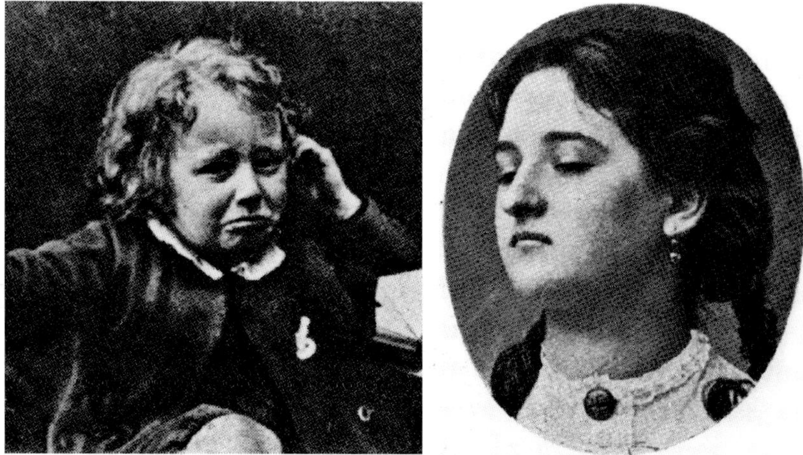

Fig. 6.3 Pictures of facial expressions from Darwin (1872): joy (*left*), grief (*centre*), contempt (*right*).

are happy', 'your child has died', 'you are angry and about to fight', or 'you see a dead pig that has been lying there a long time'. Even without the benefit of television, their expressions are the same as would be seen anywhere else in the world (Ekman 1972).

Ekman and his colleagues also tested recognition of facial expressions (see Ekman 1972). This is a tricky thing to do because there are huge cultural differences in the words and concepts people use to describe emotions. This undisputed anthropological fact was probably part of the reason that the universality thesis had not been greeted with unanimous enthusiasm—if people from different cultures can think about emotion in very different ways, why should we expect any commonality in how they recognize them? Moreover, even when members of the same culture are simply asked to describe or label facial expressions, they can give very variable responses which can themselves require considerable interpretation. How do we decide whether or not two people mean the same thing if one calls an expression 'shock' and the other calls it 'fear'?

To get round these difficulties, Ekman adopted a forced-choice procedure in which people were asked to assign photographs to one of a fixed number of categories, and they used little stories (as in the example given earlier) to make clear what each category entailed. They also took great care to ensure that their facial expression stimuli were photographs of people moving exactly the right muscles for each emotion.

With such precautions, the work of Ekman and many others has shown reasonably good recognition of a small number of basic emotional categories in nearly all cultures. These basic emotions include happiness, sadness, anger, and disgust. Fear and surprise were also included in Ekman's original set of basic emotions, but these expressions are often confused with each other in studies of non-literate cultures. This may be because surprise does not have the full

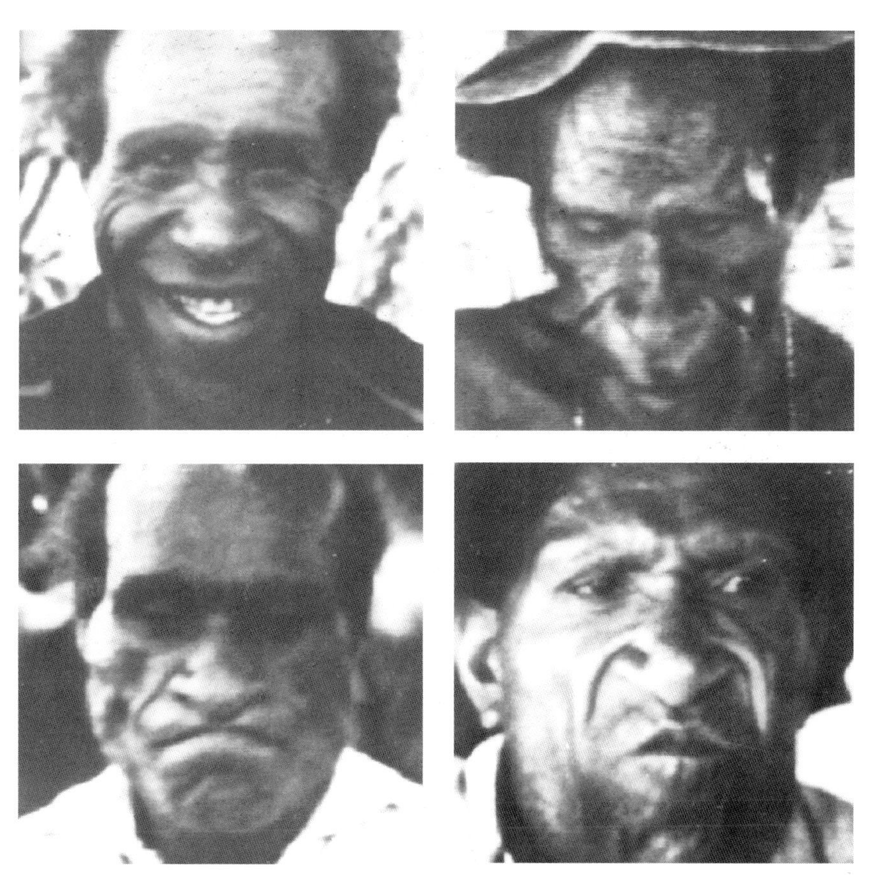

Fig. 6.4 Facial expressions posed by members of the Fore community in New Guinea in response to the stories 'your friend has come and you are happy' (*top left*), 'your child has died' (*top right*), 'you are angry and about to fight' (*bottom left*), or 'you see a dead pig that has been lying there a long time' (*bottom right*). From Ekman (1972).

status of a basic emotion. More recent studies have sometimes included contempt as another basic emotion.

It is perhaps a little surprising that static photographs of facial expressions should be so well recognized, since it is natural to think that the pattern of movement of the facial muscles will convey important information. We need to bear in mind that although photographs are recognizable, this does not mean that movement is unimportant. On the contrary, there is evidence that the timing of facial movements is carefully balanced between the needs of the sender and the intended recipient, even for a facial signal as apparently simple as raising the corners of the mouth in a smile (Leonard *et al.* 1991). However, the good recognition of static photographs of basic emotions shows that, for these emotions at least, the apex of the set of muscle contractions forms a recognizable configuration of the facial features.

Findings of universal recognition of facial expressions of emotion have not gone unchallenged, largely on the grounds that they seem to be observed most often with the particular combination of methods Ekman used (forced-choice responses, stories to back up the response categories, careful selection of target photos). To us, these challenges seem to miss the main point, which is that with appropriately careful testing there is evidence of an impressive degree of commonality across cultures in the interpretation of certain emotions. In making this point, Ekman and his colleagues have never sought to deny the richly diverse contributions from culture and upbringing.

An example of diversity involves what Ekman (1972) calls display rules. These concern the circumstances in which it is appropriate or inappropriate to display emotion, and they vary widely across the world. The important point is that although display rules concerning when it is appropriate to express certain emotions are culturally determined, the forms of the facial expressions themselves do not vary significantly.

The point about display rules applies just as much to artistic conventions. In portraiture, there are usually only a limited range of emotions commensurate with the dignity of the sitter (Stevenson 1976). This is shown amusingly in Fig. 6.5, where the portrait of the

Fig. 6.5 *William Hamilton of Bangour* (Poet, 1704–1754). Engraving by Robert Strange after Gavin Hamilton shown upper left; other pictures altered by an unknown hand. In the collection of the Royal Museum of Scotland.

poet William Hamilton of Bangour (top left picture in Fig. 6.5) has been altered by an unknown contemporary armed with scissors, paste, and a set of prints of human facial expressions.

6.3 Perceiving and recognizing facial expressions of emotion

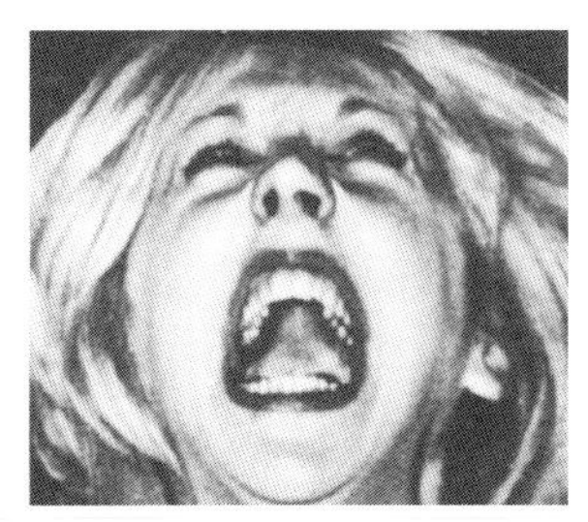

Fig. 6.6 What is the young woman feeling?

Although a small number of basic emotions seem to be recognized throughout the world, there are many facial expressions which do not correspond precisely to one or other of these categories. In everyday life, we make considerable use of the contexts in which expressions occur to assist interpretation. To demonstrate this, look at Fig. 6.6. What is the young woman feeling? Absolute terror?

The answer can be found overleaf, in Fig. 6.7. She was photographed in the audience at a pop concert. Once this context is clear, it is easy to interpret her expression as one of great excitement.

We have cheated a bit with this example, because there is a curious relationship between fear and excitement—the queues for the white-knuckle rides are always the longest at theme parks. However, less dramatic effects of context on the interpretation of expressions are common, quite possibly the norm. Indeed, if the contextual information is strongly discrepant, people will reinterpret other basic emotions as well as fear. Facial expression perception does not deal in absolute certainties. We modify how we read people's feelings on the basis of any other pertinent information available at the time.

Since even basic emotions can sometimes be misinterpreted, and because many facial expressions do not in any case correspond exactly to specific basic emotions, many people have thought that what we may be doing when we perceive facial expressions is to locate their positions along general dimensions of emotion.

This idea was made popular among psychologists in a book by Woodworth and Schlosberg (1954), which was for many years *the* textbook of experimental psychology. Woodworth and Schlosberg tried to make sense of the bewildering array of findings which had arisen from the pre-Ekman studies of expression recognition which had looked at the spontaneous labels given to highly diverse sets of faces. They noticed that while people's responses in such tasks initially seemed incredibly varied, they were not random. The key to putting them into a more systematic order was to realize that only a certain range of responses would tend to be used for each face. For example, people might say that a particular face showed fear, suffering, grief, even surprise, but no one would describe it as joy, happiness, disgust, or contempt. By grouping together some labels which seemed to them to be used almost interchangeably, and studying the patterns of responses, Woodworth and Schlosberg were able to arrange the emotions

Fig. 6.7 Facial expressions at a pop concert (from Liggett 1974).

into a circle which had the property that those expressions most likely to be confused with each other were placed in adjacent parts of its perimeter. Their scheme is shown in Fig. 6.8.

A mathematically simple way to describe locations in a two-dimensional diagram like Fig. 6.8 is in terms of their positions along two fully independent (orthogonal) axes, just as we do when plotting a graph. When they studied their facial expression circle it seemed to Woodworth and Schlosberg that two axes could be identified. These involved the difference between expressions which are associated with *pleasant* or *unpleasant* feelings, and between expressions which involve increased *attention* to the external world or *rejection* and shutting it out. They therefore suggested that our perception of facial expressions is achieved by the visual system coding them on pleasant–unpleasant and attention–rejection dimensions. Less intense expressions will fall toward the centre of the circle, and more extreme ones at the circumference.

Since Woodworth and Schlosberg, other variants of the two-factor approach have been suggested. The exact details do not matter here, what is important is that all share the perspective that facial expressions are considered to vary continuously along the dimensions proposed, each shading smoothly into the next.

This is a very different conception to Ekman's idea that facial

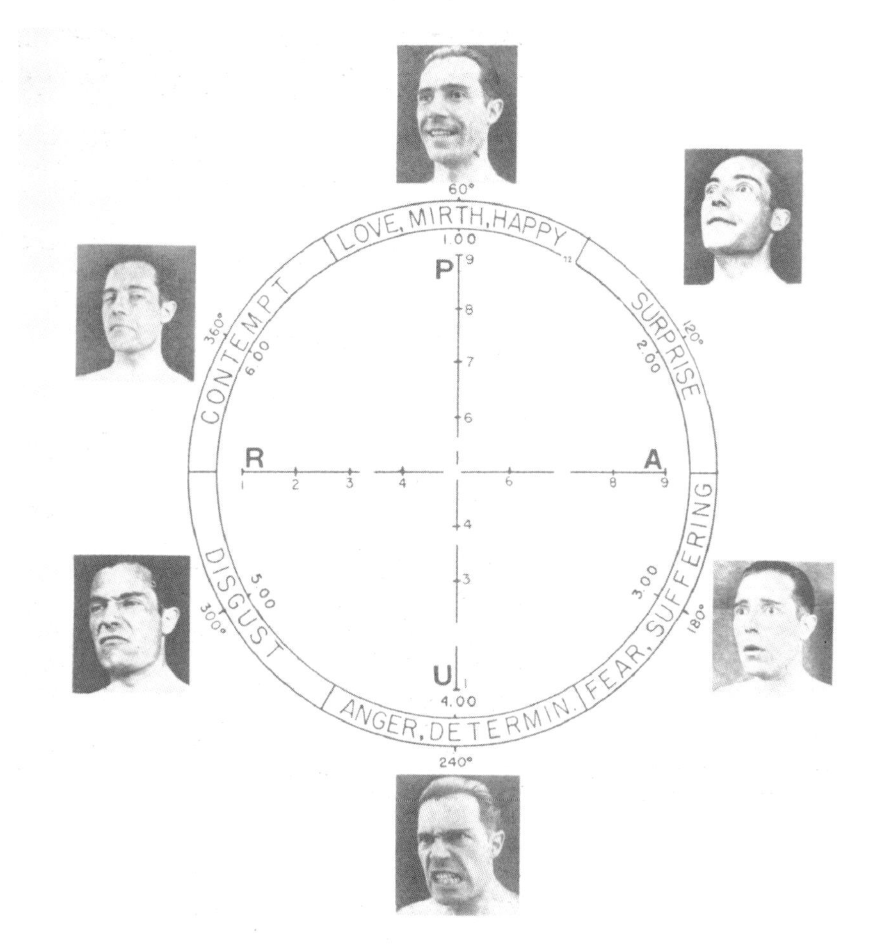

Fig. 6.8 Emotion circle proposed by Woodworth and Schlosberg (1954) as adapted by Izard (1977) with examples of facial expressions which would fall at different points around the circumference.

expressions of basic emotions belong to discrete categories, but until recently it has not been possible to determine whether dimensional or category-based accounts are correct; both were able to account for existing findings. Box 6A describes recent applications of computer image-manipulation techniques to investigate these issues. These have lent support to the categories hypothesis by demonstrating that the perceptual transitions between one emotional expression and another are not as smooth as dimensional accounts would predict.

A further insight which can be gained from computer-graphics techniques involves examining the contributions to recognition of emotion made by shape cues (raising or lowering eyebrows, corners of the lips, widening the eyes, etc.) and cues which depend more on pigmentation or texture differences (opening or closing the mouth, baring teeth, etc.). The image-manipulation techniques described in Box 6A can be used to create images which are of average shape (i.e. with mainly pigmentation cues remaining) or of average pigmentation (i.e. with shape cues remaining) to investigate how this affects recognition of each emotion.

Box 6A: Categorical perception of facial expressions

A fundamental issue concerns whether we recognize facial expressions by assigning them to discrete emotion *categories*, as Ekman suggested, or whether we interpret them by locating them on a small number of underlying *dimensions*, such as in Woodworth and Schlosberg's suggestion of pleasant–unpleasant and attention–rejection dimensions.

Computer-graphics techniques allow a novel way of testing which of these hypotheses is correct, since they allow us to make systematic changes between images depicting different expressions using a technique known as 'morphing'. Building on work with line drawings by Nancy Etcoff and John Magee (1992), Andrew Calder and his colleagues studied the perception of photographic-quality morphed images of facial expressions (Calder *et al.* 1996*a*).

The basis of the technique involves locating a number of anatomical landmarks on a photograph of a face, and using these to divide it into a pattern of small triangles, as in Fig. 6A.1. If this procedure is then repeated with a second face photograph, the shapes of the triangles in one image can now be gradually altered toward the shapes of corresponding triangles in the other image, and the brightness values from each image blended in proportion to the degree of change.

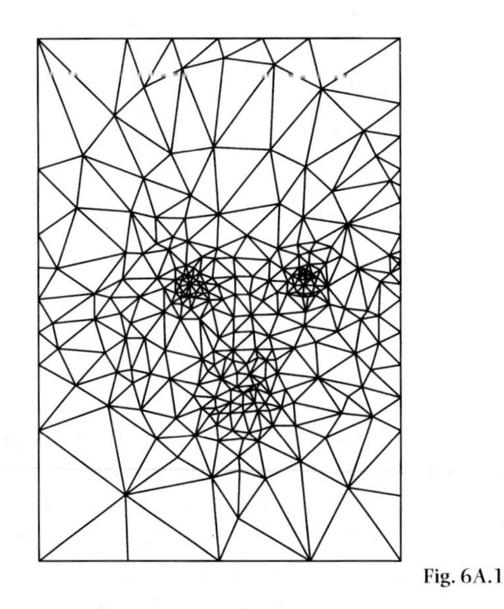

Fig. 6A.1

Figure 6A.2 shows images created in this way for continua which run from fear to happiness (first column), happiness to anger (second column), and anger to fear (third column). In each column, the images are moved 10%, 30%, 50%, 70%, and 90% of the distance from one expression to the other. The effect is to create a facial expression triangle which runs fear–happiness–anger–fear.

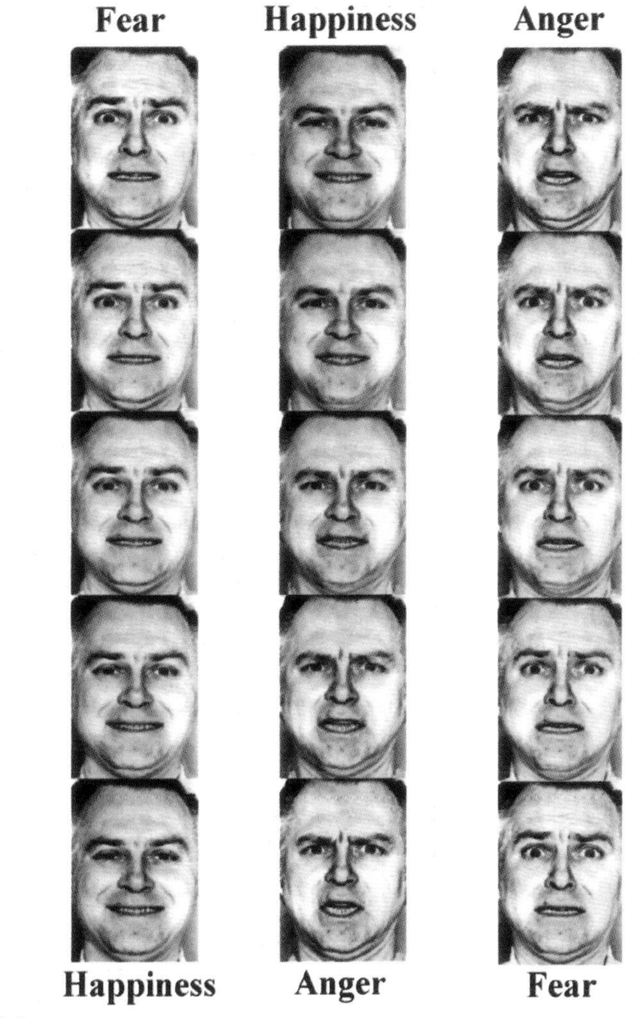

Fig. 6A.2

Identification rates for the images from Fig. 6A.2 are shown in the upper graph of Fig. 6A.3. People were asked whether the expression on each face was most like fear, happiness, or anger. It is clear that there are abrupt shifts between perceiving one emotion and the other near the mid-point of each continuum.

These abrupt discontinuities in identification rates do not really fit very well with a dimensional account, and even more persuasive evidence against the dimensional view comes from a direct test of categorical perception applied to these faces. The phenomenon of categorical perception arises when items which come from the same category seem more perceptually similar to each other than they actually are, while items from different categories seem more dissimilar. For example, in colour perception, two different wavelengths which are both considered shades of 'red' will be judged as

Box 6A: Continued

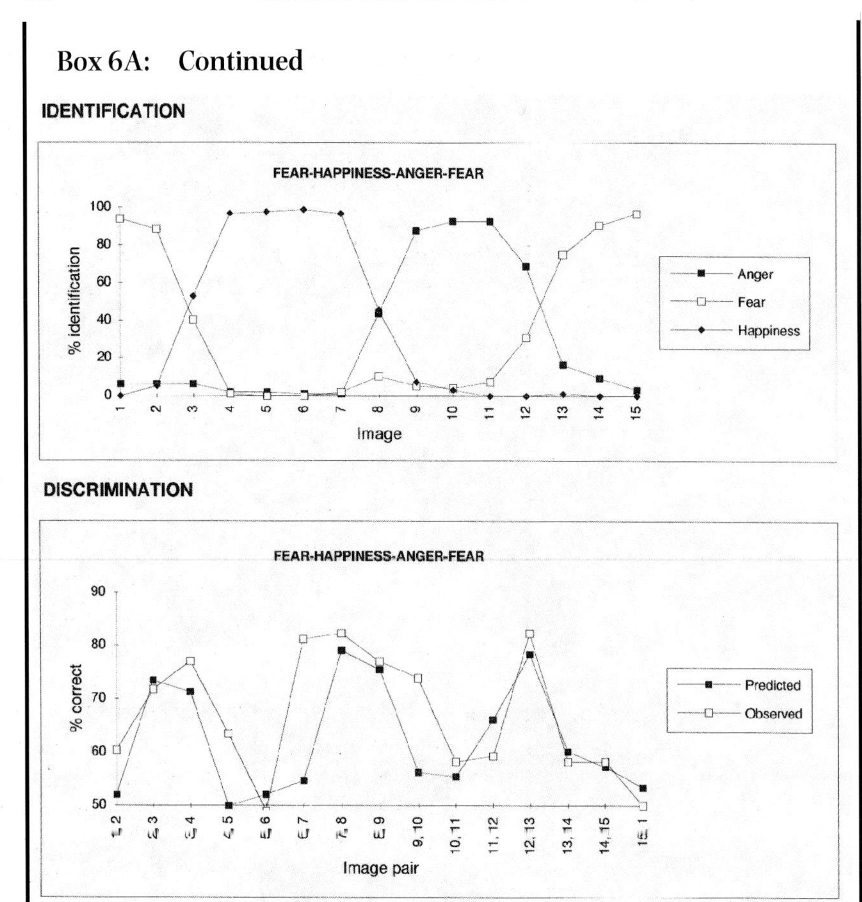

Fig. 6A.3

visually more similar to one another than two wavelengths which are physically no further apart (in terms of the difference in wavelength) but which span the boundary between a red and an orange.

If emotions are also perceived categorically, then pairs of faces which fall within the range of faces identified as 'happy' should be judged as being more visually similar to each other than equally similar pairs of faces which straddle the boundary between happiness and anger. In Fig. 6A.2, this would mean that pairs of adjacent face images from near the top or the bottom of each column will seem more similar than pairs falling nearer the centre. This was tested with a task called an 'ABX' task, in which stimuli A, B, and X were presented in sequence. A and B were adjacent images from the expression triangle shown in Fig. 6A.2, and the test was to decide whether X was exactly the same as A or exactly the same as B. If expressions are perceived as continua, this observed discrimination performance should be more or less constant (i.e. close to a horizontal line) because each continuum involves linear changes of shape and texture. Instead, though, observed discrimination performance is strikingly non-linear; it peaks at the boundaries between the perceptual categories revealed in the identification data.

These data therefore show evidence of categorical perception. Items which come from the same perceptual category seem more similar to each other than they actually are, whereas items from different categories seem more dissimilar. The curve plotted as 'predicted' discrimination performance in Fig. 6A.3 is based on the hypothesis of categorical perception of facial expressions. It uses the differences in identification rates between the members of different item pairs to predict how easily they will be discriminated. There is a reasonably good fit between predicted and observed discrimination performance, and this lends strong support to the idea that expressions are perceived as belonging to distinct categories.

As we noted, categorical perception effects are known to exist in other domains, such as colour perception (where perceived colour changes more abruptly across some wavelengths than others) and speech perception (where the perceived consonant can also shift suddenly as voice onset time is altered). It is likely that they also apply to certain other aspects of face perception (such as the perception of identity; see our observations on the recognition of identical twins in Section 4.2). However, it is important not to over-interpret such results. Although observed discrimination performance peaks near category boundaries in Fig. 6A.3, it seldom falls to the chance level of 50 per cent correct (where people would be just guessing); in other words, we can still see within-category differences even though we are more sensitive to differences between categories.

Figure 6.9 shows examples of such images. A set of 36 photographs (consisting of six women posing six emotions) from the Ekman and Friesen series (1976) was modified to retain their shapes but have the average pigmentation texture, or to retain their individual pigmentations but have the average shape. The figure shows that both types of cue make some contribution to our ability to recognize emotion, but shape seems more important since the average shape images are the most tricky.

That pigmentation plays only a limited role in the recognition of facial expressions is also demonstrated by Fig. 6.10, where expressions of the six basic emotions are shown in photographic negative. Negatives reverse the brightness values which determine pigmentation, and as we saw in Chapter 2 this has a big impact on perception of identity, making negative images of faces almost unrecognizable. However, turning the image into a negative does not really affect the shapes and positioning of the facial features very much, and the expressions can still be seen—not perfectly, but not with a great deal of difficulty either.

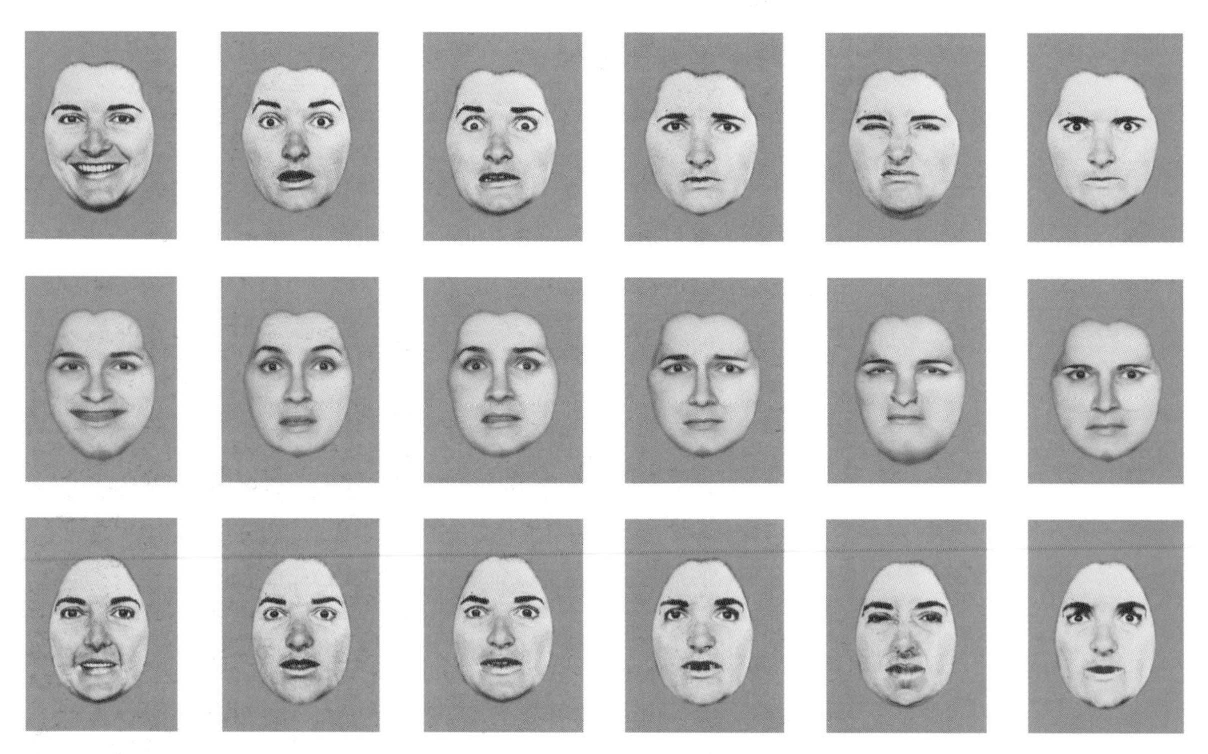

Fig. 6.9 *Top row*: facial expressions of basic emotions from the Ekman and Friesen (1976) series: from *left to right*, happiness, surprise, fear, sadness, disgust, anger. *Middle row*: images with the same feature shapes as the top row images but the average pigmentation (texture) from a set of six people posing these six expressions. *Bottom row*: images with the same pigmentation as the top row images but the average feature shapes from a set of six people posing these six expressions. Images in middle and bottom rows produced by David Perrett and Duncan Rowland, University of St Andrews.

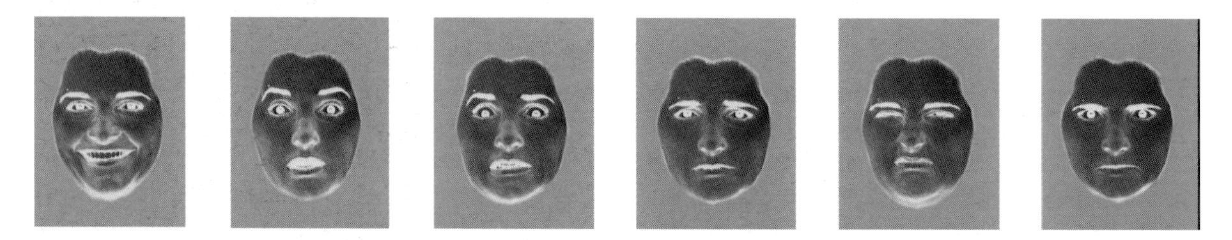

Fig. 6.10 Facial expressions in photographic negative. The images are negatives of the top row of faces from Fig. 6.9: *left to right*, happiness, surprise, fear, sadness, disgust, anger.

Shape information therefore seems critical to facial expression perception. The importance of feature-shape can also be demonstrated by using the computer-caricaturing techniques discussed in Section 5.5. The differences between the locations of features in an emotion face and a non-emotional face are measured, and image-manipulation methods are then used to increase these differences to create a caricatured expression, or reduce them to create anticaricatures (Calder *et al.* 1997). As Fig. 6.11 shows, this works well whether using a reference norm (from which the differences are calculated) based on the face's

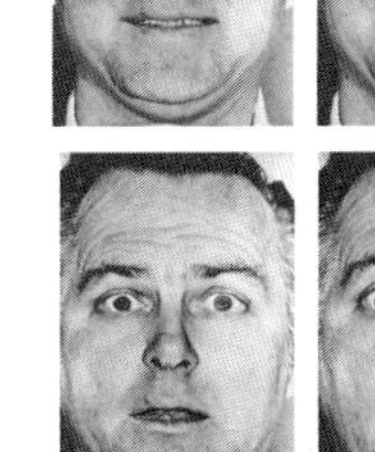
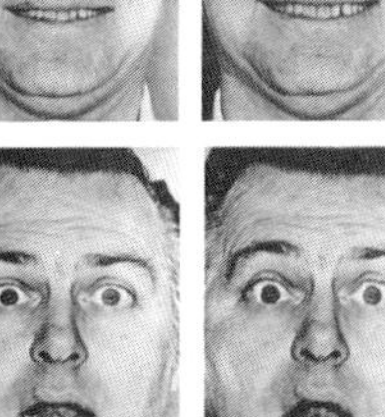
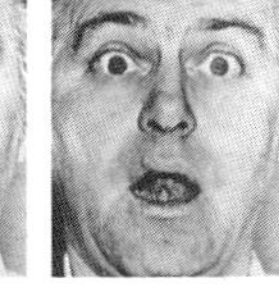
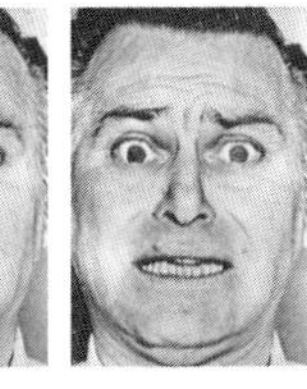
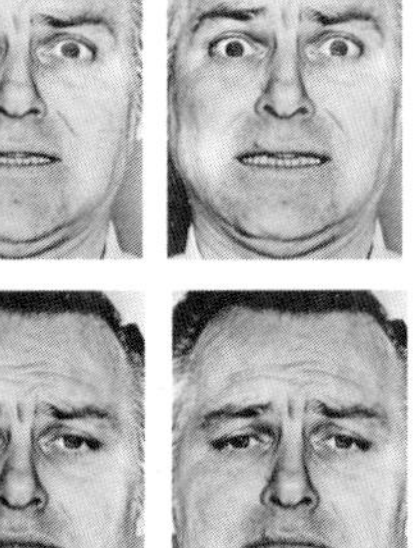
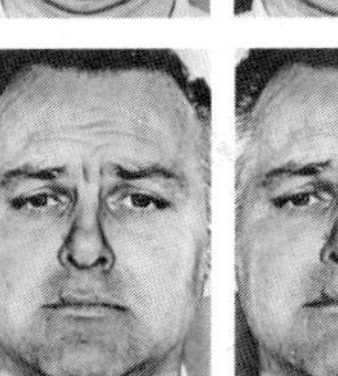
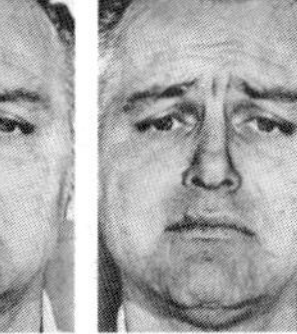
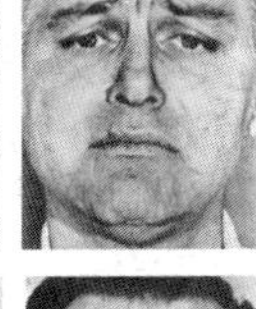
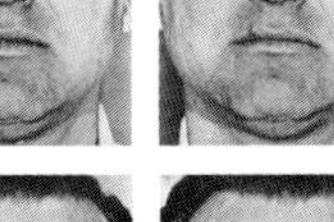
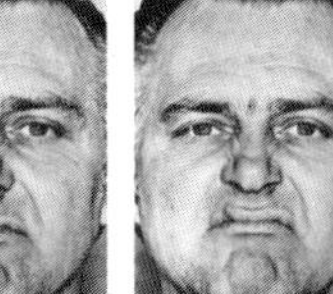
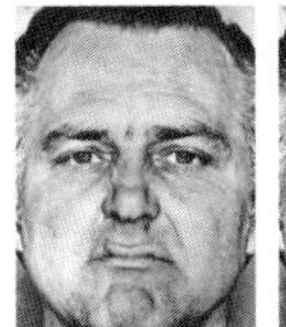
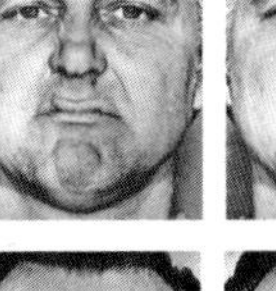
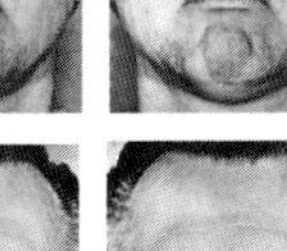
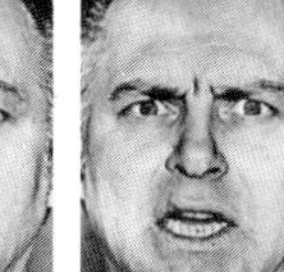
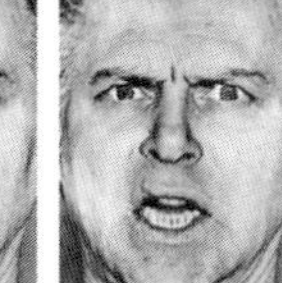
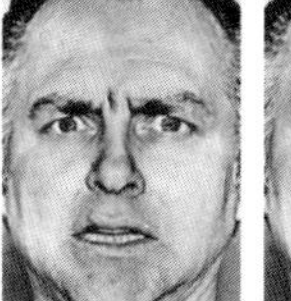
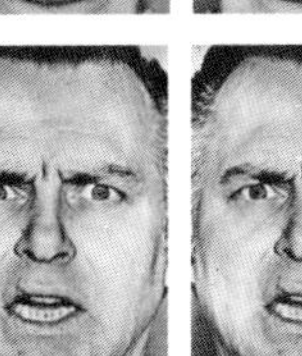
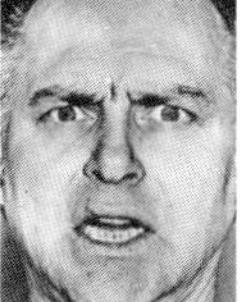

Fig. 6.11 Computer-manipulated (caricatured) emotion in facial expressions. These were created using a reference norm based on the average positions of features in the set (average-expression norm—images on the left side of the figure), or the face's muscles being at rest (neutral-expression norm—images on the right side of the figure). In each case, facial expressions of basic emotions from the Ekman and Friesen (1976) series are shown (0% images) alongside caricatured images (+50%) in which the differences in the locations of features from the reference norm are increased by 50% and anti-caricatured images (−50%) in which differences from the norm are decreased by 50%. From Calder *et al.* (1997).

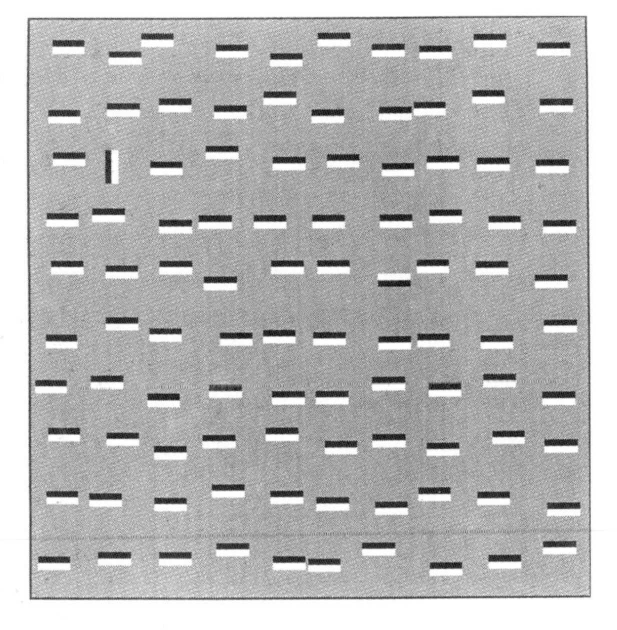

Fig. 6.12 Demonstration of the phenomenon of 'pop-out' in visual search. The differently oriented bar is almost immediately obvious. Note that there is a second target which differs from the other bars, but this (upside-down) bar has to be carefully sought. From Nothdurft (1993). Reproduced by courtesy of Pion Limited, London.

muscles being at rest (neutral-expression norm) or just on the average positions of features in the set (average-expression norm). The shape changes introduced by this process create increased intensity of perceived emotion in the caricatured images, and reduced intensity in anticaricatures.

If we put these various strands of evidence together, it seems that the way we recognize facial expressions is through combinations of features which have become communicative signals: eyes opened wide or narrowed, brows raised or lowered, mouth open or closed, corners raised or lowered, teeth showing or behind the lips, and so on. Each of these features may itself be encoded as a continuous dimension, but the effect of using several different dimensions simultaneously is to create distinct categories based on specific feature constellations.

An implication is that recognizing expressions is not dependent only on the presence or absence of a single feature; even something as apparently simple as upturned corners of the mouth need only be seen as a smile if the other features are appropriate.

When stimuli can be encoded with a single feature, they are highly noticeable when placed in a background of stimuli which differ on that feature. Look at Fig. 6.12. The misoriented stimulus is quickly detected; it seems to 'pop out' of the display.

Pop-out is not found for facial expressions (Nothdurft 1993). Try finding the sad face among the happy faces in Fig. 6.13(a), or the sad face among the happy faces in Fig. 6.13(b). It is not easy!

Now go back to Fig. 6.12. As well as the misoriented bar, there is another one which is discrepant, but finding it usually takes some time.

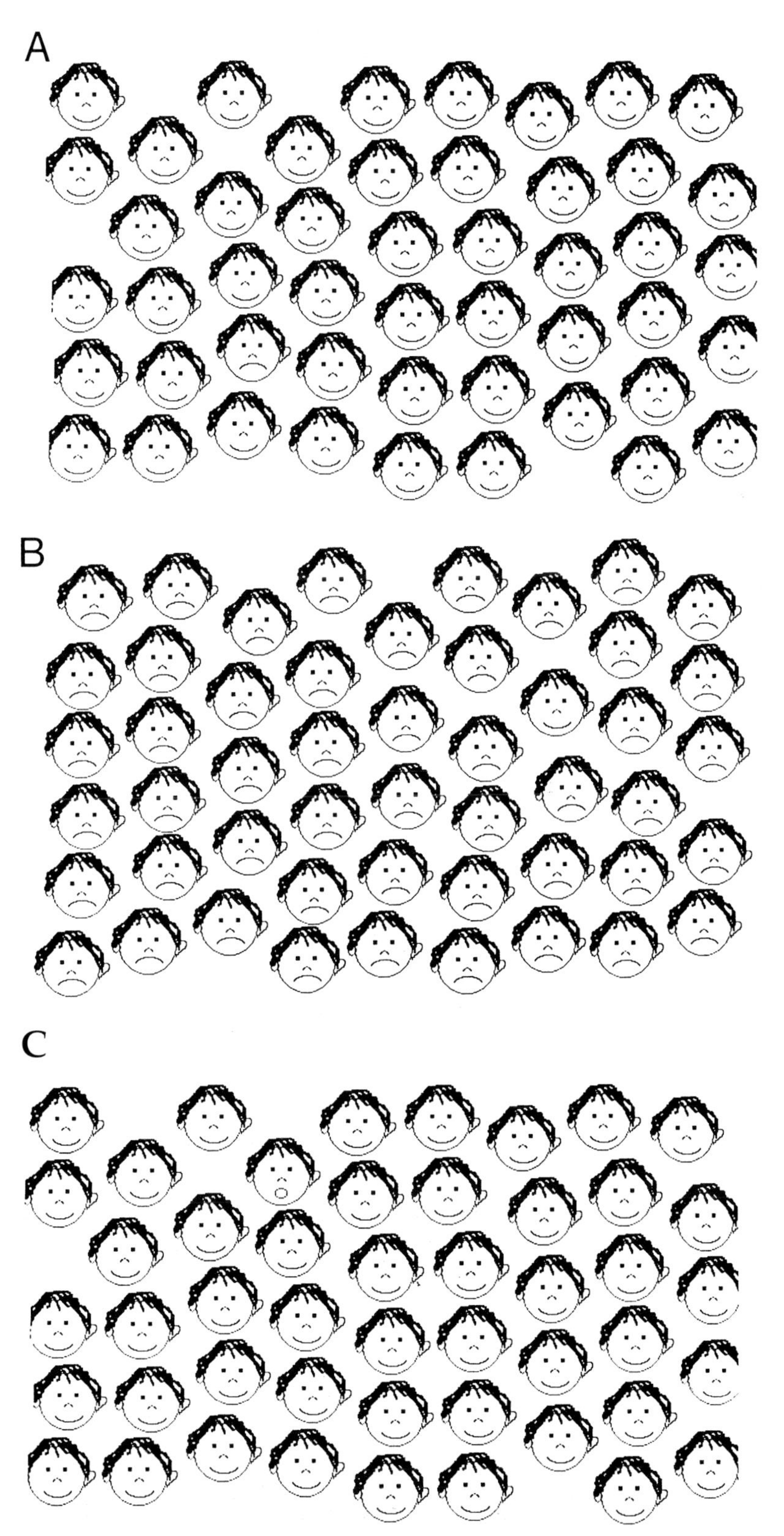

Fig. 6.13 Lack of pop-out for facial expressions. It is not easy to find a sad face among happy faces (Fig. 6.13a) or a sad face among happy faces (Fig. 6.13b). The open-mouthed expression in Fig. 6.13c is slightly easier to find, but still does not pop out. From Nothdurft (1993). Reproduced by courtesy of Pion Limited, London.

This is because it shares a common feature (horizontal orientation) with the other bars, and only differs in a secondary characteristic which requires putting the elements together (white over black, instead of black over white). Similarly, the need to put the elements together to perceive an expression may contribute to the lack of pop-out for facial expressions. This conclusion does not follow directly from Figs 6.13(a) and 6.13(b), because finding a single up-turned and down-turned mouth shape amongst opposite-shaped distractors would have caused a little difficulty even without the rest of the face. But the point is clear in Fig. 6.13(c), where one still needs to search a bit for the face with the discrepant mouth, even though the circular shape would be fairly immediately obvious among the other curves if no other facial features were present.

We have already seen that portraits usually involve a limited range of expressions (Fig. 6.5), but this restriction does not apply when an artist wants to create a sense of character and narrative. This was a speciality of David Wilkie, who was greatly influenced by his attendance at a series of special lectures for painters, given in London by fellow Scotsman the anatomist Charles Bell (see Section 1.5). It was Bell's frustrated ambition to be appointed Professor of Anatomy at the Royal Academy, but his forthright criticism of the anatomical ignorance of contemporary artists may not have helped him in this respect.

Wilkie's power in communicating character and intent had a profound impact (Errington 1985). Even Turner, who had previously concentrated on landscapes or seascapes, painted a rural interior with close-ups of the faces of arguing country people. This was a theme Turner had never used before, and he adopted it in acknowledged rivalry.

Figure 6.14 shows Wilkie's *Distraining for rent*. The main emotions expressed are despair in the father, langour and sorrow in the mother, and horror in the neighbours. In each case, they are clearly modelled on Bell's description of the appropriate facial expression (Errington 1985). The bailiffs are, of course, indifferent, which heightens the contrast.

Figure 6.15 shows a detail from the painting (centre), with enhanced (rightmost image) or reduced (leftmost image) emotion created by computer-manipulation using the same techniques as for Fig. 6.10.

6.4 Seeing speech

People with normal hearing naturally think of speech perception as a purely auditory process. We know that we can understand someone talking on the telephone, or behind our backs, and we know that blind people do not seem to have problems in following a conversation. Such facts indicate that a visual input is not necessary to speech perception. But does it follow that a visual input is irrelevant to speech perception?

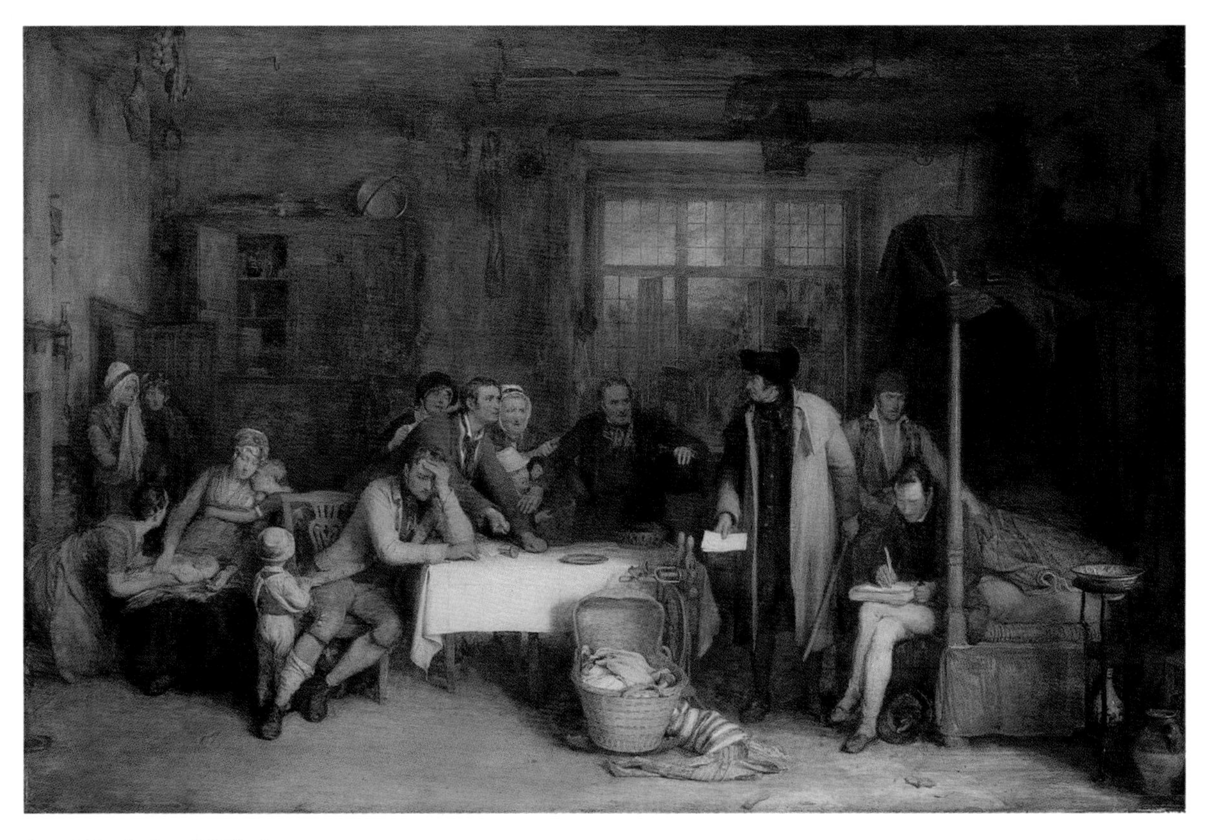

Fig. 6.14 Sir David Wilkie's, 1815, *Distraining for rent*. Courtesy of the National Gallery of Scotland.

We tend to assume so, because we know that deaf people find it very difficult to learn to lip-read, and that even the most skilled lip-readers are not always completely successful.

One of the most surprising findings of modern experimental psychology has been that this common-sense view is wrong, and that even people with entirely normal hearing make quite a lot of use of lip-reading to support speech perception. We will discuss some of the findings, and their implications.

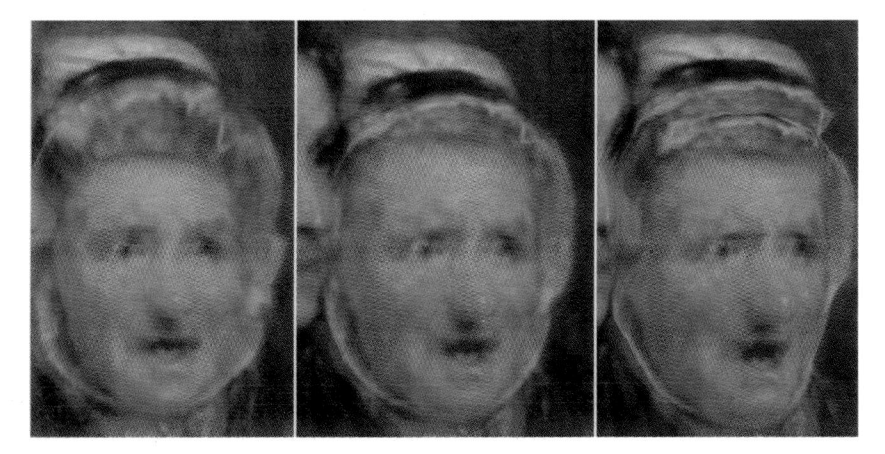

Fig. 6.15 Detail from Wilkie's *Distraining for rent* (*centre*), and versions with computer-enhanced (*right*) or reduced (*left*) emotion. Images produced by David Perrett and Duncan Rowland, University of St Andrews.

One reason for thinking that lip-reading might be of some use in speech perception is that we find it disconcerting to watch a foreign film with a dubbed soundtrack. However, this does not provide strong evidence, because it may simply be relatively gross mismatches between the soundtrack and the seen lip movements that are picked up (for example if the lips move fast when only a few different words are heard).

A more impressive result is that seeing the speaker's face helps in understanding speech heard against background noise (Miller and Niceley 1955; Sumby and Pollack 1954). Sumby and Pollack (1954) showed that the effect of seeing the speaker approximated to an increase of 15 decibels in the auditory signal to noise ratio—a substantial benefit. Hence seeing the speaker does have some role in speech perception.

However, the most dramatic evidence for involvement of visual information in speech perception has come from an effect discovered by McGurk and MacDonald (1976), now known as the 'McGurk effect'. This involves an illusion in which you see the face of a person mouthing one phoneme but a different phoneme is spoken on a synchronized soundtrack. Remarkably, the phoneme you hear is neither the auditory component (i.e. the phoneme on the soundtrack) nor the visual component (the visually mouthed phoneme), but a fusion of the two (see Box 6B for details).

Box 6B: Intermodal speech perception

The McGurk effect

McGurk and MacDonald (1976) constructed videos in which people saw the face of a person saying (mouthing) one phoneme, but heard a different phoneme on the synchronized soundtrack. Their results showed that the phoneme people actually perceived could be a fusion of the two (see Fig. 6B.1). For example, if the seen face makes the articulatory movements involved in saying 'ga', and the soundtrack is 'ba', most people perceive the fusion as 'da'. Contrary to our intuition, both auditory and visual information are used for speech perception, if available. Even if you notice or are told that the visual and auditory signals are discrepant, and are asked to report only what you hear, you will still tend to report the fusion. Like many illusions, the mechanisms are automatic and outside our conscious control.

Which parts of speech are perceived intermodally?

Are such fusion effects only found in the perception of consonants, or do they also apply to vowels? The answer is that what seem to be analogous effects can occur in the perception of vowels.

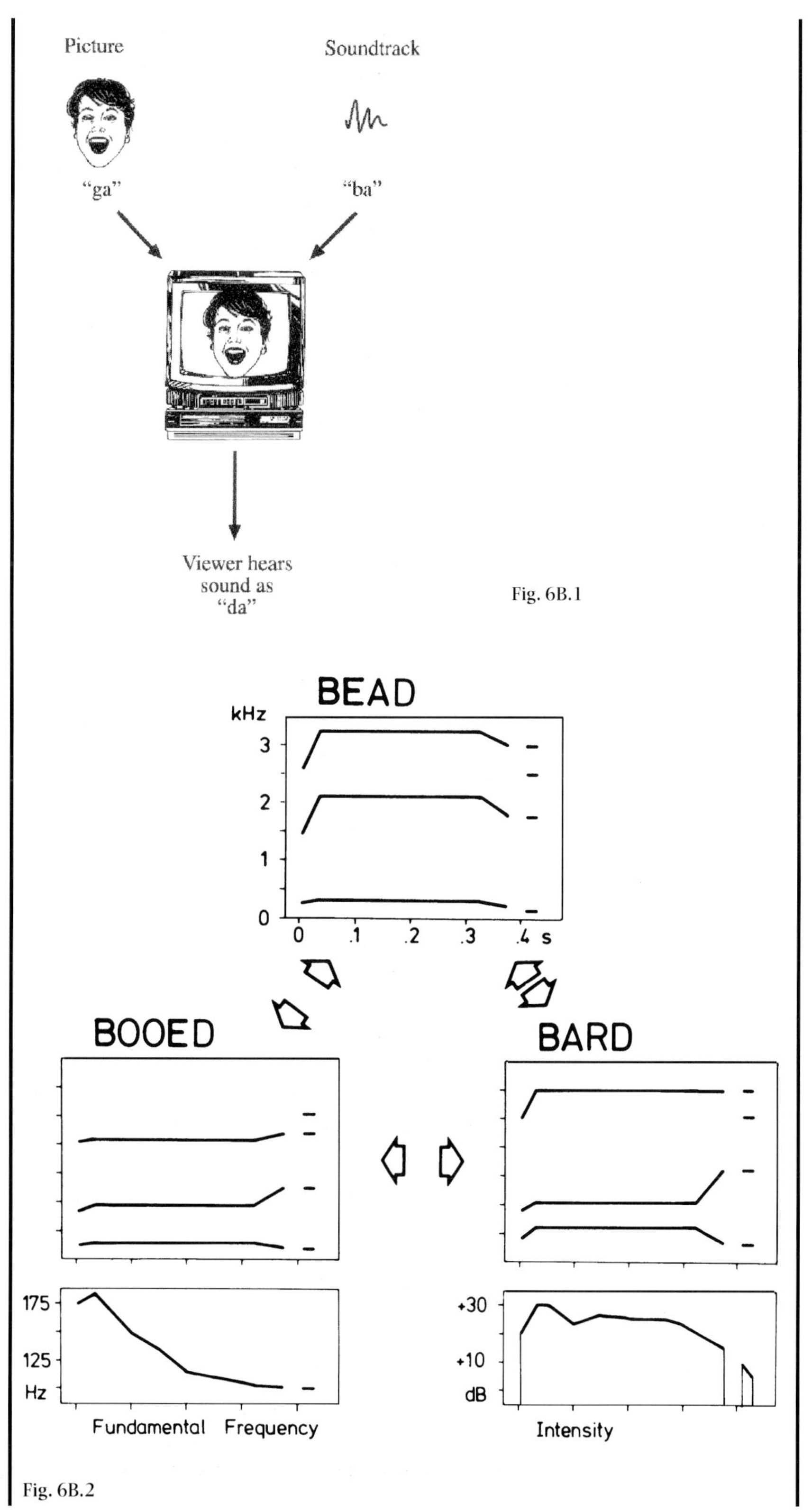

Fig. 6B.1

Fig. 6B.2

Box 6B: Continued

Summerfield and McGrath (1984) synthesized sounds to represent ten equal steps between 'booed' and 'bard', ten equal steps between 'bard' and 'bead', and ten equal steps between 'bead' and 'booed', by changing the parts of the sounds (fundamental frequencies of the first, second, and third formants) which determine the vowel, as shown in Fig. 6B.2.

These synthesized sounds were synchronized to videotapes of people saying either of the two syllables at the end of each continuum. For example, a sound lying along the booed–bard continuum would be paired with a videotape of a person saying 'booed', or with a videotape of a person saying 'bard'. Perception of the sound as 'booed' or 'bard' was then found to be influenced in the direction of the accompanying visual cues, even when observers detected the discrepancy and were instructed to report only what they heard.

This effect is clearly an analogue of that found for consonants by McGurk and MacDonald (1976), in that the observer's percept is determined by a combination of the visual and auditory inputs, but in this case the fusion simply biases perception toward one of the original inputs, rather than creating a novel percept.

When does the fusion occur?

The McGurk effect seems to be stronger for meaningless sounds or single syllables, rather than coherent, fluent speech (Easton and Basala 1982). Hence, when there is a great deal of contextual information as to what is likely to come next, the discrepant visual cue may somehow be overruled. But even so the McGurk effect still operates to some extent for meaningful speech.

The McGurk effect operates even when the face and the voice are of different sexes (for example a male face with a female voice or vice versa). It is interesting that auditory and visual information should be integrated despite such obvious incongruities, showing again the highly automatic nature of the effect (Green *et al.* 1991). However, there are limits to how far this will happen. If the face of a highly familiar person is combined with a different voice, the degree of fusion between the face and voice inputs is reduced in comparison to when the face is unfamiliar (Walker *et al.* 1995).

The McGurk effect is a powerful demonstration that speech perception is an inter-modal rather than a purely auditory phenomenon. The auditory input does not take precedence over the visual input, as one might expect. Instead, heard and seen inputs combine to form a percept that is a fusion of both, but that no longer corresponds to either input.

Although widely described by the shorthand term 'lip-reading', these effects involve more than just perceiving movements of the lips (Summerfield *et al.* 1989). Experiments have demonstrated that whilst seeing the lips alone certainly provides a lot of useful information, there are also contributions from seeing the teeth and the way that the tongue and jaws are moving as well.

A key finding in understanding lip-reading concerns the order of difficulty for lip-read or heard phonemes. To understand its implications, we need to think about how speech sounds are produced (see Chapter 1, Box 1A for details). These depend mainly on the place of articulation (at the lips, or within the mouth), the manner of articulation (the way in which the airflow from the mouth is obstructed across time), and voicing (the involvement or non-involvement of the vocal chords).

A demonstration may help. To understand the importance of place of articulation, try saying 'b' or 'p', which are articulated at the lips (bilabial consonants); contrast these with 'd' or 't', formed by placing the tongue against the alveolar ridge near the front of the roof of the mouth (alveolar consonants). To understand manner of articulation, try 'b' and 'p' again, noting that these are stop consonants in which the airflow from the mouth is completely obstructed for a period of time and then released; contrast these with 'f' or 's', which are fricatives, in which the airflow is obstructed but not completely stopped. For voicing, try saying 'b' or 'z', in which the vocal chords are kept together so that the air passing through vibrates them, making a voiced sound; contrast these with unvoiced sounds in which the vocal chords are kept apart, as in 'p' or 'f'.

When you have completed this little self-tutorial, you are ready to grasp why lip-reading is useful. Sounds that are easy to lip-read are those whose distinctiveness rests heavily on place of articulation (since articulation at the lips is easily visible), whereas manner of articulation (how the airflow is obstructed across time) is harder to see. It turns out that the sounds that are easiest to lip-read tend to be those where the differences are the most difficult to hear, because they generally involve rapid acoustic changes of relatively low intensity. This point had been grasped by Miller and Niceley (1955, p. 352) from their work on perception of speech in noise, 'The place of articulation, which was hardest to hear in our tests, is the easiest of features to see on a talker's lips. The other features are hard to see, but easy to hear.' This suggests a simple hypothesis as to why these lip-reading skills have been developed by people with normal hearing; they are probably of great value in infancy, when the baby is learning the sounds of its native language. Much communication is then face to face, and the use of lip-read information will assist in disambiguating the sounds that are auditorily the more difficult. To check the plausibility of this hypothesis, we need to know whether or not infants can lip-read.

A technique for investigating this was developed by Patricia Kuhl and Andrew Meltzoff (1982). Babies were shown simultaneous videos of two faces, and heard one of two possible sounds ('a' or 'i'). One face articulated the sound the baby was hearing, and the other face articulated the unheard sound in synchrony with it. Kuhl and Meltzoff (1982) measured whether babies showed any preference for looking at one face rather than the other. They found that 4–5-month-old infants looked longer at the face articulating the vowel they heard ('i' or 'a') than they looked at the same face articulating the other vowel ('a' or 'i'). Hence infants are sensitive to the mismatch between the seen and heard information, even though at this age they cannot produce speech themselves.

A further impressive demonstration of the integration of lip-reading into speech perception comes from the work of Calvert *et al.* (1997). They used modern imaging techniques to measure blood flow in different regions of the brain when people looked at a speaking face (without sound) or listened to a voice (without seeing the face). The reason for measuring blood flow is that the more active an area of the brain is, the more oxygen (and therefore blood) it requires. As Fig. 6.16 shows, the part of the brain which was activated both by seen speech and by heard speech was in the auditory cortex; in other words merely seeing a person speaking activates brain regions which are involved in hearing.

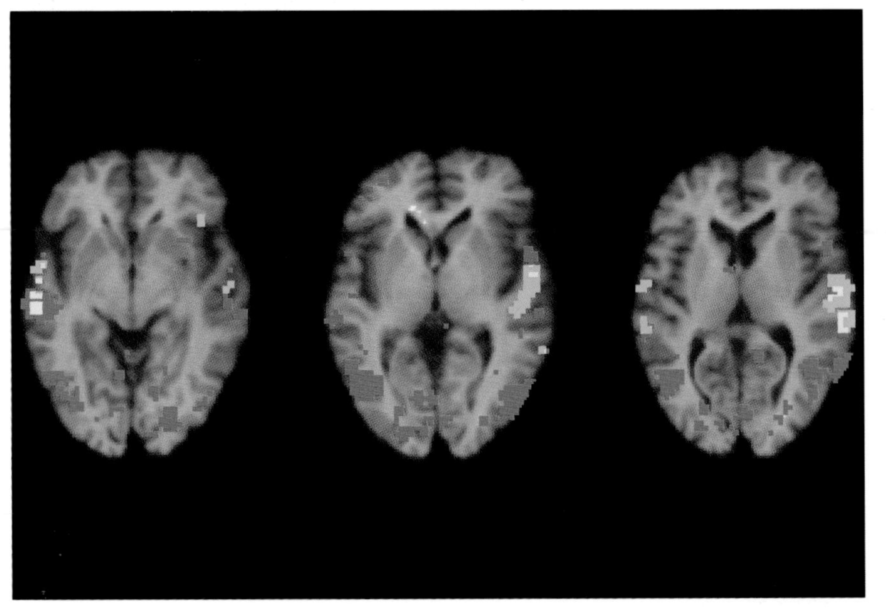

Fig. 6.16 Brain areas activated by silent lip-reading (coloured pink), auditory speech perception (blue), or by both tasks (yellow) as recorded in an experiment by Calvert *et al.* (1997). The region of overlap (yellow) involves auditory cortex. In this figure, brain structures are seen in a series of near-horizontal sections viewed as if from below. Adapted from Calvert *et al.* (1997). Courtesy of Gemma Calvert, University of Oxford, and Michael Brammer, Institute of Psychiatry, London. Copyright, 1997, American Association for the Advancement of Science.

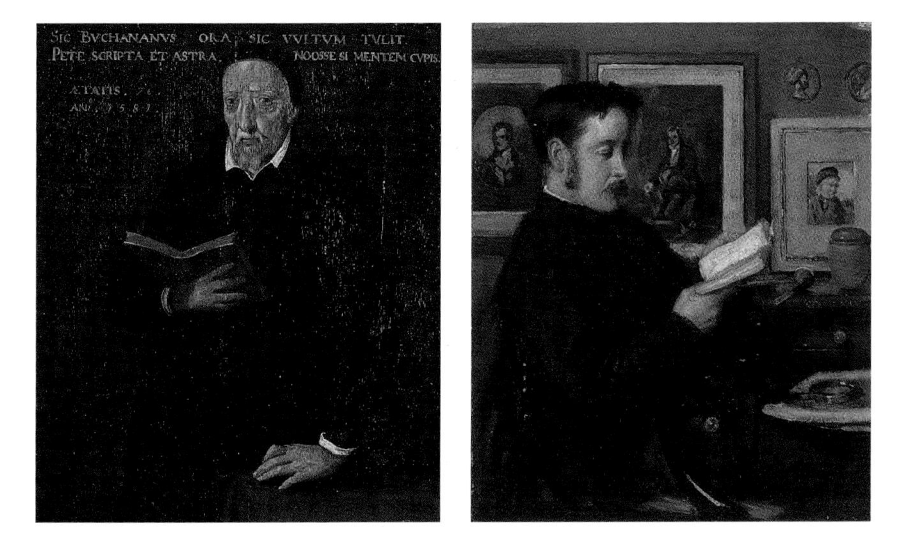

Fig. 6.17 Portraits of George Buchanan (sixteenth-century historian, poet and reformer; attributed to Arnold Bronckorst) and John Miller Gray (the first curator of the Scottish National Portrait Gallery; painted by Patrick William Adam). Courtesy of the Scottish National Portrait Gallery.

6.5 Gaze

In Section 1.2 we noted that the whites of human eyes are unusually prominent, with the consequence that the pupil and iris stand out, making the direction of gaze potentially easy to see. This is likely to be more than a coincidence, because gaze direction is a powerful social signal.

Compare the portraits of George Buchanan and John Miller Gray, shown in Fig. 6.17. It is immediately obvious that Gray is reading, whereas the book held by Buchanan is just a prop—an emblem that he is a cultured and literary man. We can see this because we are skilled at detecting the direction of another's gaze. Gray is looking at his book, whereas Buchanan, in a device often used in portraits, looks back at the viewer.

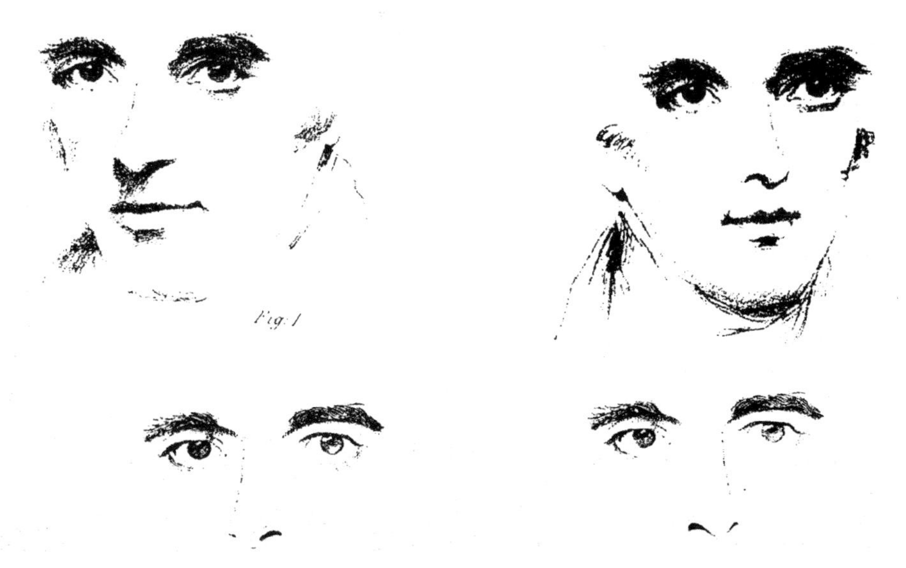

Fig. 6.18 Some of Wollaston's (1824) demonstrations that head orientation influences perceived direction of gaze.

Even though our ability to perceive gaze direction is highly skilled, it is not infallible. A fascinating study was made by William Wollaston (1824), and this was discussed by Sir David Brewster (1832) in his *Letters on natural magic*, written at the suggestion of Sir Walter Scott.

Figure 6.18 shows illustrations taken from Wollaston's paper. In the upper two pictures, the face on the left seems to look directly at the viewer, whereas the face on the right seems to look slightly to one side. In fact, however, the eye regions of both pictures are identical—only the lower part of each drawing has been changed. The same effect is apparent in the lower two pictures in Fig. 6.18, which show that the presence of a slight change of direction of the nose is sufficient to induce a change in apparent direction of the person's gaze.

Wollaston (1824) pointed out that this phenomenon demonstrates that we do not base our judgement of gaze direction *solely* on the position of the iris and pupil relative to the whites of the eyes; instead, this is combined with information about head direction. The effect would repay further investigation to establish just why this happens.

Another anomaly of gaze perception is the feeling that a portrait's eyes will follow you round a room—people say that the *Mona Lisa* seems to gaze at them from wherever she is viewed. This was also discussed by Wollaston (1824) and by Brewster (1832). Wollaston reasoned that the phenomenon may relate to the way in which we judge the orientations of things relative to other frames of reference. In Fig. 6.19, a compass needle is oriented in the same direction as the face's eyes, and the sides of the compass box are oriented as is the head. As for the eyes, the apparent direction of the needle is little affected across a fairly wide range of changes in the angle at which it is viewed.

Simon Baron-Cohen and his colleagues (Baron-Cohen *et al.* 1995) have pointed out that gaze direction can be used to draw conclusions about other people's mental states. It is easy for us to infer which person is *thinking* in Fig. 6.20, or which sweets the carton character *wants* in Fig. 6.21. This ability to infer mental states from gaze direction has been found to be very poorly developed in children with autism, a developmental disorder characterized by marked social abnormalities, leading to the suggestion that failures to 'mindread' from gaze cues are intimately linked to the genesis of this very disabling problem (Baron-Cohen and Ring 1994).

Fig. 6.19 The phenomenon of invariance of apparent gaze direction despite changes in the viewer's position (with the eyes seeming to follow you about) can also apply to the apparent direction of a compass needle.

In social interaction, gaze and eye contact (looking at another person's eyes) serve a number of different functions (Kleinke 1986). These include:

- Regulating turn-taking—for example establishing slightly longer than usual eye contact at the end of an utterance is one of the signals that it is the listener's turn to speak.
- Expressing intimacy—eye contact increases as a function of positive attraction, and romantic relationships are one of the few circumstances in which prolonged gazing into another person's eyes is considered acceptable.

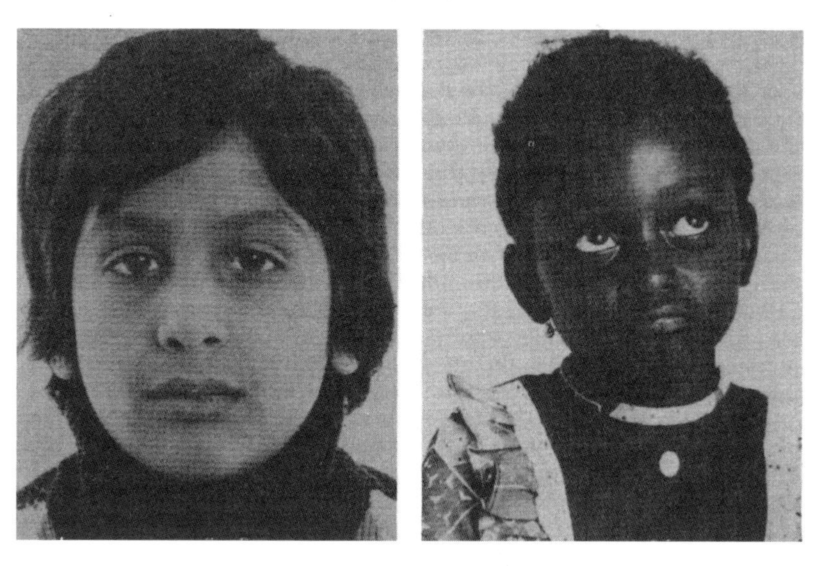

Fig. 6.20 Examples of faces used by Baron-Cohen and Cross (1992). Which one is thinking?

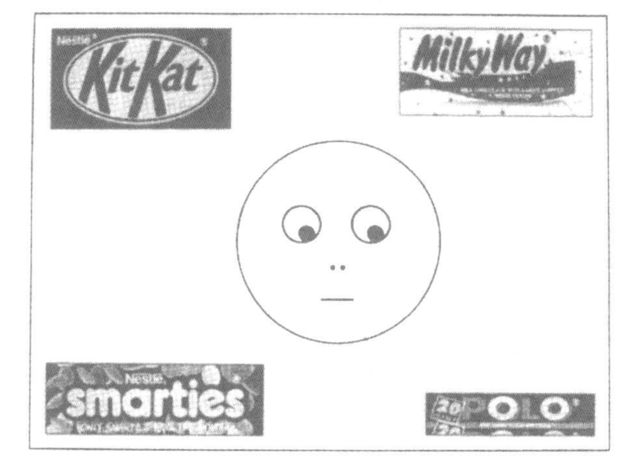

Fig. 6.21 Task used by Baron-Cohen *et al.* (1995). Which sweets does he want? © The British Psychological Society.

Fig. 6.22 *Princess Elizabeth* and *Princess Anne* (Daughters of Charles I) by Sir Anthony van Dyck. Courtesy of the Scottish National Portrait Gallery.

- Exercising social control—through staring to intimidate, or by increasing the amount of eye contact when attempting to persuade or to deceive.
- Facilitating service and task goals—such as looking at another person to check their understanding or approval, or looking at them to seek information or clarification.

People therefore become very adept at using gaze and eye contact to draw inferences about liking and attraction, attentiveness, competence, credibility, and even the mental health of others. The lovely picture of the daughters of Charles I (Fig. 6.22) illustrates how readily we use gaze to draw inferences about personal intimacy. The basic skill of using gaze direction to infer what a person is looking at is developed in infancy, and used to establish shared foci of attention between babies and their carers; most 9-month-old babies can do it (Scaife and Bruner 1975).

Box 6C: Eye-centring in portraits

The great importance of the eyes in portraiture is demonstrated in Christopher Tyler's (1997) recent observation that artists will very often place one of the sitter's eyes on the vertical centre line of the canvas. This is illustrated with several famous portraits in Fig. 6C.1. These portraits include examples of several different styles and a good range of head orientations. They are by Rogier van der Weyden (c. 1460), Sandro Botticelli (c. 1480), Leonardo da Vinci (1505), Titian (1512), Peter-Paul Rubens (1622), Rembrandt (1659), Gilbert Stuart (c. 1796, as reproduced on the U.S. $1 bill), Graham Sutherland (1977), and Pablo Picasso (1937). The white stripe marks the vertical midline of each image.

Figure 6C.2 summarizes this tendency across portraits painted by 170 different artists in the last 500 years. The graph shows that the locations of the best-centred of the two eyes (plotted as a blue solid line) fall in a tight band around the vertical midline. In contrast, if one plots the location of the centre-point between the eyes (that is, the midline of the face itself—plotted as a dashed black line) the majority of placements of this facial midpoint are slightly left or slightly right of the centre of the picture, which is exactly what one would expect to find if artists are actually aligning one of the eyes to the centre of the picture, rather than the face itself.

The importance of the centre of the canvas has long been appreciated in art theory, with rules of artistic composition that highlight the importance of the central vertical line. Equally, in portrait paintings, emphasis has been placed on the axis defined by the centre of the face. But there are no formal compositional rules about the placement of the eyes in relation to the frame of the canvas. Despite this, Tyler's findings reveal a highly consistent tendency of artists to place one eye close to the picture's vertical midline. This must be based on unwritten rules about what feels right when one looks at a picture; placing one of the eyes at the centre may reflect the great importance of gaze and eye contact for human beings.

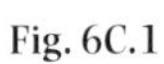
Fig. 6C.1

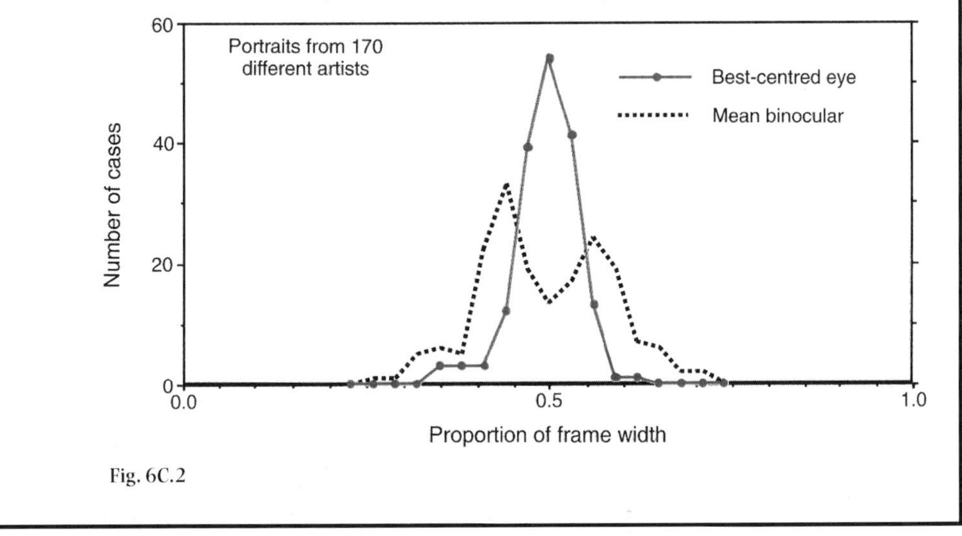

Fig. 6C.2

6.6 Face to face

Face to face communication involves the integrated use of all the cues we have described so far: facial expressions, gaze, lip-reading, and then some more.

Other facial gestures include nodding or shaking the head to indicate agreement or disagreement, frowning for puzzlement, raised eyebrows for questioning, and so on. Some of these gestures, like certain facial expressions, seem to be universally understood, others vary across cultures. Eibl-Eibesfeldt (1989) has given an especially thorough analysis of the functions of the apparently simple gesture of raising the eyebrows (see Fig. 6.23).

Such gestures are often called 'paralanguage', because they supplement and complement the information conveyed by speech. The linguist David McNeill (1985) has argued forcefully, by analysing the

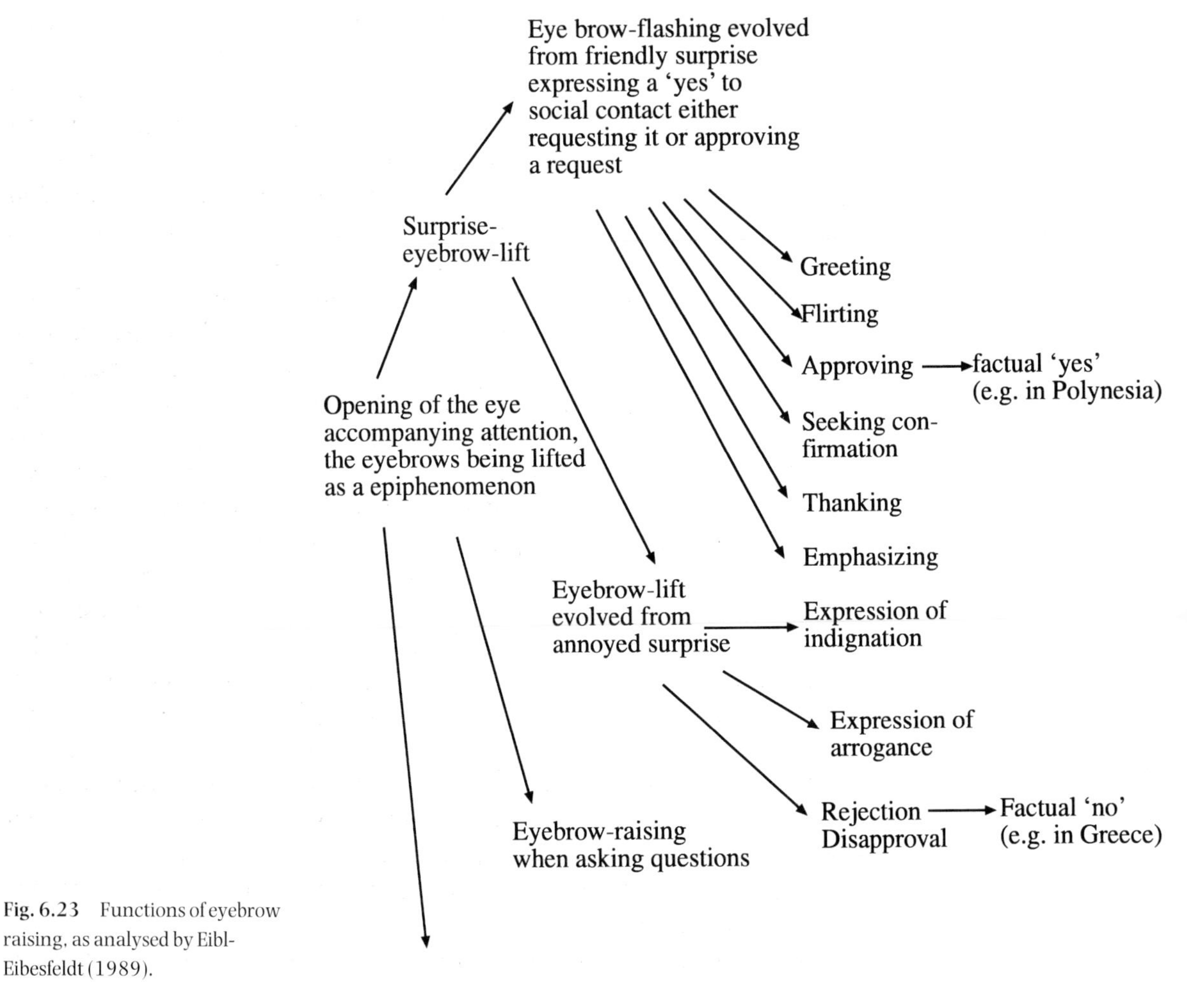

Fig. 6.23 Functions of eyebrow raising, as analysed by Eibl-Eibesfeldt (1989).

speech and gestures of people describing cartoon movies, that gestures are not just redundant rhythmic accompaniments to speech, but modify our interpretation of its content. These gestures include a range of arm, head, and facial movements.

In order to study the way in which non-verbal signals can complement speech in everyday face-to-face communication, researchers need to examine both verbal and non-verbal behaviour in a task which produces naturalistic dialogue, but which also allows performance and behaviour to be measured. One task that has been used to explore how visual and verbal signals are used together is the 'Map Task' (Brown *et al.* 1984).

In this task, one participant is given a map with a series of landmarks and a route drawn on it, and their task is to describe the route to another participant so that they can reproduce it on their own version of the map, which shows the landmarks but not the route (see Fig. 6.24). However, there are some crucial differences between the landmarks shown on the two maps. Some are absent, others in different locations, and the two partners must negotiate the correct route by trying to work out the differences between the maps. The task gives a measure of performance (how close the drawn route was to the target route) as well as producing natural, though task-directed, dialogue.

Gwyneth Doherty-Sneddon and her colleagues (Doherty-Sneddon *et al.* 1997) compared the structure of the dialogues produced when participants in the Map Task could see each other with the dialogues obtained when they communicated only with words. When participants were hidden from each other, more words were needed to complete the task successfully, and these words were used particularly to provide or elicit certain forms of feedback. For example if the instruction giver says 'You know that wee curve?' they are seeking confirmatory feedback from their partner. If their partner says 'yes', they are providing the requested feedback. Both of these kinds of speech activities were more common in the audio-only group. When participants interact face to face, they can substitute glances, nods, or other non-verbal behaviours for these speech acts. Interestingly, however, when dialogues were conducted via high quality video links, only some of these features of face-to-face interaction were preserved. Video-links, particularly where eye contact is made possible through careful arrangements of cameras and mirrors, seem to provoke, at least initially, some behaviour which is not characteristic of normal face-to-face interaction. One possible reason for this is that participants linked by video can see each other, but they cannot see each other's environments and therefore cannot see to which objects participants are paying attention.

Video-mediated communication technologies are increasingly important, allowing teaching and counselling services, for example, to

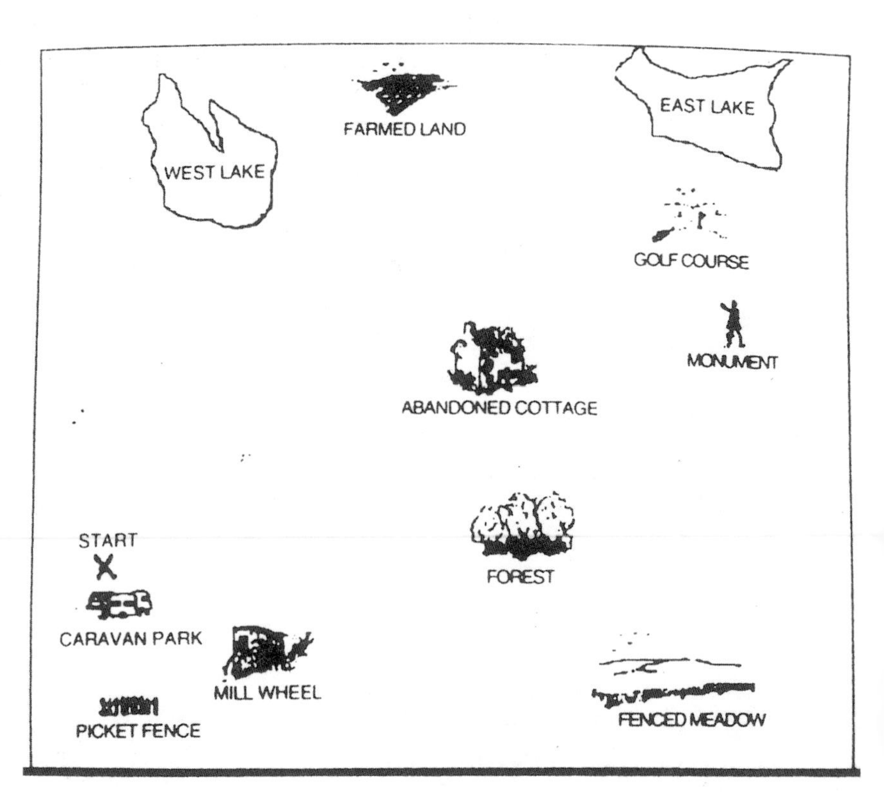

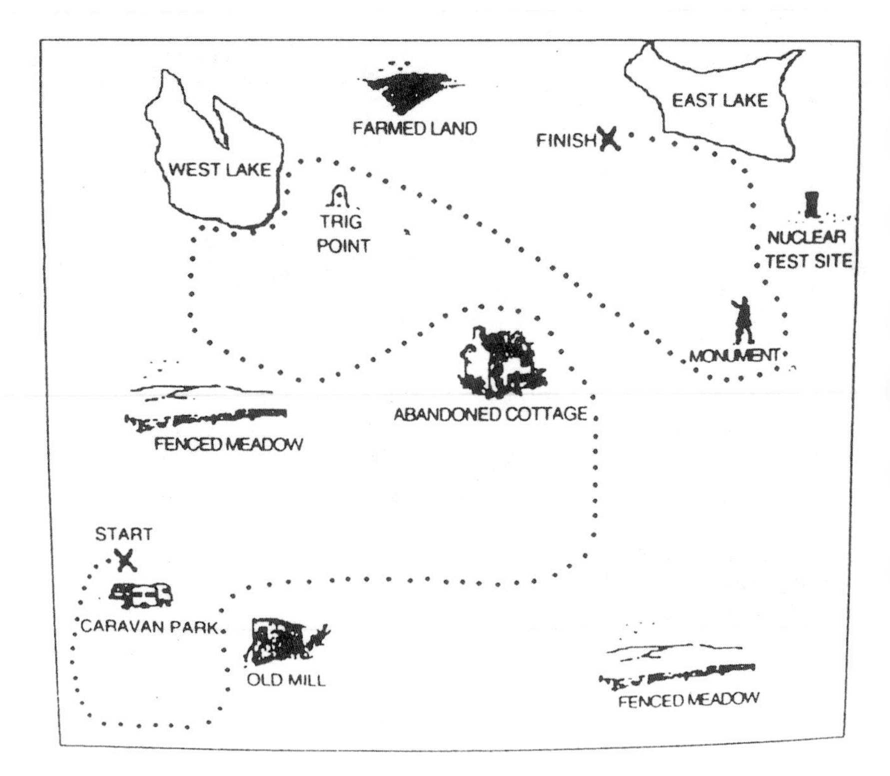

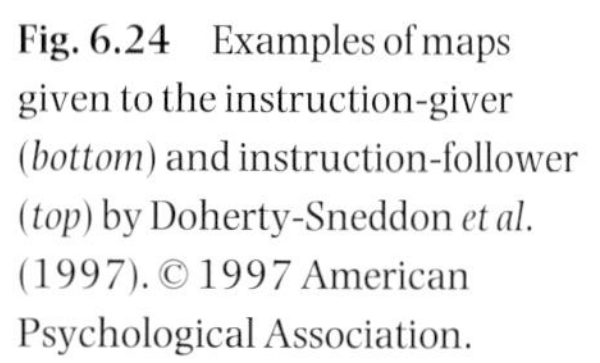

Fig. 6.24 Examples of maps given to the instruction-giver (*bottom*) and instruction-follower (*top*) by Doherty-Sneddon *et al.* (1997). © 1997 American Psychological Association.

be delivered to geographically remote locations. It is important that we understand the rather subtle interactions that there may be between people's natural interpersonal behaviours and particular technological configurations, so that we can exploit the advantages of these technologies and minimize their disadvantages.

In our everyday lives, then, we are able almost effortlessly to combine the analysis of many different social signals to support two-way and even multi-party interactions. In art, it is possible to capitalize on these abilities to create complex yet readily interpretable social scenes, as in Fig. 6.25.

Fig. 6.25 Alexander Moffat, 1980, *Poets' pub*. Courtesy of the Scottish National Portrait Gallery.

7 In the brain of the beholder: the neuroscience of face perception

7.1 Overview

Because faces are of such fundamental social importance to a creature that lives in a complex society, extensive areas of the brain are involved in their perception. The functions of these areas are being revealed in studies of the effects of different types of brain injury, and in studies using modern neuro-imaging methods to picture face processing in the normal brain. One of the most interesting points that emerges is that the brain seems to 'farm out' different aspects of the task to different specialized areas; for example some regions of the brain are more closely involved in determining an individual's identity from their facial appearance, others in interpreting facial expressions of emotion. The brain seems to be wired up to be responsive to faces from birth, as studies with newborn babies have shown.

To discuss different regions of the brain, it helps to have an idea where these are located. Figure 7.1 shows the positions of structures mentioned in this chapter. The wrinkly grey surface of the brain (the cerebral cortex) is divided into anatomically separate left and right cerebral hemispheres. Within each cerebral hemisphere, anatomical landmarks can be used to make a subdivision of what is actually one continuous (albeit very wrinkled and folded) sheet of tissue into subdivisions known as frontal, parietal, temporal, and occipital lobes. The optic nerves from the eyes project to the primary visual cortex, which forms part of the occipital lobes. The other structure we will mention is the amygdala, which lies beneath the temporal lobes.

7.2 Phrenology and the beginnings of neuropsychology

Though always controversial, phrenology enjoyed a considerable vogue from the late-eighteenth to the mid-nineteenth centuries. Its originator was Franz Joseph Gall, a Viennese physician who thought that character traits could be read from the shape of the skull. The

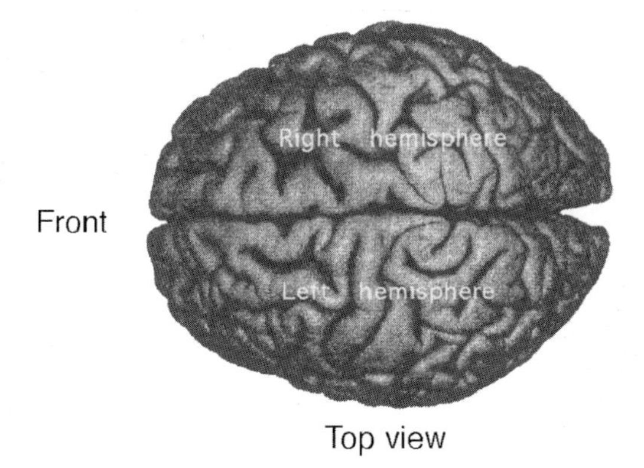

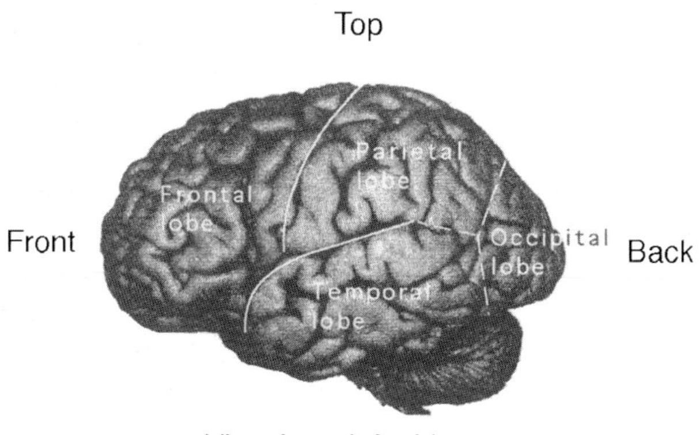

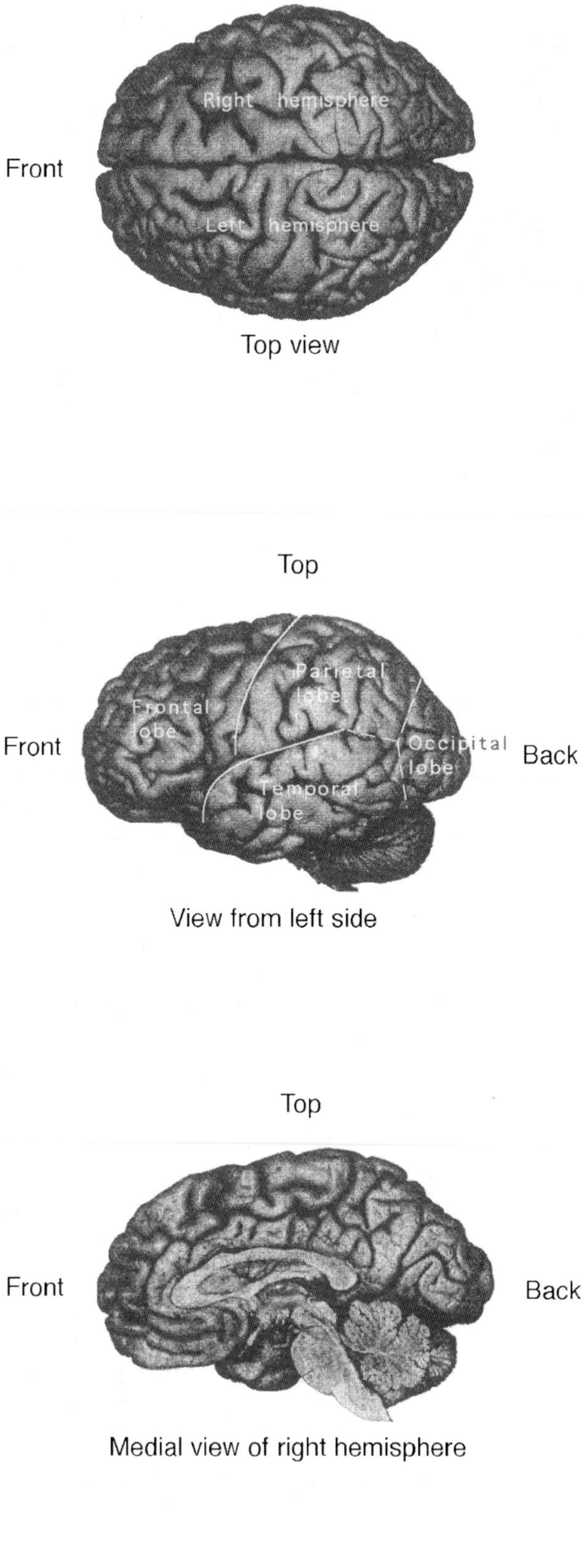

View from left side

Top

Front

Back

Medial view of right hemisphere

Fig. 7.1 Anatomical loci mentioned in this chapter. The upper illustration shows the brain as seen from above, to indicate the left and right cerebral hemispheres. The middle illustration shows a side view of the left hemisphere, indicating how each cerebral hemisphere can be subdivided into frontal, temporal, parietal and occipital lobes. The lower illustration shows a medial (brain midline) view of the right cerebral hemisphere. Adapted from Rose (1995).

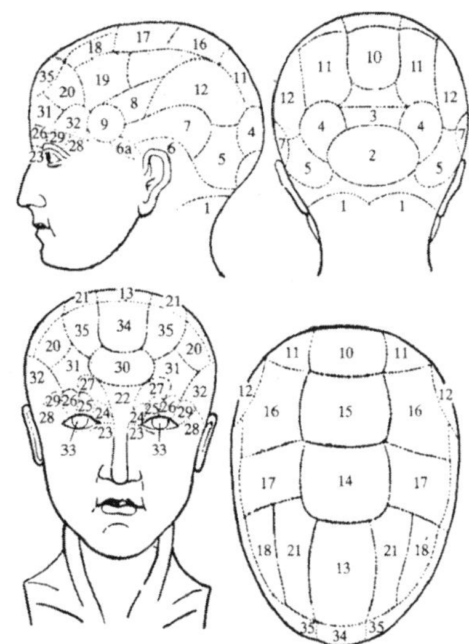

1. Amativeness
2. Philoprogenitiveness
3. Concentrativeness
4. Adhesiveness
5. Combativeness
6. Destructiveness
6a. Alimentiveness
7. Secretiveness
8. Acquisitiveness
9. Constructiveness
10. Self-esteem
11. Love of approbation
12. Cautiousness
13. Benevolence
14. Veneration
15. Conscientiousness
16. Firmness
17. Hope
18. Wonder
19. Ideality
20. Wit
21. Imitation
22. Individuality
23. Form
24. Size
25. Weight
26. Colour
27. Locality
28. Number
29. Order
30. Eventuality
31. Time
32. Tune
33. Language
34. Comparison
35. Causality

Fig. 7.2 Gall demonstrating his method (from Kaufman 1988), and a diagram of the mental organs from a phrenology text (from Gregory and Zangwill 1987).

underlying idea was that, like the rest of the body, the brain could be divided into component organs, each dedicated to a particular mental faculty. It was assumed that the skull would take on a shape reflecting the sizes of these underlying organs.

Gall is shown demonstrating the technique in Fig. 7.2, together with one of the many charts of the different faculties which resulted. The organ for recognizing things, which presumably included recognizing people's faces, involved the faculty of *individuality*. It was located immediately above the nose on the grounds that this region was large in Michelangelo and small in the Scots!

Phrenology caught on fast; by 1832 there were twenty-nine phrenological societies in Britain alone. Two of the people who were most influential in the spread of phrenology were the Combe brothers. George Combe (Fig. 5.11) was an Edinburgh lawyer and social reformer. In 1815, his attention was reputedly drawn by an article in the Edinburgh *Review* denouncing phrenology as 'a piece of thorough quackery from beginning to end'. Combe became convinced otherwise when he attended a party where he witnessed Gall's pupil and protégé Spurzheim dissecting a brain which he had brought along in a paper bag, to show how its parts had influenced the actions of its late owner. Together with his younger brother Andrew, who despite persistent ill-health became a prominent physician, George Combe edited the *Phrenological Journal* from Edinburgh.

Life and death masks of Gall, Spurzheim, and the brothers Combe are shown in Fig. 7.3 for tyro phrenologists to test their observational skills. When you have done this, try comparing them to the life masks of Burke and Hare (Fig. 4.19), whose heads were of course carefully studied in Edinburgh phrenological circles.

After the 1850s phrenology went into rapid decline. Many text-

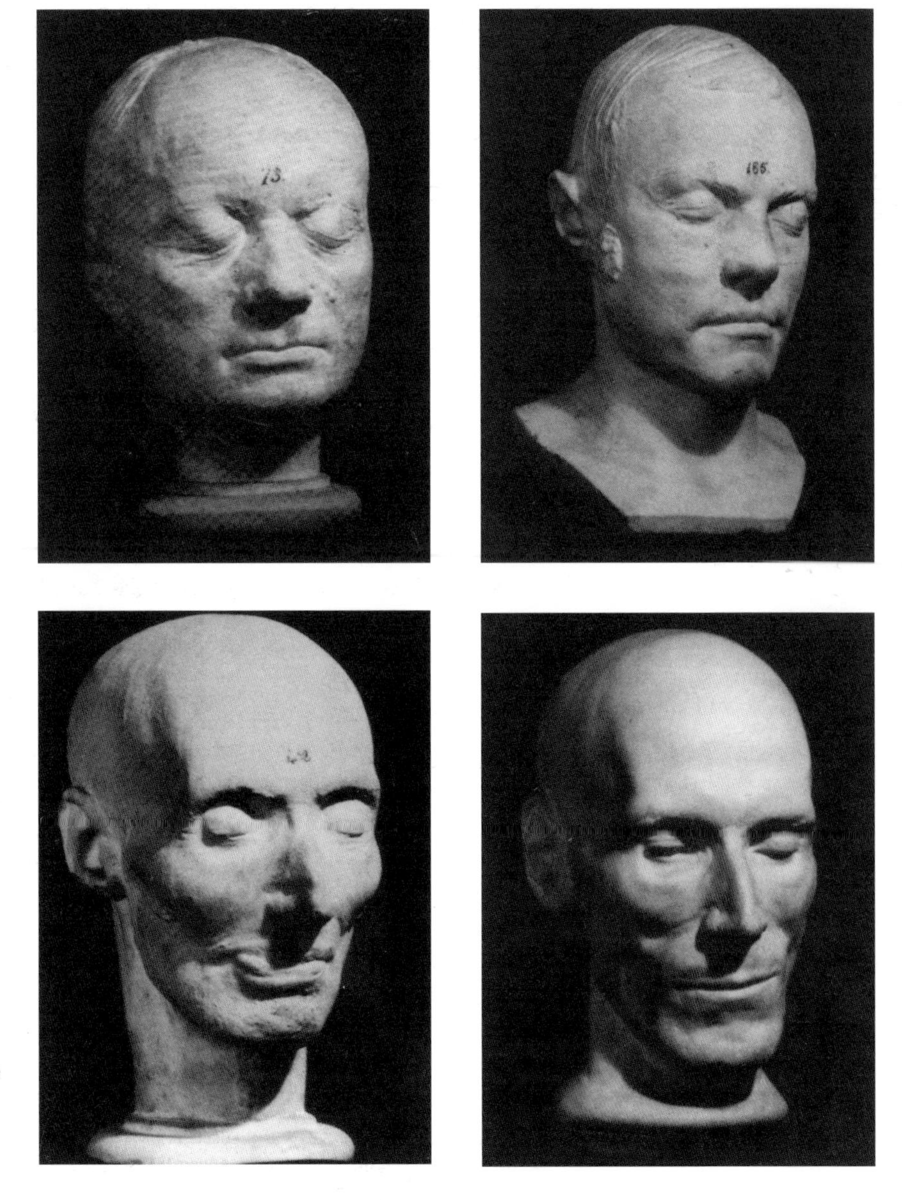

Fig. 7.3 Life masks for Gall (*upper left*), Spurzheim (*upper right*), and death masks of George Combe (*lower left*) and Andrew Combe (*lower right*). From Kaufman (1988).

books now dismiss it as a pseudo-science, akin to physiognomy (see Section 4.5). What went wrong? It is easy to ridicule some of the faculties proposed, which have not stood the test of time. The identification of bumps and protuberances on the skull undoubtedly owed a great deal to wishful thinking, and there were disagreements over the number of faculties and their locations.

What was interesting about phrenology, though, was the idea that different parts of the brain might have different functions, and that these could be determined scientifically, through careful investigation. This bold conception deserves our admiration, not contempt. Gall himself was a talented anatomist, and he had done important work on the differences between grey matter (neurons or brain cells) and white matter (axons, interconnecting the neurons with each other and with the body) in the brain. What let phrenology down was Gall's indirect method of inferring the sizes of brain organs from small variations in the shape of the skull, and his erroneous but entirely plausible assumption that in the brain bigger always means better. Phrenologists were careful to qualify this claim by using relative rather than absolute sizes of the various organs, thus allowing for the fact that some people's brains just are bigger than others, but even so it had little empirical justification.

Rather than seeing it as a pseudo-science, we can think of phrenology as an attempt at a genuine science which was simply ahead of contemporary technical capabilities. The task of demonstrating the fruitfulness of the basic phrenological concept of localization of function in the brain fell to others.

A key figure was the neurologist Sir David Ferrier, who was born near Aberdeen in 1843. Ferrier devised a method of electrical stimulation of the brain which allowed him to establish localization of function to certain regions. Brain neurons work by causing tiny electrical impulses to travel along the nerve fibres (axons) attached to them. Because the brain itself has no pain receptors, mild electric stimulation is not itself felt, but it produces activity of the neurons in that region. So, for example, stimulation of the part of the brain involved in moving a leg will cause the leg to move when the small electrical current is applied.

Figure 7.4 is Ferrier's (1876) summary of his findings. It shows a side view of the left cerebral hemisphere of the human brain, with numbers

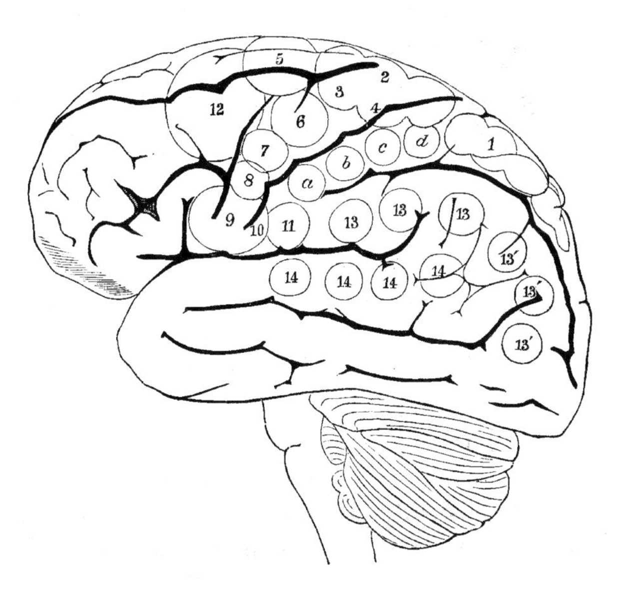

Fig. 7.4 Regions of the brain with functional specialization, according to Ferrier (1876). Numbers 1–12 were identified as being concerned with various types of movement, 13 with vision, 14 with hearing.

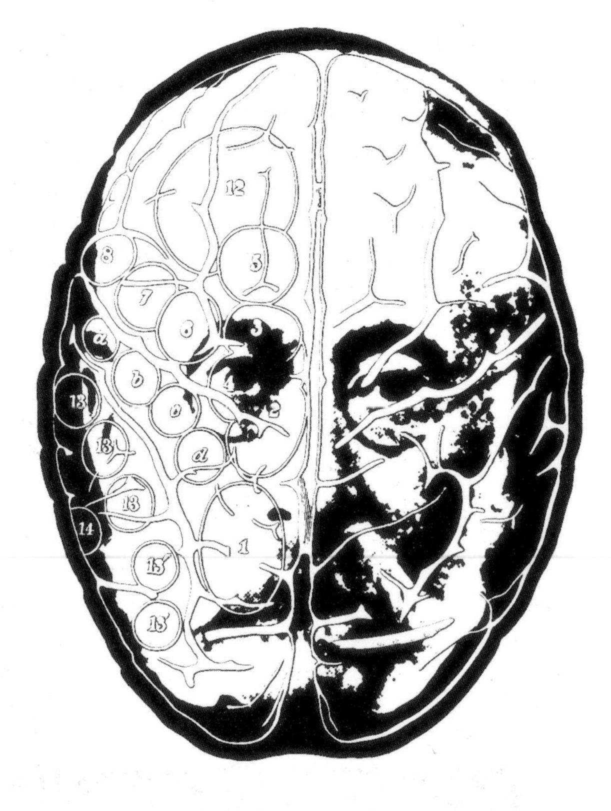

Fig. 7.5 Nicholas Wade, *Ferrier*. © Nicholas Wade.

denoting which functions Ferrier identified for each region. Whilst modern knowledge is not in entire agreement with all of the findings, they are not that far from the mark. Nicholas Wade's picture (Fig. 7.5) shows Ferrier peering out from his diagram of the same sites viewed from above.

The kind of work done by Ferrier relied on technical capabilities which would not have been available when phrenology was taken seriously. But another method was also devised in the nineteenth century which required less technical sophistication, but great patience. This involved careful examination of the brains of people who had suffered brain injuries and subsequently died (sometimes many years later). The sites of the brain lesions (regions of damage) found at autopsy could then be compared to the problems these people had experienced following their brain injuries.

This was a variant of Spurzheim's party piece, but looking at the effects of actual damage to the brain, rather than the sizes of putative mental organs. The method of lesion analysis was painstakingly perfected in the late nineteenth and early twentieth centuries, and it has formed the cornerstone of neuropsychology ever since. It has confirmed the main thrust of Ferrier's findings, and also revealed a great wealth of fascinating information about the brain's functions.

Some of the most important studies have involved examining the effects of brain injuries sustained by soldiers in the terrible wars which have scarred the twentieth century. Freda Newcombe and her colleagues worked with many of the ex-servicemen injured by shrapnel during the D-day landings and subsequent Normandy campaign. These men were injured in 1944–1945, when they were young and fit, and many of those who survived were able to recover and lead normal lives. Yet Newcombe's work showed that even forty years after they were injured, some of these ex-servicemen were still performing poorly on certain tasks (Newcombe *et al.* 1987; Newcombe and Russell 1969).

Examples are shown in Fig. 7.6. The upper stimuli are faces in which the contrast has been exaggerated to create very light or very dark regions. Ex-servicemen with injuries affecting the brain's right temporal lobe found it very hard to see whether such images showed young or old men or women. The lower part of Fig. 7.6 shows a stylus maze, in which a pointer has to be traced along a route between two markers. If the path taken is incorrect, a buzzer sounds, until the correct route has been learnt. Ex-servicemen with injuries affecting the right parietal lobe found this task most difficult. Importantly, however, the people with right parietal injuries were able to perform the face task quite normally, and the people with right temporal lobe injuries

Fig. 7.6 Face and stylus maze tasks used by Newcombe *et al.* (1987). Copyright 1987, with kind permission from Elsevier Science Ltd.

performed the maze task normally. The impairments were highly selective, forming the pattern neuropsychologists call a double dissociation, in which one type of brain injury affects task A but not task B, whereas another type of injury affects task B but not task A.

These features of Newcombe's work highlight that the brain uses separable mechanisms for learning to navigate spatial routes, which requires an intact right parietal lobe, and for the analysis of faces, which requires an intact right temporal lobe.

The finding of abnormalities of face perception after temporal lobe lesions is especially interesting because the nerves from the eye do not project to this region. The optic nerves go to the primary visual cortex, in the occipital lobes. Studies such as Newcombe's have helped to show that the primary visual cortex is simply an early stage in a complex process in which information is fed forward from the occipital lobes along pathways passing into the parietal lobes for the analysis of space and spatially-directed action (Milner and Goodale 1996) and into the temporal lobes for the analysis of significant visual stimuli, of which faces form the most important class.

7.3 Hemispheric specialization and face perception

In December 1911 the painter Lovis Corinth suffered a stroke which severely affected his right cerebral hemisphere. Before this stroke, Corinth had developed a considerable reputation in the fields of landscape and portraiture. The stroke left him unable to work for several months, but eventually he resumed his artistic career. Figure 7.7 shows self-portraits and portraits of his wife Charlotte drawn by Corinth before (left column in Fig. 7.7) and after (right column) his brain injury.

Art critics commented on the heightened intensity of Corinth's work after the stroke. They attributed this to the psychological effects of his illness, such as preoccupation with the fragility of existence after his brush with death. Gardner (1975, p. 323) quotes the writings of Alfred Kuhn:

> When Corinth arose from his sickbed, he was a new person. He had become prescient for the hidden facets of appearance... The contours disappear, the bodies are often as if ript asunder, deformed, disappeared into textures... Corinth always seems to be painting a picture behind the picture, one which he alone sees... at this point Corinth shifted from the ranks of the great painters into the circle of the great artists.

We agree that the portraits in the rightmost column of Fig. 7.7 can be considered the more expressive. But we doubt that was Corinth's intention. Instead, the pictures he painted after his stroke show a phenomenon that is now well-known to neuropsychologists—unilateral neglect.

Fig. 7.7 Self-portraits and portraits of his wife by the artist Lovis Corinth. Those in the left column date from before his right hemisphere stroke, those in the right column from after his stroke. From Gardner (1975).

Notice that it is the left side of each image which is the more sketchy and unfocused. Now compare them to the faces shown in Fig. 7.8. These were made by Keith, a man who had also suffered a right hemisphere stroke (Young *et al.* 1990). Unlike Corinth, Keith had no artistic experience, so he was given the face outline, eyes, nose and mouth already drawn on separate transparent sheets, and asked to arrange these into the best faces he could manage. He professed himself quite satisfied with his efforts, and did not think them in any way unusual.

This problem of left-sided neglect is often noted following right hemisphere strokes, and it has considerable impact on the patients' daily lives since they may fail to shave one side of their face or only comb one side of their hair. Usually, the neglect affects many other objects as well as faces, but in rare cases (such as Keith's) it may present as face-specific. Figure 7.8 shows a comparable task in which Keith was asked to place the headlights, radiator, and window of a car into their correct positions; unlike the faces, he was near-perfect at doing this.

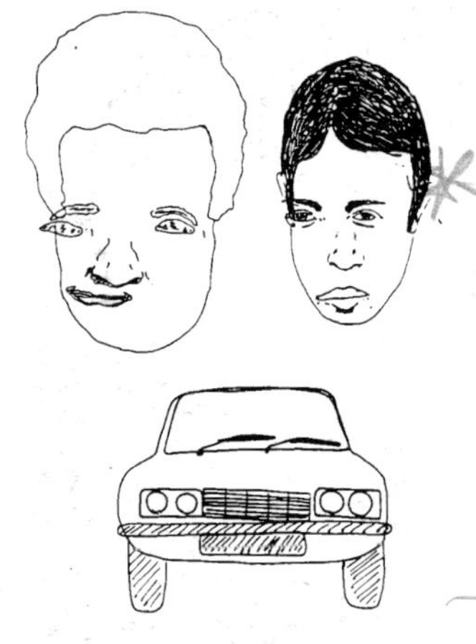

Fig. 7.8 Face and car pictures constructed by Keith after his right hemisphere stroke. The task was to arrange already-drawn fragments of pictures into the correct positions. From Young *et al.* (1990). Copyright 1990, with kind permission from Elsevier Science Ltd.

In Section 7.2 we noted that Freda Newcombe had found that it was ex-servicemen with right temporal lobe injuries who performed poorly at face perception. However, their problems affected all of the seen face, whereas unilateral neglect primarily affects the left side. The critical site for a brain lesion within the right hemisphere to cause neglect of the left sides of faces is not yet known, but a likely possibility is that it will affect the right parietal lobe, since this is damaged in other forms of unilateral neglect. The parietal lobe is involved in attention and in the perception of the spatial layout of the world, and especially in the visual abilities we need to control our actions (Milner and Goodale 1996).

In evolutionary terms, the key task of vision has been to construct representations of the external environment which will permit effective actions, and many parts of the visual system are tightly integrated with the mechanisms that control our movements. To some extent, the left-sided neglect evident in Figs 7.7 and 7.8 might therefore reflect the fact that both involve motor activity, in the form of drawing or putting things in their appropriate positions.

However, neglect is not simply a failure to move the limbs correctly. When chimaeric stimuli consisting of halves of two different photographs of faces joined at the midline are presented to people with left-sided neglect, they will often identify only the half-face falling to their right.

Part of the reason for this seems to lie in a different form of motor ability—eye movements. Figure 7.9 shows examples of the eye movements made by another person with unilateral neglect, Robert (Walker *et al.* 1996). When we look at things, our eyes make sharp movements (saccades) to get the things we are especially interested in onto the central part of the retina, which has the most acute vision. In Fig. 7.9 Robert's eye movements have been superimposed onto the pictures he was looking at by marking the series of points which he fixated and then joining these to show the pattern of his saccades. In each case, Robert began at the midline of the face, because the procedure involved asking him to look at a central fixation square before the face was presented, but after that he only explored the right side of each image. This makes it a little less surprising that Robert only recognized the chimaeric's right side.

Even so, it is not clear whether it is failure to scan the left side of a

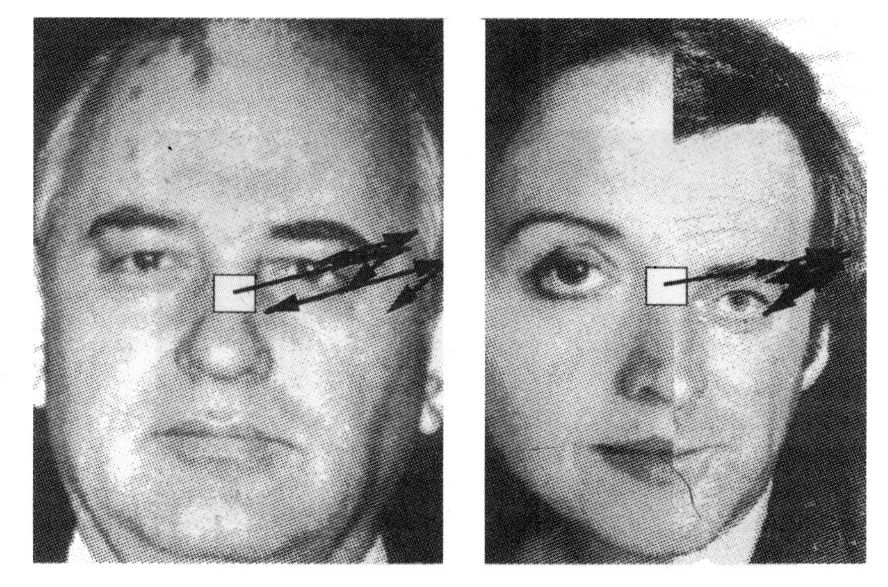

Fig. 7.9 Eye movements to a face (Mikhail Gorbachev) and a chimaeric face (Anna Ford and Terry Wogan) by Robert, who had suffered a right hemisphere stroke. The white squares show Robert's initial fixation position, black lines show his saccades, with arrows indicating regions of fixation. Despite being warned that many stimuli would be made from two people joined together, he identified the faces as those of Mikhail Gorbachev and Terry Wogan. From Walker *et al.* (1996).

chimaeric figure which causes the failure to recognize it as different, or failure to be aware of the left side which causes the failure to scan it.

This is something of a chicken and egg question, but there are grounds for thinking that simple inability to look to the left is not the answer. Young *et al.* (1992) investigated another person with neglect, Barbara, in some detail. When they pointed to parts of the chimaeric on the left, Barbara could describe them accurately, yet she nearly always identified the chimaeric as the right half-face. With a chimaeric in which the left half came from Mick Jagger's face and the right half from Roger Moore, Barbara described accurately the shape of the left eye, the long hair over the left ear, the full lips on the left, and the stubble on the left chin when these features were pointed out. None of them resembled the parts of Roger Moore's face on the other side of the chimaeric, yet even after describing all of these features of Mick Jagger's face Barbara identified the chimaeric only as 'Roger Moore', and did not accept that part of another person's face was present. On another occasion she showed partial insight; when made to describe parts of a face chimaeric with the left half of Michael Parkinson and the right half of Terry Wogan, Barbara commented that 'It's Terry Wogan ... but there's a touch of a Picasso about him'.

These examples show that even when a person with left neglect has carefully inspected the left side of a face, it may still not be recognized! Neglect is a very puzzling clinical phenomenon, whose exact nature has yet to be determined, but current thinking is that it seems to involve some kind of difficulty in forming an adequate representation of the face's left side or an inability or disinclination to attend properly to it.

What seems to be a related phenomenon is evident in neurologically normal perceivers. The anatomical arrangement of the optic nerves (see Fig. 7.10) means that when we look at the world, information falling to the left of our fixation point is initially projected to the primary visual cortex of the right cerebral hemisphere, whereas information falling to the right of fixation is projected to the visual cortex of the left hemisphere. Of course, as we move our eyes about to take in the whole scene, what is projected to the right or to the left hemisphere will

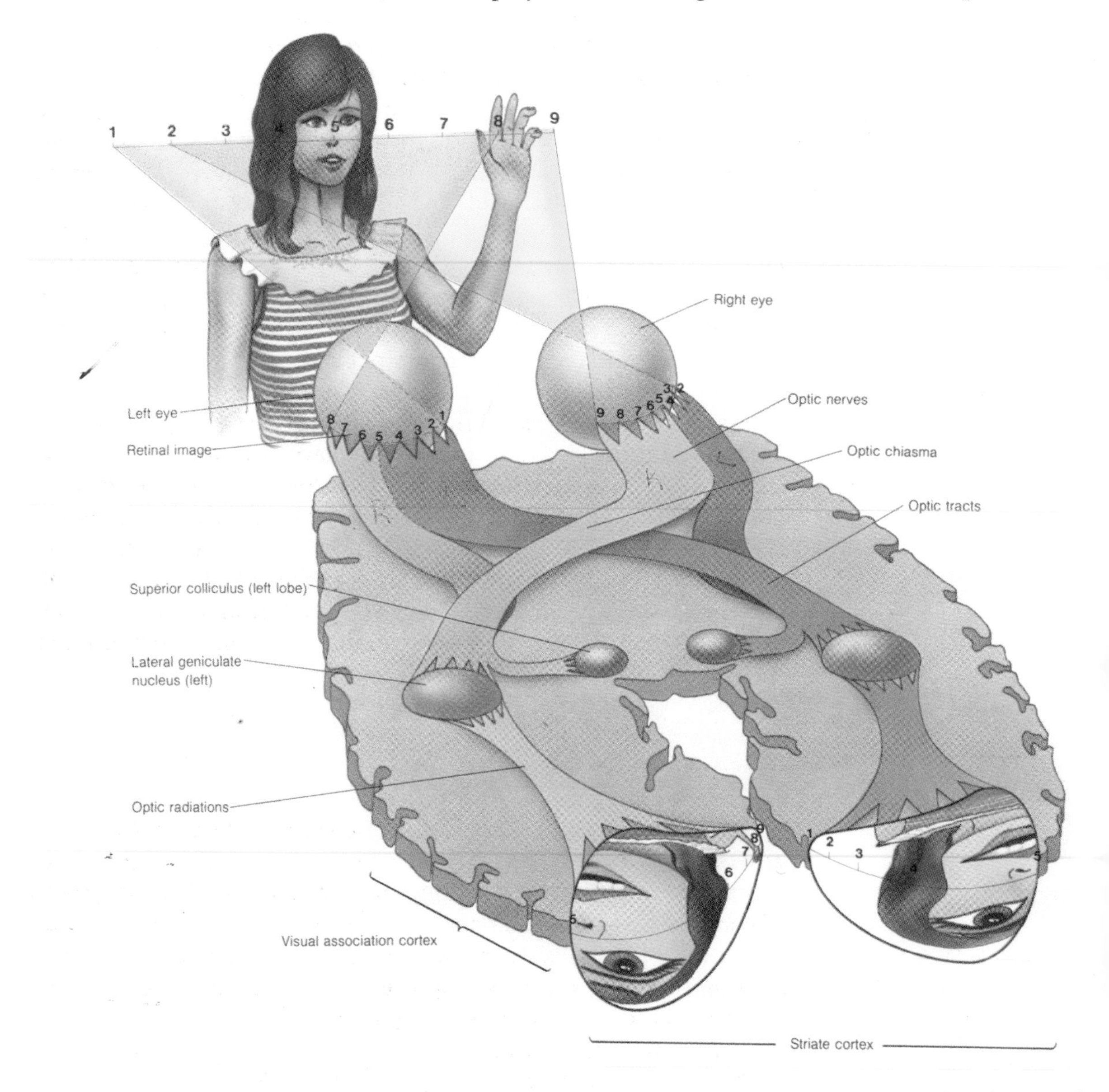

Fig. 7.10 Diagram of the principal visual pathways from the retina to the brain. Note how the retinal image is split at the point of fixation (the woman's nose). The pictures superimposed on the primary visual cortex (striate cortex) show how the retinal image is projected in topographic form, but divided at the midline and crossed, so that the left side of the image goes initially to right visual cortex, and vice versa. From Frisby (1979).

change from moment to moment according to where we are fixating at the time, but if the average position of our fixations tends to fall near the midline of a display (as it usually does for a normal person looking at a face), then on balance the right hemisphere is mainly getting its initial visual input from the side of the face falling to our left.

In consequence, because the right hemisphere is so important to face perception, neurologically normal people show almost the opposite to the left neglect found after right brain injury; instead, the normal observer unknowingly tends to overestimate the importance of information from the left side of the face.

Look at the images in the first column of Figure 7.11. The centre face is a portrait of the photographer James Craig Annan. In the upper image, the left side of the portrait is combined with a mirror-reversed version of the same side, and in the lower image the right side of the portrait is combined with its mirror-reversed version. To most right-handed people, the top image will look more like the original (centre) image than does the bottom image (for left-handed people, there is generally no particular direction of preference, since cerebral asymmetry for face perception does not seem so pronounced in left-handers).

To show that this does not simply reflect an idiosyncrasy of Annan's face, the centre column in Figure 7.11 shows the same phenomenon using Sir Donald Currie as a model. Finally, to really convince you that the effect is in your brain, not in the face before you, the middle image in the last column has itself been mirror-reversed. Yet it is still the top image (based on what is now the left side) which looks more like the original to most right-handed perceivers (Gilbert and Bakan 1973). This is of course the opposite of the face they would have chosen when the original image was the right way round!

7.4 The phrenologists' revenge

There is a remarkable degree of functional specialization of the brain to deal with faces. In Section 7.2 we saw how damage to the right temporal lobe could cause problems in face perception. The temporal lobe forms the target for a neurological pathway from primary visual cortex whose job includes recognizing familiar visual stimuli. If this pathway is severely affected, especially by lesions affecting the parts that lie close to the brain's midline, an almost complete inability to recognize familiar faces can result.

This problem is known as prosopagnosia. People with prosopagnosia know when they are looking at a face, but lose all sense of who it might be. The most familiar faces may go unrecognized—friends, family, and even the patient's own face when seen in a mirror.

Fig. 7.11 Composites of portrait faces made from two left halves (*top row*) and two right halves (*bottom row*) of the centre row images. Faces are those of the photographer James Craig Annan (*left column*) and the shipping magnate and educational benefactor Sir Donald Currie (*centre column*). In the right column, Annan's face is itself mirror-reversed. Original portraits courtesy of the Scottish National Portrait Gallery. Computer-manipulated images courtesy Gary Jobe, Medical Research Council Applied Psychology Unit, Cambridge.

Failure to recognize faces is indeed a social handicap. Newcombe *et al.* (1994, p. 108) described some of the problems experienced by a retired mathematician who had been prosopagnosic all his life:

> At the age of 16 years, while on holiday, he stood in a queue for 30 min next to a family friend whom he had known all his life, without any experience of familiarity. The friend complained—mildly—to his parents and his father reacted strongly to this apparent discourtesy. Soon after, cycling from the village to his home, he saw a man walking toward him. In his own words: 'Mindful of my father's recent forceful comments, I decided to play it safe. As we passed, I said "Good morning, Sir." My father said later that I had never addressed him as politely before or since.

Fortunately, recognition from other cues (such as voices) may remain possible, allowing some scope for compensatory strategies. These are seldom entirely successful, however. A particularly unfortunate example is given by Pevzner and his colleagues (Pevzner *et al.* 1962, p. 336). Their patient

> was accustomed to frequent meetings with his lawyer for business discussions but in court faced with two lawyers, both dressed in a similar fashion, he could not distinguish between the two and discussed the proceedings of the case with his opponent's attorney, thinking him to be his own—with disastrous consequences. When his physician afterwards asked him what had happened he replied that he had always met his lawyer in the latter's office and had come to recognize the furnishings and surroundings, these serving as the means of identifying the man. But in the courtroom, presented with two persons wearing identical gowns and not in their usual setting, he was at a loss.

Figure 7.12 shows a magnetic resonance brain scan image of Valerie, a lady with a degenerative disease affecting the right temporal lobe. This disease led to a progressive prosopagnosia, in which Valerie gradually became poorer and poorer at recognizing familiar faces. Eventually she began to lose her memories of who people were, so that she would fail to recognize familiar names as well as familiar faces (Evans *et al.* 1995).

The development of techniques for making images of the structure of the brain of a living person, as shown in Fig. 7.12, has revolutionized contemporary neuropsychology, because it is now possible to know with reasonable precision where a brain lesion is sited whilst the person is still alive. Further exciting advances have come from techniques which measure blood flow in the brain. When neurons are highly active, they need lots of oxygen, which means that extra blood must pass through the arteries feeding that region. This allows us to examine the activity of different parts of the normal brain when a person is doing different tasks.

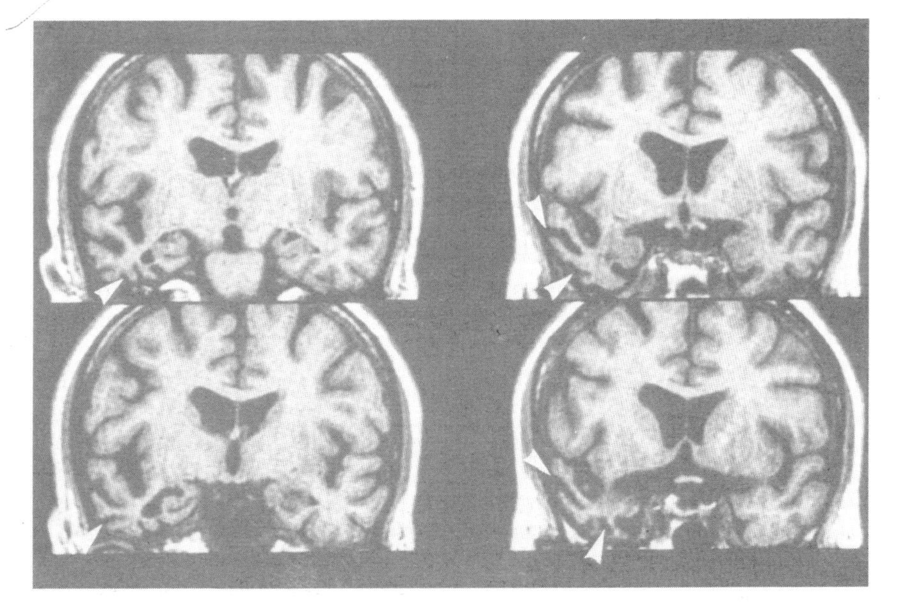

Fig. 7.12 Magnetic resonance brain scan image of a person with a progressive prosopagnosia, showing degeneration of the right temporal lobe. The brain scans show a series of vertical (coronal) sections through the temporal lobes. The right temporal lobe (marked with arrows) is noticeably atrophied in comparison to the left temporal lobe. From Evans *et al.* (1995).

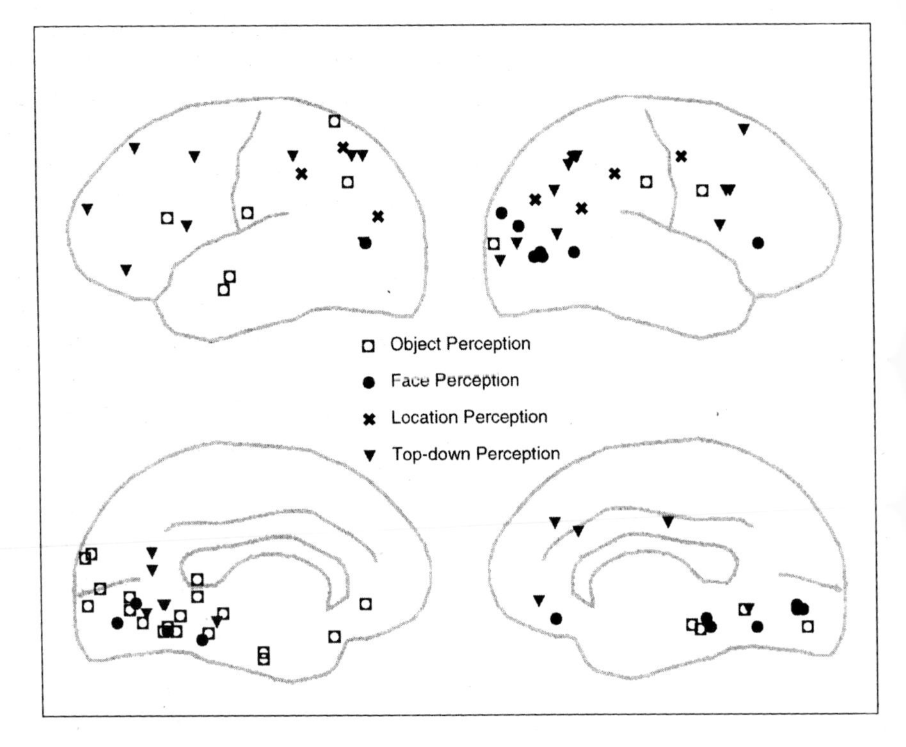

Fig. 7.13 Summary of brain locations where activation has been found in brain imaging studies of object perception, face perception, location perception, and top-down perception (attention). The upper diagrams show the outer surfaces of the left and right cerebral hemispheres, with inner (midline) surfaces below. From Cabeza and Nyberg (1997). © 1997 Massachusetts Institute of Technology.

Cabeza and Nyberg (1997) have reviewed the findings of many such studies. Figure 7.13 shows their summary of the parts of the brain which have been found to be the most active when people perceive objects, faces, and spatial locations, and for top-down perception (roughly, a measure of attention). As the figure shows, attention and spatial perception involve extensive regions of parietal and frontal lobes, whereas object and face perception involve occipital and temporal regions, and especially those parts of the occipital and temporal lobes nearest the midline of the brain (the lower diagrams in Fig. 7.13). It is a beautiful fit to the neuropsychological evidence.

A question often asked about prosopagnosia concerns whether it is *only* faces which are affected? Usually, the answer is that there are also problems in recognizing the individual members of other categories with high inter-item similarity such as cars, flowers, and birds. The problem with faces is just more likely to receive the most attention because it is the most distressing. In very rare cases, however, it does seem that the deficit can be face-specific (De Renzi 1986).

This is as might be expected from the pattern of findings summarized in Fig. 7.13, where it is clear that there is substantial overlap between the neurological pathways involved in face and object perception, making it likely that brain injury will usually compromise both abilities to some extent.

One of the astonishing things about prosopagnosia is that people who can recognize virtually no familiar faces at all may remain able to see the faces sufficiently well to be able to recognize their emotional expressions or to match photographs accurately to say whether they show the same or different people. The possibility that familiar face recognition, unfamiliar face matching, and recognition of facial expressions may each involve partially separable pathways in the brain has been investigated in Freda Newcombe's group of ex-servicemen with shrapnel wounds sustained in the Second World War (Young *et al.* 1993*a*), with results that favour this hypothesis (see Box 7A).

The picture we are piecing together is one of very rich specialization. We can see that in effect the brain farms out the task of perceiving faces to specialist regions or specialist pathways, each with its own distinct purposes. In part this may be driven by the different visual requirements of these tasks, which we have discussed in Chapters 2–6. If age and sex perception require us to analyse differences in 3D craniofacial structure which arise slowly, over many years, it may make sense to keep these separate from something like expression perception, for which the relevant time-scale can be of the order of fractions of a second. But it is also likely that at least some of the pressures creating this specialization come not just from visual factors, but from the different social purposes for which the information is used.

A good example comes from work on the contribution of the amygdala to recognition of emotion. The amygdala is a small structure lying beneath the temporal lobe, which has long been recognized by neurophysiologists as involved in emotion, and especially in the emotion of fear (LeDoux 1995).

Box 7B shows data for recognition of facial expressions of emotion after amygdala damage. Some emotions are recognized quite normally, whereas the recognition of anger and (especially) fear is compromised. The amygdala seems to make an important contribution to the recognition of these emotions.

The same point can be demonstrated in studies of blood flow when normal people look at facial expressions. Figure 7.14 shows computer-manipulated facial expressions with increasing intensities of happiness or fear; the images were created using techniques described in Chapter 6. People were asked to look at a series of such faces and classify them as men or women (Morris *et al.* 1996). When cerebral blood flow was analysed to find the brain structure which required more blood as the face became more afraid, it was the left amygdala. The result is shown in Fig. 7.15, in which the region of increased blood flow is marked with an increasingly bright colour. Note that in this experiment nothing was actually asked about emotion, the task was just to decide male or female. This shows that the amygdala is automatically responsive to displays of fear by others; it is not something we can switch on or off at will.

Box 7A: Recognition of identity and expression after brain injury

Freda Newcombe and her colleagues studied the face perception abilities of 34 ex-servicemen who had suffered injuries affecting the left or the right cerebral hemisphere during the last years of the Second World War (32 cases) or in the Korean War (2 cases). These ex-servicemen form a uniquely important group for several reasons; they were injured when young and fit, they have since had plenty of time to recover (most of the people tested were again leading normal lives), and the shrapnel wounds produced injuries which affected discrete regions of brain tissue.

The face perception tests were carried out in the 1980s, which would generally be around forty years after these men were injured. Six different tests were used (Young *et al.* 1993*a*). Two tests examined ability to recognize familiar faces; these involved deciding whether or not faces were of famous people (Test A) or giving the occupations of several famous faces (Test B). A further two tests involved ability to match views of unfamiliar faces. In one test (Test C) a target face had to be matched to six simultaneously presented choices (Benton *et al.* 1983), and in the other test slides of unfamiliar faces were presented one after the other (Test D), to be judged as those of same or different people. The final pair of tests examined ability to recognize facial expressions of emotion; in one test by matching a target expression to one of four simultaneously presented expressions (Test E), and in the other by recognizing facial expressions of six basic emotions (Test F).

The logic of using two tests of familiar face recognition, two tests of unfamiliar face matching, and two tests of facial expression recognition was to look for cases of selective impairment, defined as poor performance on both tests of a particular ability and normal performance on the other four tests.

Several cases of selective impairment were found among the ex-servicemen, and these included cases of selective impairment of each ability. Three examples are shown in Fig. 7A.1. Each chart represents the performance of a different person across the six tests, and in each case this is expressed as a *z*-score indicating how far each person's accuracy at performing a test differs from the performance of an age-matched comparison group of neurologically normal people; higher *z*-scores indicate increasing degrees of difficulty, with values greater than 2.00 being severely impaired.

As Fig. 7A.1 shows, Peter was impaired for both familiar face recognition tasks, Stephen for both unfamiliar face matching tasks, and Geoffrey for both facial expression tasks.

Such findings support the idea that these abilities each involve at least partially separable neurological pathways in the brain. A consequence of this would be that brain injury can compromise one

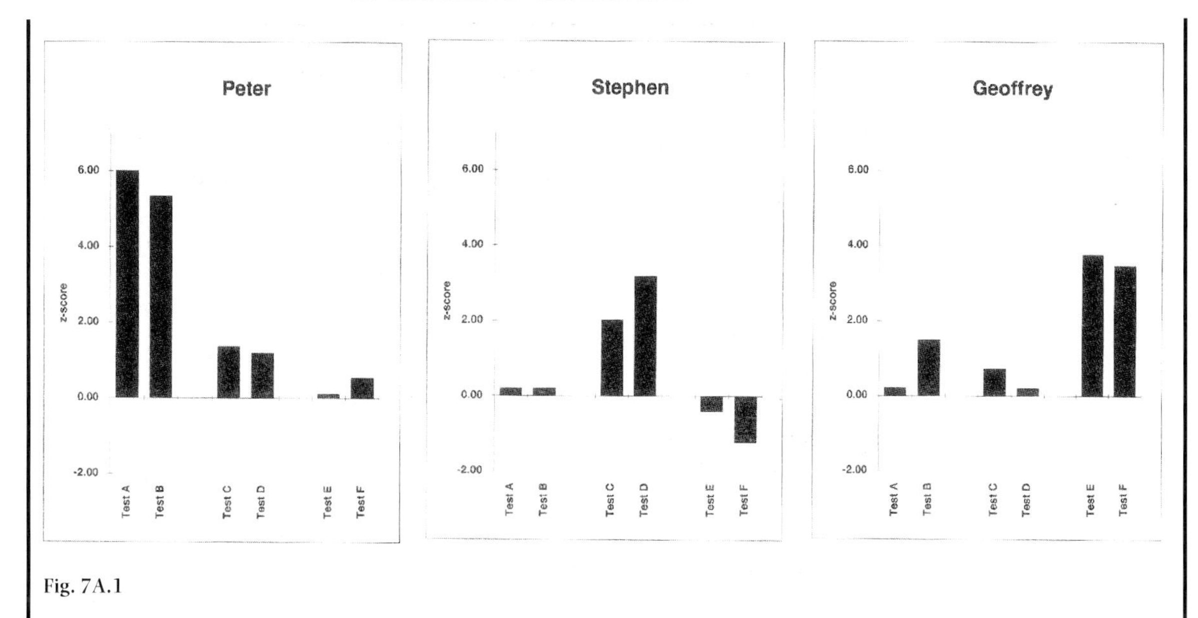

Fig. 7A.1

type of ability whilst sparing the others. Because of the considerable importance attached to the various forms of social information we derive from the faces we see, the most efficient method for analysing the different types of social signal may be to farm out parts of the task to specialist sub-components, each of which can then be optimally tuned to analyse a particular kind of information.

Box 7B: Emotion recognition after damage to the amygdala

The amygdala has often been thought to be involved in emotion, but studies of its contribution to our ability to recognize the emotions of other people have only been carried out recently.

Andrew Calder and his colleagues studied the ability of two people with amygdala damage to recognize facial expressions of emotion (Calder *et al.* 1996*b*). The stimuli for one of their tests are shown in Fig. 7B.1. Facial expressions from the Ekman and Friesen (1976) series were computer-manipulated (using the technique described in Box 6A) to create a hexagon of faces whose expressions wereevenly graded between happiness and surprise (top row), surprise and fear (second row), fear and sadness (third row), sadness and disgust (fourth row), disgust and anger (fifth row), and anger and happiness (bottom row).

Images from this emotion hexagon were presented one at a time—each had to be recognized as most like happiness, surprise, fear, sadness, disgust, or anger. In total, each of the 30 images from Fig. 7B.1 was presented five times, in random order.

Box 7B: Continued

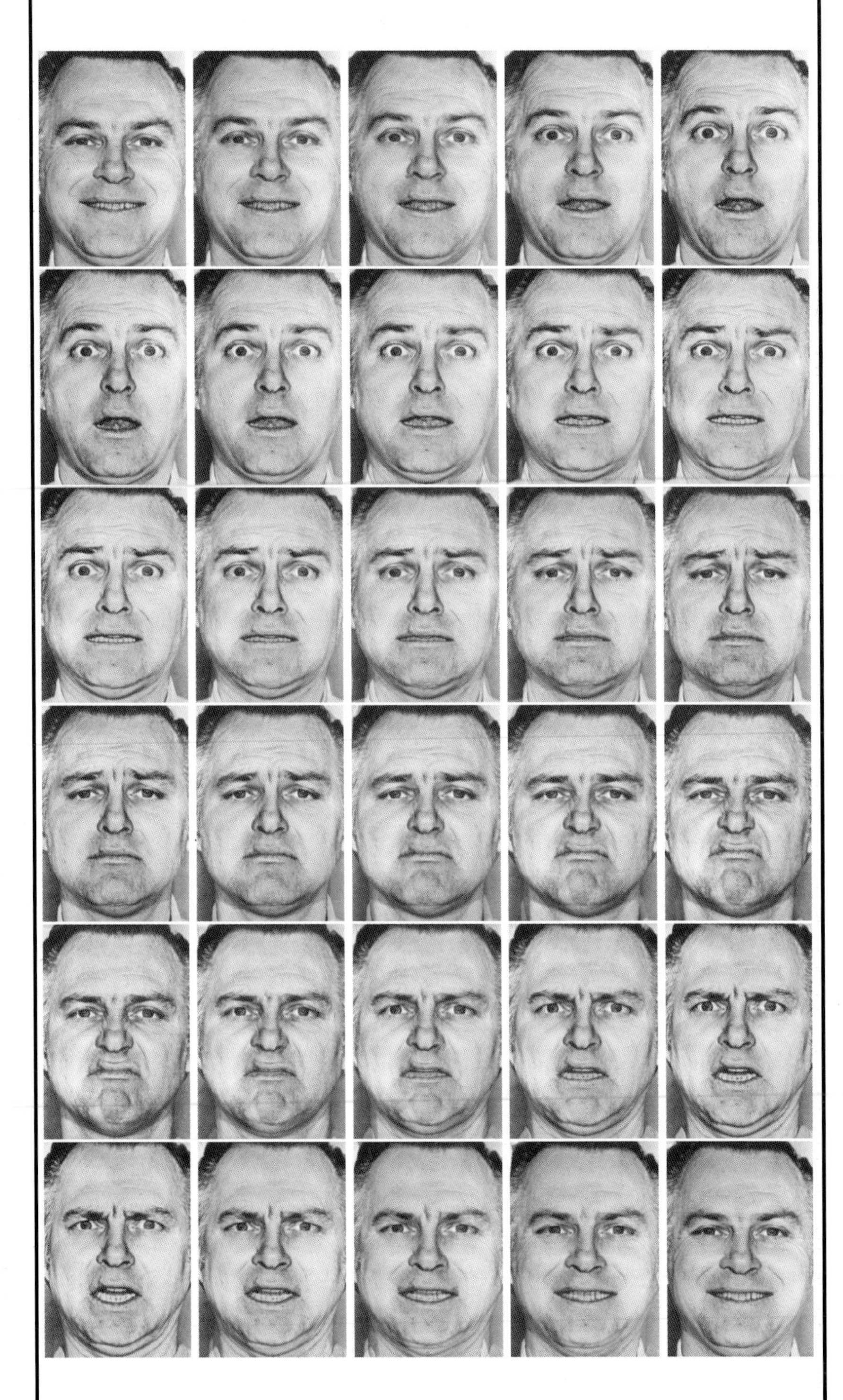

Fig. 7B.1

Figure 7B.2 shows the rate at which each image was recognized as each emotion. The top graph shows recognition by neurologically normal people. There is a clear sequence of abrupt transitions between regions which are consistently perceived as each of the six emotions.

The lower graph in Fig. 7B.2 shows the average recognition rates for the two people with amygdala damage. The recognition of happiness, surprise, and sadness is not too different from normal performance, but there are signs of some difficulty with disgust and severe problems in recognizing anger and fear.

Such findings demonstrate that the amygdala may be more involved in the evaluation of some emotions than others, especially fear and anger. Facial expressions of fear and anger are important indications of the presence of danger in the immediate environment, and the problems in recognizing fear or anger after damage to the amygdala may form part of a wider problem in the appraisal of danger.

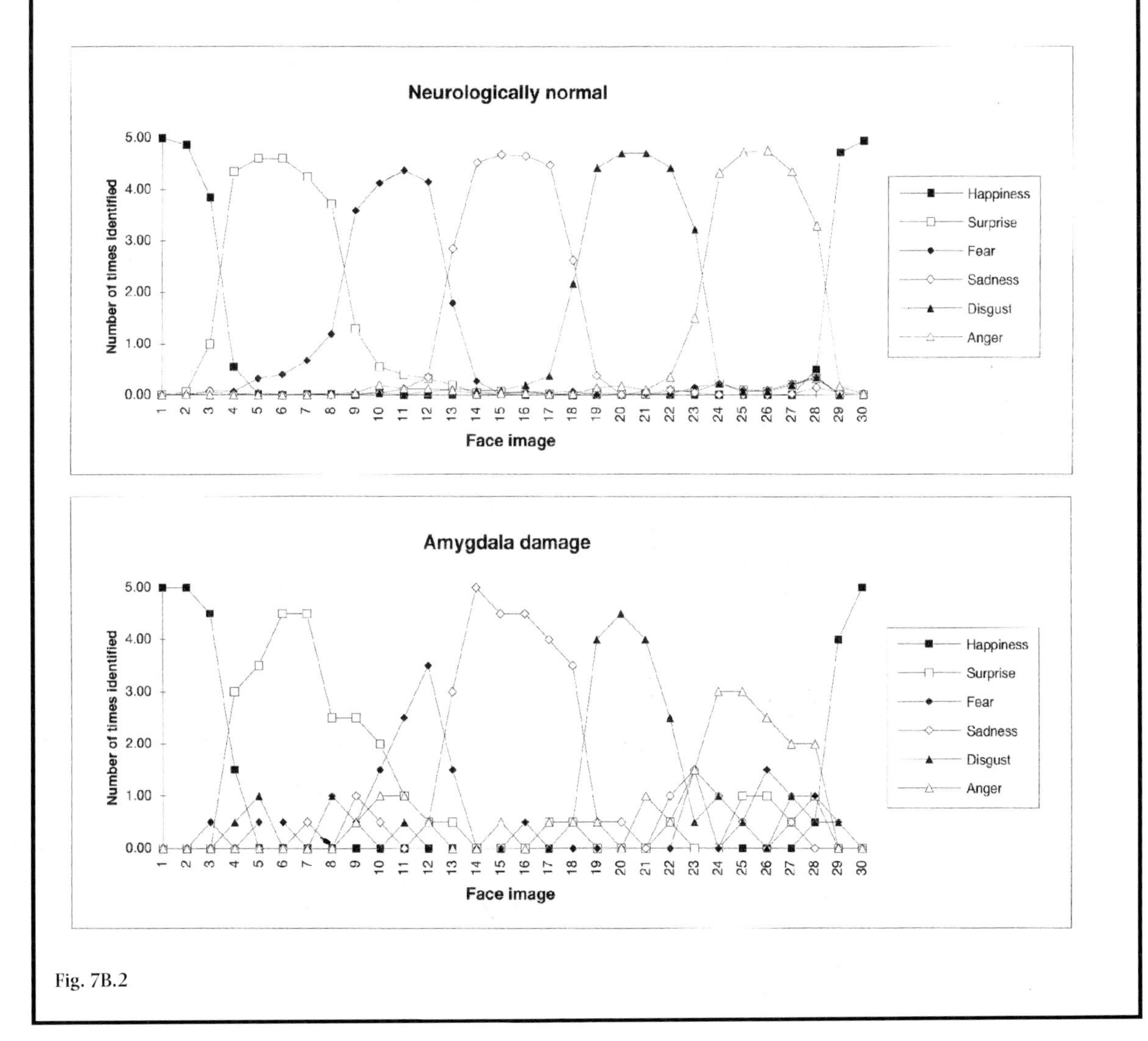

Fig. 7B.2

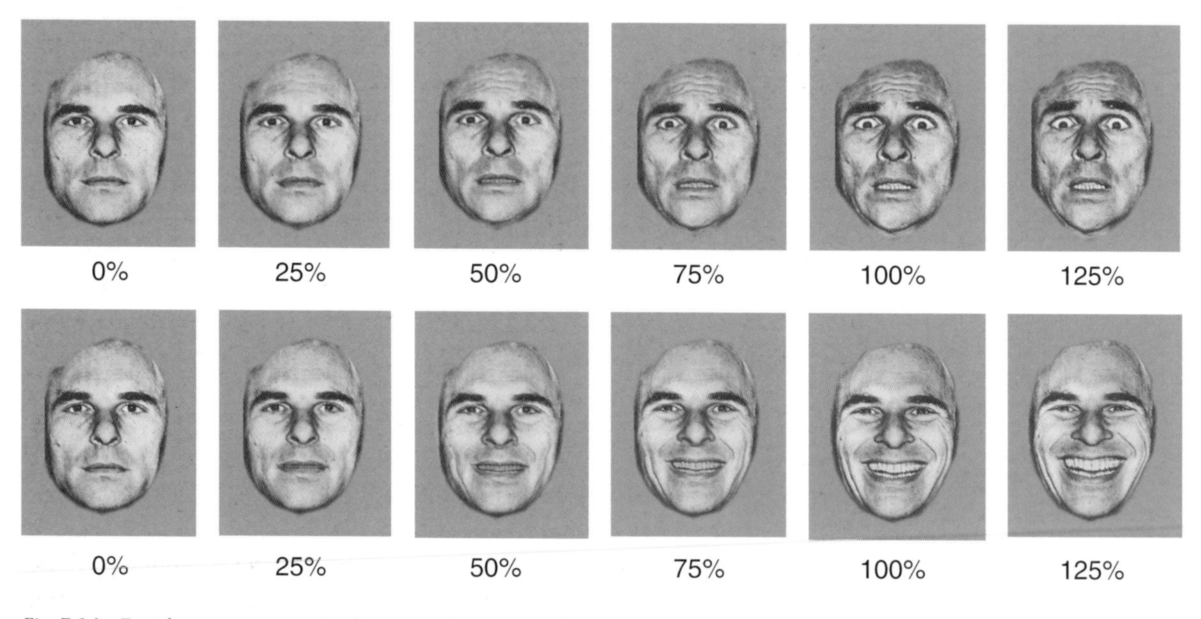

Fig. 7.14 Facial expressions ranging from neutral to extreme fear (*top row*) or extreme happiness (*bottom row*). The 0% and 100% images are prototype expressions from the Ekman and Friesen (1976) series, and the other images were created using computer-graphics techniques described in Chapter 6. Computer manipulation by Duncan Rowland and David Perrett, University of St. Andrews.

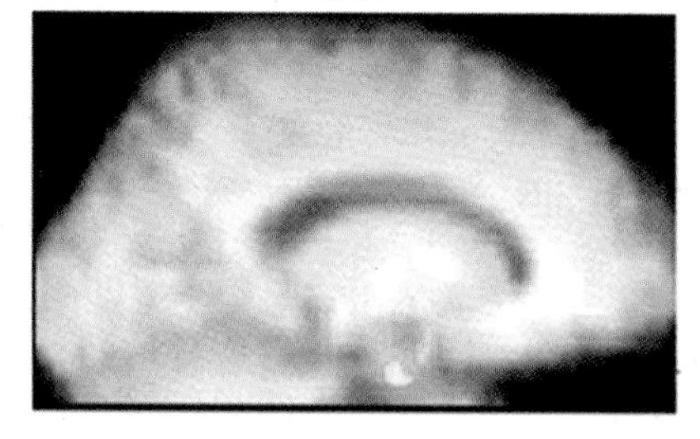

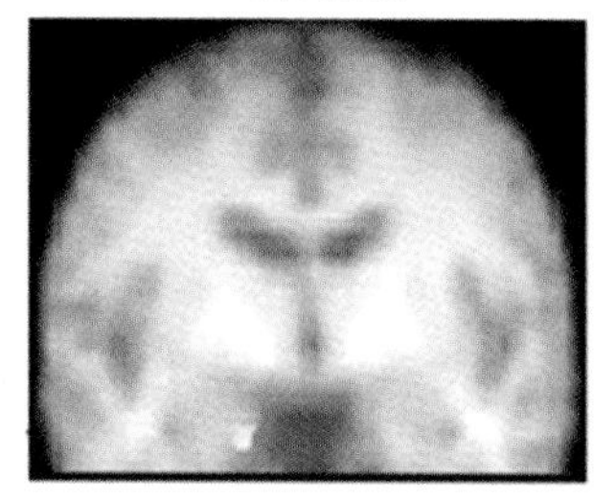

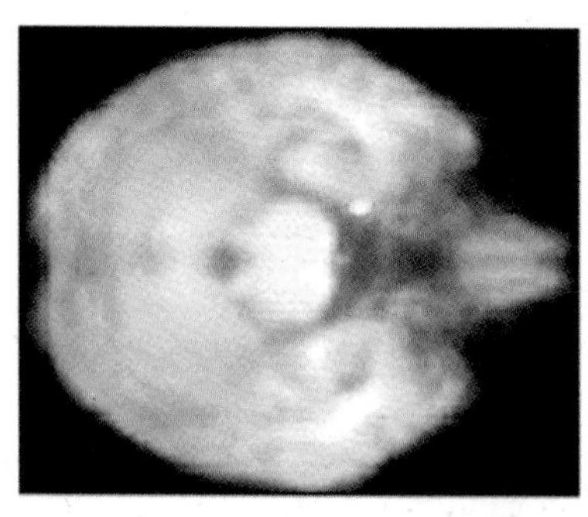

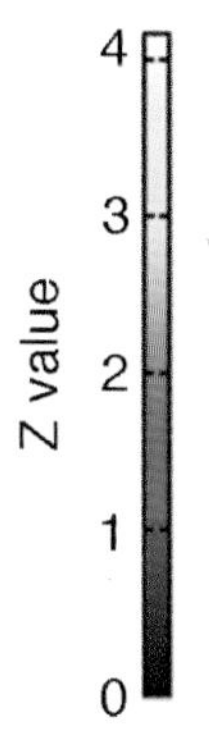

Fig. 7.15 Brain response when people look at the type of images shown in Fig. 7.14. Vertical (coronal), horizontal (transverse) and front to back (saggital) sections are shown. The region of increasingly intense colour is centred on the left amygdala, which becomes more active as the face becomes less happy and more afraid. Adapted from Morris *et al.* (1996).

Further testing of a person with brain lesions confined to the amygdala region has shown that she was also poor at recognizing vocal expressions of fear and anger (Scott *et al.* 1997), implying that the amygdala is involved in a general way in the recognition of these emotions regardless of the type of input (face or voice).

This may happen because the amygdala plays a more general role in the appraisal of danger (LeDoux 1995). As Darwin (1872) had noted, basic emotions represent important evolutionary adaptations; Ekman (1992) characterized them as facilitating rapid responses to fundamental life-tasks in ways which have enhanced our fitness to survive. For example if an animal comes into contact with something dangerous it may become frightened and flee, or it may aggressively stand its ground; danger induces a basic emotion, fear, and a set of preparatory responses that aid in the preservation of life. One can suppose, then, that any animal that can experience fear (or, at least, mobilize a fear response) under the appropriate circumstances will be better equipped to survive than one that cannot. Similarly, an animal that is sensitive to perceiving fear or anger in other animals is in a better position to decide when a rapid exit or a display of aggression could pay off. This is not to deny that emotions like fear will also involve learning and cultural influences, but these are likely to be built upon mechanisms honed by evolution.

Reports of the behaviour of a person with amygdala damage, Yvette, are consistent with this view (Broks *et al.* 1998). Yvette showed impaired recognition of facial expressions of fear in a laboratory test, but we will focus here on her real-life behaviour. While on holiday, Yvette and her husband were returning to their hotel after an evening out when they were mugged by some youths. In the ensuing struggle (during which, at one stage, Yvette was physically accosted) her husband's wallet was seized. He resisted and managed to retrieve it, and the youths eventually made off empty-handed. As well as the physical struggle, there was a loud and aggressive verbal exchange. Although closely involved, Yvette, according to her husband, showed no sign of concern, 'She seemed to think they were just larking around'. In contrast to this, a couple of months later Yvette became terrified when watching a mildly aggressive exchange between two female characters in a fictional television programme, and in the days that followed she displayed episodes of inappropriate fear. Once when her husband innocently entered the room Yvette cowered, hands over face, pleading 'Don't hit me', and she responded in a similar fashion on other occasions involving her son and her care assistant.

These incidents reinforce the view that appraisal of danger is a key function of the amygdala. The episodes of inappropriate fear show that fear responses can be initiated and sustained under certain circumstances without the involvement of the amygdala, whereas what characterizes all of the observations is that the fear response (absent to a mugging, excessive to a TV programme and some everyday behaviours) was out of line with an appropriate evaluation of the dangers inherent in each incident. The idea that the amygdala plays an important role in the appraisal of danger thus provides a perspective which can integrate observations of real-life behaviour after amygdala damage with findings of impaired recognition of the emotions expressed by others.

7.5 Neuropsychiatry

The striking portrait of the alienist (psychiatrist) Sir Alexander Morison (Fig. 7.16) was painted by Richard Dadd. Dadd became famous especially for his fairy pictures, but behind their quaint exterior lay a tragic history (Allderidge 1974). Dadd had been an artist of great promise, but he spent most of his life in Bethlem and Broadmoor hospitals after he killed his father in 1843. This was a premeditated murder of someone who, because of his devotion to his son, was trying to care for him after he had shown clear signs of mental instability. His father did this against medical advice that Richard should be put under restraint. This recommendation had been made because Richard was at the time the victim of many delusions, a number of which centred on the belief that he was an envoy of the god Osiris, but also included the idea that his father was the devil. In Dadd's own words, these beliefs

> induced me to put a period to the existence of him whom I had always regarded as a parent, but whom the secret admonishings I had, counselled me was the author of the ruin of my race. I inveigled him, by false pretences, into Cobham Park, and slew him with a knife, with which I stabbed him, after having vainly endeavoured to cut his throat. (Allderidge 1974, p. 24)

Dadd's most famous work was done when he was a mental patient, with some of that time spent under the care of Sir Alexander Morison. His younger brother George went insane at the same time, and was also in Bethlem Hospital. Figure 7.17 shows one of Dadd's fairy pictures, painted for Dr William Charles Hood who had greatly improved conditions for the patients at Bethlem (widely known as 'bedlam' at the time). It depicts a scene from *A Midsummer Night's Dream* in which Oberon and Titania quarrel over her refusal to give up an Indian changeling to be Oberon's page.

When he killed him, Dadd thought his father was someone else (the devil). This is a form of delusional misidentification—a psychiatric symptom which involves thinking that other people are not who they appear to be. Psychiatrists have identified a number of distinct forms of

Fig. 7.16 The alienist (psychiatrist) *Sir Alexander Morison*, painted by Richard Dadd. Courtesy of the Scottish National Portrait Gallery.

delusional misidentification, and recognize that some of them present a small but significant risk of violence (de Pauw and Szulecka 1988).

One of the most extensively investigated forms of delusional misidentification has been the Capgras delusion; the claim that one or more close relatives has been replaced by near-identical impostors. Joseph Capgras was a French psychiatrist who gave one of the first descriptions of this bizarre phenomenon. It used to be considered extremely rare, but studies of people with dementing illnesses show that it is not as uncommon as was supposed.

Cases of Capgras delusion have been found in many cultures throughout the world, and they show a noticeable consistency of certain features. Capgras delusion patients can be otherwise rational and lucid, able to appreciate that they are making an extraordinary claim. If you ask 'what would you think if I told you my family had been replaced by impostors?', you will often get answers to the effect that it would be unbelievable, absurd, an indication that you had gone mad. Yet the same patients will claim that, none the less, this is exactly what has happened to their own relatives. If you ask for evidence that it is an impostor, the patients often tell you that they can *see* the difference, yet they find it hard to express this difference in words. Further probing will sometimes reveal more pervasive feelings that many things seem strange, unfamiliar, almost unreal.

Fig. 7.17 *Contradiction. Oberon and Titania* by Richard Dadd.

Many clinicians have seen this as a psychodynamic problem. All of us find things we like and things we dislike in our loved ones, but acknowledging the existence of the things we dislike about them can make us feel uneasy. A much discussed possibility has therefore been that the Capgras delusion is a pathological way of resolving chronic uneasiness; by splitting the relative into a good original and a bad double, the double can be hated without guilt.

Like many psychodynamic hypotheses, this is ingenious but not grounded in evidence. Behaviour to the alleged impostor is more variable than would be expected on the psychodynamic account, and sometimes quite friendly. More importantly, many recent reports show that the Capgras delusion is associated with certain types of abnormal brain activity or can follow brain injury.

Figure 7.18 shows the distribution of blood flow in the brain of a person with Capgras delusion (Lebert *et al.* 1994). Abnormalities were

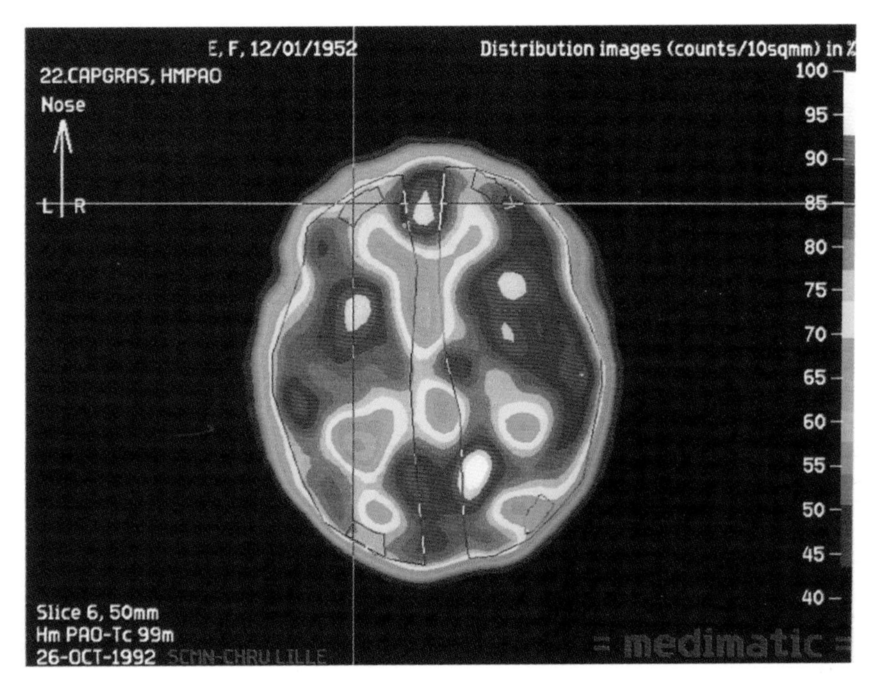

Fig. 7.18 Image showing blood flow in the brain of a person with Capgras delusion. There is a 20% reduction of blood flow in the right parietal region. From Lebert *et al.* (1994). Reproduced with permission of S. Karger AG, Basel.

noted in parts of the right parietal lobe considered likely to be involved in our 'emotional' reactions to visual stimuli, and faces in particular.

This is exactly as would be expected from a neuropsychiatric hypothesis of the basis of Capgras delusion suggested by Hadyn Ellis and Andy Young (1990). When we look at faces of people we know, we recognize who they are and parts of our brains set up preparatory reactions for the type of interaction that is likely to follow. Russian psychologists named these preparatory reactions the *orienting response*. In Ellis and Young's (1990) account, recognizing who it is and preparing for what you are likely to do (orienting responses) involve separable neurological pathways (see Box 7C), and the Capgras delusion can happen when the pathway responsible for the orienting response is affected. The consequence will be that faces can be recognized, but seem somehow odd because they do not provoke the usual reactions. The impostor claim is a rationalization of this highly disquieting sense of strangeness.

7.6 Perception of faces by newborn babies

Given the involvement of several brain regions in face perception, a natural question concerns how this comes about. In particular, one can ask to what extent the human infant comes into the world prepared to see faces?

Box 7C: The Capgras delusion and prosopagnosia

In prosopagnosia nearly all familiar faces are not recognized (see Section 7.4), whereas in Capgras delusion certain familiar people are alleged to have been replaced by impostors (Section 7.5). Because both types of problem affect the recognition of familiar people, there has been speculation that they might be related to each other. This speculation has been fuelled by findings that people who experience the Capgras delusion often perform quite poorly on face perception and face recognition tasks (Young *et al.* 1993*b*), and that they may show neurological abnormalities in brain regions similar to those involved in prosopagnosia (Lewis 1987).

Even so, the relation between these conditions must be relatively subtle. The face recognition impairments noted to accompany Capgras delusion are nothing like as severe as those found in prosopagnosia whereas, conversely, people with prosopagnosia do not usually make delusional claims about their relatives.

An extraordinary finding made by Russell Bauer (1984) may provide a clue. Bauer tested the skin conductance response (SCR) of a person with prosopagnosia when he was shown a familiar face and a series of names was read out to him. The SCR is usually measured by recording electrical conductivity from the finger or the palm of the hand. When we have an emotional response to something, activity of the autonomic nervous system causes secretions from sweat glands and alters skin conductance; even very small degrees of emotional arousal can be measured in this way.

Bauer's astonishing finding was that his prosopagnosic patient showed a greater SCR change when the correct name for a face was read out than when an incorrect name was given. Yet he could not recognize the face if asked to do so, and he could not even make an accurate verbal guess as to which was the correct name. The SCR was therefore picking up some form of non-conscious, covert recognition of the face's identity. Bauer considered that this was some form of non-conscious orienting response. He argued that emotional orienting responses and conscious recognition of identity are mediated by different neurological pathways, and that it is mainly the pathway to overt recognition which is compromised in prosopagnosia. This idea is shown in schematic form in Fig. 7C.1, using a face which still produces a strong emotional orienting response in many people.

From Bauer's account of his findings in prosopagnosia, Hadyn Ellis and Andy Young (1990) reasoned that the opposite pattern of relatively preserved conscious recognition but loss of emotional orienting responses might form the basis of the Capgras delusion. Lack of an orienting response might give social interactions an odd, awkward, and emotionless tone, like interacting with a stranger.

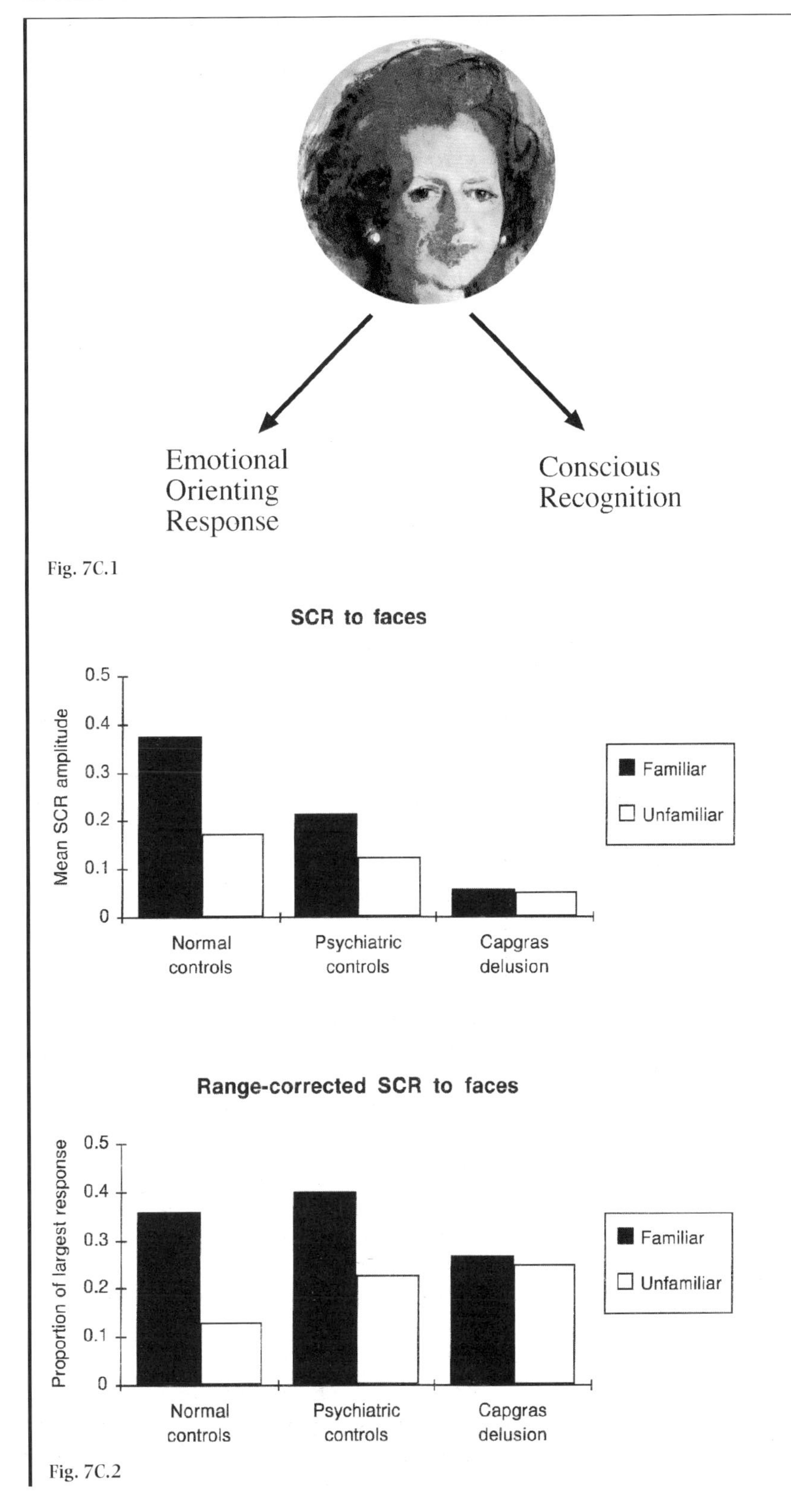

Fig. 7C.1

Fig. 7C.2

Box 7C: Continued

Ellis and Young's theory predicts that people with Capgras delusion will not show any SCR to familiar faces. Figure 7C.2 shows this is correct; normal people have a larger SCR (measured in micro-Siemens) to familiar than to unfamiliar faces, and this is also shown by people with other types of psychiatric problem, but not by people experiencing the Capgras delusion (Ellis *et al.* 1997). This conclusion holds even when SCRs are range-transformed to allow for the generally lower responsiveness of the people with Capgras delusion.

In contrast when responses to non-visual stimuli are measured, people with Capgras delusion show normal SCRs across a series of presentations of an auditory tone (see Fig. 7C.3). Their lack of response is therefore limited to certain kinds of stimuli, as might be expected from Ellis and Young's (1990) account.

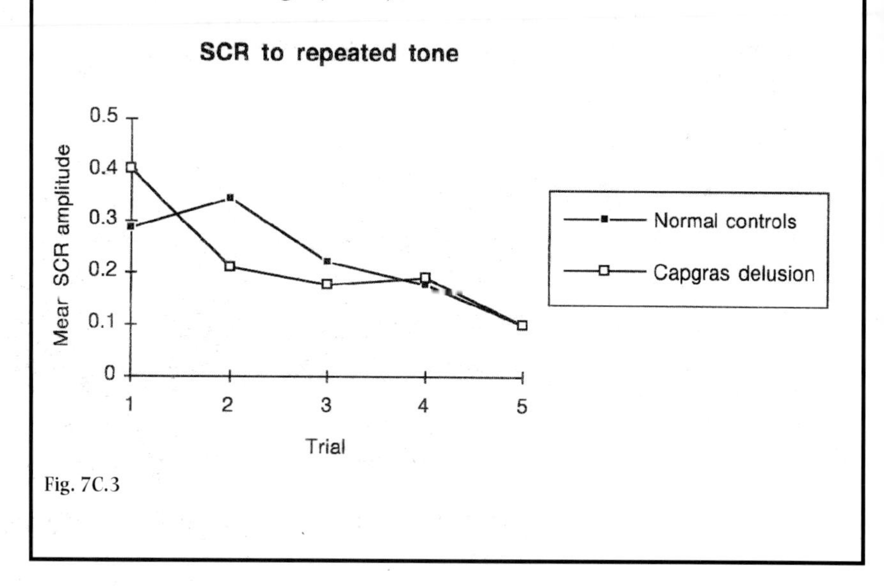

Fig. 7C.3

Such questions are very difficult to answer, because newborn babies have a limited repertoire of behaviour. Psychologists therefore have to devise ingenious methods to test what babies can see.

In humans, the visual system is fairly immature at birth. This places definite limits on what the baby can see, especially in terms of its ability to resolve fine detail and gradations of shading. Figure 7.19 gives an impression of what a face might look like to a 1-month-old and a 3-month-old infant, demonstrating both the initial immaturity of the baby's vision and its rapid development.

Although vision is clearly limited in the first month of life, infants none the less show some extraordinarily precocious abilities. Work by Andrew Meltzoff and Keith Moore (1977) has shown that infants who

(a)

(b)

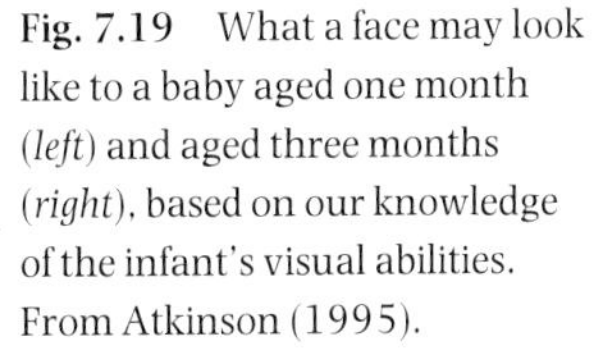
Fig. 7.19 What a face may look like to a baby aged one month (*left*) and aged three months (*right*), based on our knowledge of the infant's visual abilities. From Atkinson (1995).

Fig. 7.20 Photographs (from videotaped recordings) of 2- to 3-week-old infants imitating tongue protrusion, mouth opening, and lip protrusion. From Meltzoff and Moore (1977). © 1977 American Association for the Advancement of Science.

are only 2–3 weeks old can imitate facial movements including tongue protrusion, mouth opening, and lip protrusion (see Fig. 7.20). An important aspect of this finding is that it shows the baby must have some kind of 'map' to indicate which of its own facial muscles corresponds to those of another human being, even though it has never experimented with a mirror. This inborn mapping may well provide an important contribution to learning to make the movements necessary for speech (see Section 6.4).

There is also evidence for innate knowledge of faces which leads to the following of face-like patterns by newborn infants. Figure 7.21 illustrates an experiment reported by Mark Johnson and his colleagues (Johnson *et al.* 1991). The infant is lying down and one of the four patterned boards is moved in an arc in front of its face.

Using this technique, Johnson *et al.* (1991) found that the infant's eyes would follow the most face-like pattern longer than the un-face-like patterns shown at the bottom of the right-hand-panel, with tracking of the slightly less face-like pattern (top right in Fig. 7.21) falling in between. The average age of these babies was 43 minutes from delivery when the test began—there was therefore virtually no chance that they could have learnt this preference for faces!

Even the most face-like of the stimuli shown in Fig. 7.21 is a schematic representation, with only moderate realism. But it contains intensity changes which approximate (or even exaggerate) those of real faces in the infant's environment, and for an innate mechanism that is all that is required. Studies of many animal species show that innate behaviours can often be elicited by simple triggers, and it makes good biological sense to keep them underspecified in this way.

Inborn attentiveness to faces probably serves several functions, not the least of which is to make the baby seem interested in its caregivers, thereby encouraging them to nurture it. In addition, it means that babies can begin the task of learning about the particular faces of their parents and siblings. This is a lengthy business, but experiments by Ian

Fig. 7.21 Newborn infant being tested. It is lying down and one of four different types of pattern is moved in an arc in front of its head. It is then possible to measure which pattern the baby will most readily follow with its eyes. Courtesy of Mark Johnson, MRC Cognitive Development Unit, London.

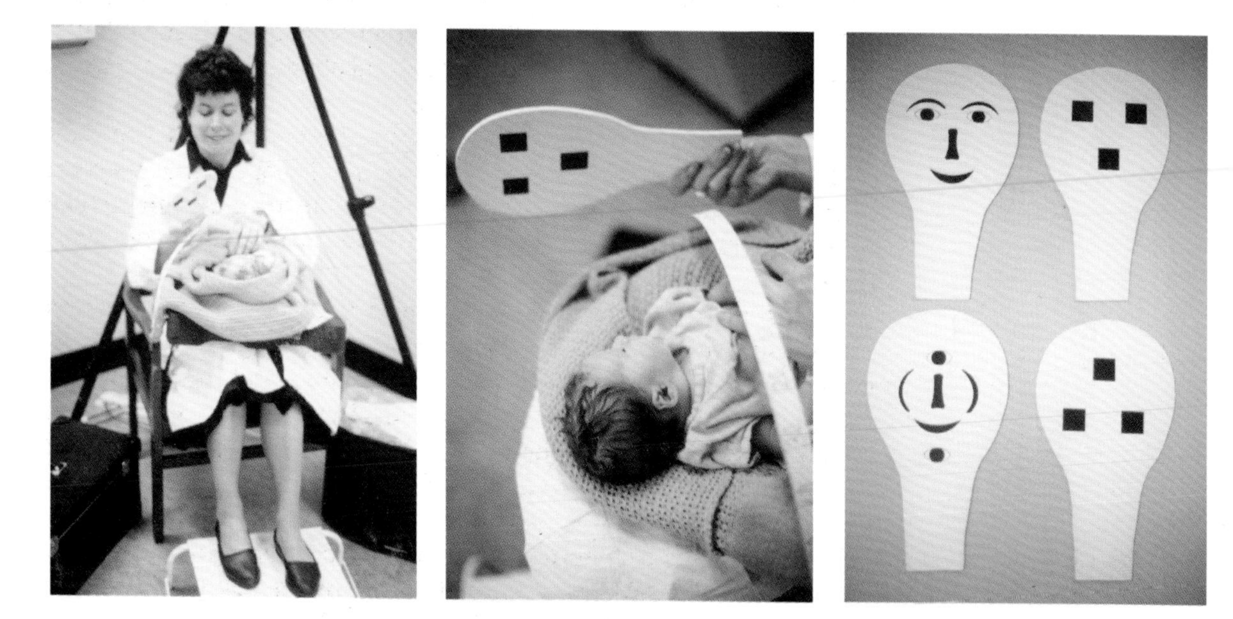

Bushnell and his colleagues (Bushnell *et al.* 1989) showed that it gets off to a prompt start. They found that two-day-old infants looked longer at their mother's face than a stranger's face, indicating that within two days babies have learnt the rudiments of recognizing their own mothers.

These studies of infants allow us to glimpse the intricate interplay between the innate organization of the brain and its astonishing capacity for perceptual learning. The importance of learning should never be underestimated, and is clear from many of the topics we have discussed in this book. There is a sense in which the infant's brain is highly plastic—ready to be moulded by the experiences it encounters. But it also contains crafty mechanisms (such as attention-capturing properties for face-like stimuli) which keep the odds high that these experiences will be optimal for what the baby will need to learn.

We have seen, then, that psychologists have shown great ingenuity in finding ways that babies can inform us about what they see, and this has led to spectacular advances which have shown that the infant's visual world is infinitely richer than was once assumed. The strong evidence of precocious face perception abilities in human infants is consistent with the remarkable discovery by Charles Gross in the 1960s of cells in the primate brain which are responsive to biologically significant stimuli, including hands and faces. Such observations have subsequently been confirmed and extended by many research groups (see reviews in Bruce *et al.* 1992*a*).

Even so, many questions remain unanswered. One of the key issues concerns the parts of the baby's brain which are active in vision. We now know a great deal about specialization of function in the adult's brain, but next to nothing about how this comes about. Some researchers think that in the first month or so after birth the infant's cerebral cortex is so immature that its visual abilities may be mediated by subcortical parts of the visual system.

To answer such questions directly, we need to find ways of recording and imaging the activity of the baby's brain. It is not easy! For several technical reasons, the methods used with adults to create images like those shown in Fig. 7.15 and Fig. 7.18 are unsuitable for use with infants.

A new approach is to measure the electrical activity of the baby's brain, using a technique similar to an EEG recording. The activity of nerve cells generates tiny electrical currents, and these can be measured by placing little current sensors (electrodes) on the surface of the head (Johnson 1996). Figure 7.22 shows researcher Mark Johnson with a 'geodesic hair-net' consisting of 128 such electrodes, and a baby wearing it.

The hair-net is easy to put on, and safe and comfortable to wear, making it ideal for work with infants. The sensors rest gently on the

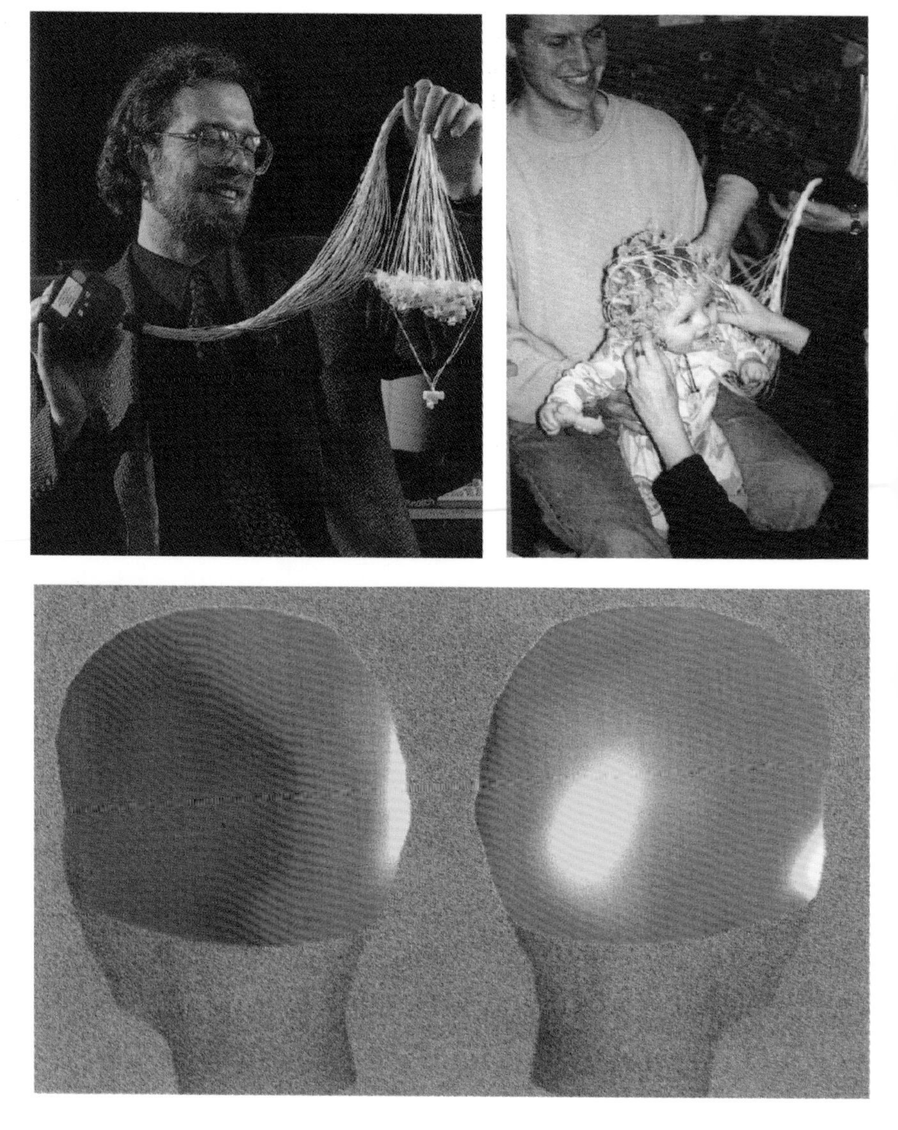

Fig. 7.22 Mapping infant brain activity. The 'geodesic hair-net', a baby wearing it, and a picture of the relative activity of different brain regions. From Johnson (1996).

head, and transmit information to a computer about the voltage changes which result as groups of brain cells become active. It is then possible to construct a picture of the relative activity of different brain regions, as in the bottom panel of Fig. 7.22.

This technique has considerable promise for improving our understanding of how the baby's brain works. The pictures obtained do not have the very high spatial accuracy of some of the techniques used with adults—they can only measure brain activity across fairly broad regions. On the other hand, this limitation may be offset by the fact that the technique allows changes in activity across time to be measured with good precision, and this may allow 'the study of brain events at the speed of thought' (Johnson 1996, p. 33).

In this book we have covered many phenomena of face perception from life and art, and examined how some of these phenomena can be understood scientifically and what they reveal about the workings of our eyes, brains, and minds. However, many things remain mysterious. It seems appropriate that we end with one of the deepest mysteries of all—how the human brain develops so rapidly over the first weeks and months of life the intricate skills which allow the effortless appreciation by eye (and brain) of so many different messages and meanings from a human face.

References

Allderidge, P. (1974). *The late Richard Dadd.* Tate Gallery, London.

Atkinson, J. (1995).Through the eyes of an infant. In R. Gregory, J. Harris, P. Heard, and D. Rose (Eds), *The artful eye.* Oxford: Oxford University Press.

Baraitser, M. and Winter, R. M. (1983). *A colour atlas of clinical genetics.* Wolfe Medical, London.

Barkowitz, P. and Brigham, J. C. (1982). Recognition of faces: own-race bias, incentive and time delay. *Journal of Applied Social Psychology*, **12**, 255–68.

Baron-Cohen, S. and Cross, P. (1992). Reading the eyes: evidence for the role of perception in the development of a theory of mind. *Mind and Language*, **7**, 182–6.

Baron-Cohen, S. and Ring, H. (1994). A model of the mindreading system: neuropsychological and neurobiological perspectives. In *Children's early understanding of mind: origins and development* (ed. C. Lewis and P. Mitchell), pp. 183–207. Lawrence Erlbaum, Hove, East Sussex.

Baron-Cohen, S., Campbell, R., Karmiloff-Smith, A., Grant, J., and Walker, J. (1995). Are children with autism blind to the mentalistic significance of the eyes? *British Journal of Developmental Psychology*, **13**, 379–98.

Bartlett, J. C. and Searcy, J. (1993). Inversion and configuration of faces. *Cognitive Psychology*, **25**, 281–316.

Bassili, J. M. (1978). Facial motion in the perception of faces and of emotional expression. *Journal of Experimental Psychology: Human Perception and Performance*, **4**, 373–9.

Bauer, R. M. (1984). Autonomic recognition of names and faces in prosopagnosia: a neuropsychological application of the guilty knowledge test. *Neuropsychologia*, **22**, 457–69.

Bell, C. (1844). *The anatomy and philosophy of expression* (3rd edn) George Bell, London.

Benson, P. J. and Perrett, D. I. (1991*a*). Synthesising continuous-tone caricatures. *Image and Vision Computing*, **9**, 123–9.

Benson, P. J. and Perrett, D. I. (1991*b*). Perception and recognition of photographic quality facial caricatures: implications for the recognition of natural images. *European Journal of Cognitive Psychology*, **3**, 105–35.

Benson, P. J., Perrett, D. I., and Davis, D. N. (1992). Towards a quantitative understanding of facial caricatures. In *Processing images of faces* (ed. V. Bruce and M. Burton), pp. 69–87. Ablex, New Jersey.

Benton, A. L., Hamsher, K. S., Varney, N., and Spreen, O. (1983). *Contributions to neuropsychological assessment: a clinical manual.* Oxford University Press, Oxford.

Berry, D. S. and McArthur, L. Z. (1985). Some components and consequences of a babyface. *Journal of Personality and Social Psychology*, **48**, 312–23.

Biederman, I. (1987). Recognition by components: A theory of human image understanding. *Psychological Review*, **94**, 115–145.

Berry, D. S. and Zebrowitz-McArthur, L. (1988). What's in a face. The impact of facial maturity and defendent intent on the attribution of legal responsibility. *Personality and Social Psychology Bulletin*, **14**, 23–33.

Blakemore, C. B. and Campbell, F. W. (1969). On the existence of neurones in the human visual system selectively sensitive to the size and orientation of retinal images. *Journal of Physiology*, **203**, 237–60.

Bornstein, R. F. (1989). Exposure and affect: overview and meta-analysis of research, 1968–1987. *Psychological Bulletin*, **106**, 265–89.

Braddick, O. (1995). The many faces of motion perception. In *The artful eye* (ed. R. Gregory, J. Harris, P. Heard and D. Rose), pp. 205–31. Oxford University Press.

Brennan, S. E. (1985). The caricature generator. *Leonardo*, **18**, 170–8.

Brewster, D. (1832). *Letters on natural magic, addressed to Sir Walter Scott, Bart.* John Murray, London.

Brigham, J. C. (1986). The influence of race on face recognition. In *Aspects of face processing* (ed. H. D. Ellis, M. A. Jeeves, F. Newcombe and A. Young), pp. 170–7. Martinus Nijhoff, Dordrecht.

Brilliant, R. (1991). *Portraiture*. Reaktion Books, London.

Broks, P., Young, A. W., Maratos, E. J., Coffey, P. J., Calder, A. J., Isaac, C., Mayes, A. R., Hodges, J. R., Montaldi, D., Cezayirli, E., Roberts, N., and Hadley, D. (1998). Face processing impairments after encephalitis: amygdala damage and recognition of fear. *Neuropsychologia*.

Brown, G., Anderson, A. H., Yule, G., and Shillcock, R. (1984). *Teaching talk*. Cambridge University Press, Cambridge.

Bruce, V. (1982). Changing faces: visual and non-visual coding processes in face recognition. *British Journal of Psychology*, **73**, 105–16.

Bruce, V. and Valentine, T. (1988). When a nod's as good as a wink. The role of dynamic information in face recognition. In *Practical aspects of memory: current research and issues* (Vol 1) (ed. M. Gruneberg *et al.*), pp. 169–74.

Bruce, V. and Langton, S. (1994). The use of pigmentation and shading information in recognising the sex and identities of faces. *Perception*, **23**, 803–22.

Bruce, V., Valentine, T., and Baddeley, A. D. (1987). The basis of the three-quarter view advantage in face recognition. *Applied Cognitive Psychology*, **1**, 109–20.

Bruce, V., Burton, M., Doyle, T., and Dench, N. (1989). Further experiments on the perception of growth in three dimensions. *Perception and Psychophysics*, **46**, 528–36.

Bruce, V., Healey, P., Burton, A. M., Doyle, T., Coombes, A., and Linney, A. (1991). Recognising facial surfaces. *Perception*, **20**, 755–69.

Bruce, V., Cowey, A., Ellis, A. W., and Perrett, D. I. (ed.) (1992*a*). *Processing the facial image*. Oxford University Press, Oxford.

Bruce, V., Hanna, E., Dench, N., Healey, P., and Burton, M. (1992*b*). The importance of 'mass' in line-drawings of faces. *Applied Cognitive Psychology*, **6**, 619–28.

Bruce, V., Burton, A. M., Hanna, E., Healey, P., Mason, O. Coombes, A., Fright, R., and Linney, A. (1993). Sex discrimination: how do we tell the difference between male and female faces? *Perception*, **22**, 131–52.

Bruce, V., Green, P. R., and Georgeson, M. A. (1996). *Visual perception: physiology, psychology and ecology*. Psychology Press, Hove.

Bruck, M., Cavanagh, P., and Ceci, S. (1991). Fortysomething: Recognizing faces at one's 25th reunion. *Memory and Cognition*, **19**, 221–8.

Bull, R. and Hawkes, C. (1982). Judging politicians by their faces. *Political Studies*, **30**, 95–101.

Bull, R. and Rumsey, N. (1988). *The social psychology of facial appearance*. Springer-Verlag, New York.

Bull, R., Jenkins, M., and Stevens, J. (1983). Evaluations of politicians' faces. *Political Psychology*, **4**, 713–16.

Burt, D. M. and Perrett, D. I. (1995). Perception of age in adult Caucasian male faces: computer graphic manipulation of shape and colour information. *Proceedings of the Royal Society of London*, **B259**, 137–43.

Burton, A. M. and Vokey, J. (1998). The face-space typicality paradox. Understanding the face-space metaphor. *Quarterly Journal of Experimental Psychology*, in press.

Burton, A. M., Bruce, V., and Dench, N. (1993). What's the difference between men and women? Evidence from facial measurement. *Perception*, **22**, 153–76.

Bushnell, I. W. R., Sai, F., and Mullin, J. T. (1989). Neonatal recognition of the mother's face. *British Journal of Developmental Psychology*, **7**, 3–15.

Cabeza, R. and Nyberg, L. (1997). Imaging cognition: an empirical review of PET studies with normal subjects. *Journal of Cognitive Neuroscience*, **9**, 1–26.

Calder, A. J., Young, A. W., Perrett, D. I., Etcoff, N. L., and Rowland, D. (1996*a*). Categorical perception of morphed facial expressions. *Visual Cognition*, **3**, 81–117.

Calder, A. J., Young, A. W., Rowland, D., Perrett, D. I., Hodges, J. R., and Etcoff, N. L. (1996*b*). Facial emotion recognition after bilateral amygdala damage: differentially severe impairment of fear. *Cognitive Neuropsychology*, **13**, 699–745.

Calder, A. J., Young, A. W., Rowland, D., and Perrett, D. I. (1997). Computer-enhanced emotion in facial expressions. *Proceedings of the Royal Society: Biological Sciences*, **B264**, 919–25.

Calvert, G. A., Bullmore, E. T., Brammer, M. J., Campbell, R., Williams, S. C. R., McGuire, P. K., Woodruff, P. W. R., Iversen, S. D., and David, A. S. (1997). Activation of auditory cortex during silent lipreading. *Science*, **276**, 593–6.

Campbell, F. W. and Robson, J. G. (1968). Application of Fourier analysis to the visibility of gratings. *Journal of Physiology*, **197**, 551–66.

Cavanagh, P. and Leclerc, Y. G. (1989). Shape from shadows. *Journal of Experimental Psychology: Human Perception and Performance*, **15**, 3–27.

Charles-Dominique, P. (1977). *Ecology and behaviour of nocturnal primates*. Duckworth.

Chevalier-Skolnikoff (1973). Facial expressions in non-human primates. In P. Ekman (Ed). *Darwin and facial expression. A century of research in review*. New York: Academic Press.

Chiroro, P. and Valentine, T. (1995). An investigation of the contact hypothesis of the own-race bias in face recognition. *Quarterly Journal of Experimental Psychology*, **48A**, 879–94.

Combe, G. (1855). *Phrenology applied to painting and sculpture*. Simpkin, Marshall, London.

Cook, S. W. (1939). The judgment of intelligence from photographs. *Journal of Abnormal and Social Psychology*, **34**, 384–9.

Coombes, A. M., Moss, J. P., Linney, A. D., Richards, R., and James, D. R. (1991). A mathematical method for the comparison of 3-dimensional changes in the facial surface. *European Journal of Orthodontics*, **13**, 95–110.

Cooper, H. and Cooper, P. (1983). *Heads, or the art of phrenology*. London Phrenology Company, London.

Costen, N. P., Parker, D. M., and Craw, I. (1996). Effects of high-pass and low-pass spatial filtering on face identification. *Perception and Psychophysics*, **58**, 602–12.

Darwin, C. (1872). *The expression of the emotions in man and animals*. John Murray, London.

Davies, G., Ellis, H., and Shepherd, J. (1978). Face recognition accuracy as a function of mode of representation. *Journal of Applied Psychology*, **63**, 180–7.

Dawkins, R. (1976). *The selfish gene*. Oxford University Press, Oxford.

de Pauw, K. W., and Szulecka, T. K. (1988). Dangerous delusions: violence and the misidentification syndromes. *British Journal of Psychiatry*, **152**, 91–7.

De Renzi, E. (1986). Current issues in prosopagnosia. In *Aspects of face processing* (ed. H. D. Ellis, M. A. Jeeves, F. Newcombe and A. Young), pp. 243–52. Martinus Nijhoff, Dordrecht.

Devlin, Lord (1976). *Report to the Secretary of State for the Home Department of the Departmental Committee on Evidence of Identification in Criminal Cases*. HMSO, London.

Dion, K. (1972). Physical attractiveness and evaluation of children's transgressions. *Journal of Personality and Social Psychology*, **24**, 207–13.

Dion, K., Berscheid, E., and Walster, E. (1972). What is beautiful is good. *Journal of Personality and Social Psychology*, **24**, 285–90.

Doherty-Sneddon, G., O'Malley, C., Garrod, S., Anderson, A., Langton, S., and Bruce, V. (1997). Face-to-face and video-mediated communication: a comparison of dialogue structure and task performance. *Journal of Experimental Psychology: Applied*, **3**, 1–21.

Dowling, J. E. (1968). Synaptic organisation of the frog retina. An electron microscopic analysis comparing the retinas of frogs and primates. *Proceedings of the Royal Society of London*, B, **170**, 205–228.

Duchenne (de Boulogne), G.-B. (1862). *Mécanisme de la physionomie humaine ou analyse électro-physiologique de l'expression des passions applicable a la pratique des arts plastiques*. Renouard, Paris.

Dunbar, R. (1996). *Grooming, gossip and the evolution of language*. Faber and Faber, London.

Easton, R. D. and Basala, M. (1982). Perceptual dominance during lipreading. *Perception and Psychophysics*, **32**, 562–70.

Edwards, O. D. (1980). *Burke and Hare*. Polygon, Edinburgh.

Eibl-Eibesfeldt, I. (1989). *Human ethology*. de Gruyter, New York.

Ekman, P. (1972). Universals and cultural differences in facial expressions of emotion. In *Nebraska symposium on motivation*, **1971** (ed. J. K. Cole), pp. 207–83. University of Nebraska Press, Lincoln, Nebraska.

Ekman, P. (1979). About brows: emotional and conversational signals. In M. Von Cranach, K. Foppa, W. Lepenies, and D. Ploog (Ed.). *Human Ethology*. Cambridge University Press.

Ekman, P. (1992). An argument for basic emotions. *Cognition and Emotion*, **6**, 169–200.

Ekman, P., and Friesen, W. V. (1976). *Pictures of facial affect*. Consulting Psychologists Press, Palo Alto, California.

Ekman, P., and Friesen, W. V. (1978). *Facial action coding system*. Consulting Psychologists Press, Palo Alto, CA.

Enlow, D. H. (1982). *Handbook of facial growth*, 2nd edn. W. B. Saunders Company, Philadelphia.

Ellis, H. D. (1986). Face recall: A psychological perspective. *Human Learning*, **5**, 189–96.

Ellis, H. D. and Young, A. W. (1990). Accounting for delusional misidentifications. *British Journal of Psychiatry*, **157**, 239–48.

Ellis, H. D., Davies, G. M., and Shepherd, J. W. (1978). A critical examination of the Photofit system for recalling faces. *Ergonomics*, **21**, 297–307.

Ellis, H. D., Young, A. W., Quayle, A. H., and de Pauw, K. W. (1997). Reduced autonomic responses to faces in Capgras delusion. *Proceedings of the Royal Society: Biological Sciences*, **B264**, 1085–92.

Errington, L. (1985). *Tribute to Wilkie*. National Galleries of Scotland, Edinburgh.

Etcoff, N. L. and Magee, J. J. (1992). Categorical perception of facial expressions. *Cognition*, **44**, 227–40.

Evans, J. J., Heggs, A. J., Antoun, N., and Hodges, J. R. (1995). Progressive prosopagnosia associated with selective right temporal lobe atrophy: a new syndrome? *Brain*, **118**, 1–13.

Ferrier, D. (1876). *The functions of the brain*. Smith, Elder, London.

Fletcher, D. J. C. and Michener, C. D. (ed.) (1987). *Kin recognition in animals*. Wiley, Chichester.

Fogden, M., and Fogden P. (1974). *Animals and their colours*. Peter Lowe. Copyright Eurobook Ltd. ISBN 0 8564 605 4. Page 149.

Fridlund, A. J. (1994). *Human facial expression: an evolutionary view*. Academic Press, San Diego.

Fright, W. R. and Linney, A. D. (1993). Registration of 3-D head surfaces using multiple landmarks. *IEEE Transactions on Medical Imaging*, **12**, 515–20.

Frisby, J. P. (1979). *Seeing: illusion, brain and mind*. Oxford: Oxford University Press.

Fromkin, V. A. and Rodman, R. (1974). *An introduction to language*. Holt, Rinehart and Winston Inc.

Galton, F. (1883). *Inquiries into human faculty and its development*. Macmillan, London.

Gardner, H. (1975). *The shattered mind*. Knopf, New York.

Gardner, M. (1967). *The ambidextrous universe*. Penguin, London.

Gilbert, C. and Bakan, P. (1973). Visual asymmetry in perception of faces. *Neuropsychologia*, **11**, 355–62.

Goldman, M. and Hagen, M. (1978). The forms of caricature: physiognomy and political bias. *Studies in the Anthropology of Visual Communication*, **5**, 30–6.

Gombrich, E. H. (1976). *The heritage of Apelles: studies in the art of the renaissance*. Phaidon Press Ltd, Oxford.

Gombrich, E. H. (1982). *The image and the eye: Further studies in the psychology of pictorial representation*. Phaidon Press Ltd, Oxford.

Graham, N. and Nachmias, J. (1971). Detection of grating patterns containing two spatial frequencies: A comparison of single-channel and multiple-channel models. *Vision Research*, **11**, 251–9.

Green, K. P., Kuhl, P. K., Meltzoff, A. N., and Stevens, E. B. (1991). Integrating speech information across talkers, gender, and sensory modality: female faces and male voices in the McGurk effect. *Perception and Psychophysics*, **50**, 524–36.

Gregory, R. L. and Zangwill, O. L. (Eds.). (1987). *The Oxford Companion to the Mind*. Oxford: Oxford University Press.

Hancock, P. J. B., Burton, A. M., and Bruce, V. (1996). Face processing: human perception and principal components analysis. *Memory and Cognition*, **24**, 26–40.

Hancock, P. J. B., Bruce, V., and Burton, A. M. (1998). A comparison of two computer-based face identification systems with human perceptions of faces. *Vision Research*, in press.

Harmon, L. D. (1973). The recognition of faces. *Scientific American*, **227**, Nov, 71–82.

Harmon, L. D. and Julesz, B. (1973). Masking in visual recognition: Effects of two-dimensional filtered noise. *Science*, **180**, 1194.

Hartline, H. K., Wagner, H. G., and Ratliff, F. (1956). Inhibition in the eye of Limulus. *Journal of General Physiology*, **39**, 651–73.

Hay, D. C., Young, A. W., and Ellis, A. W. (1991). Routes through the face recognition system. *Quarterly Journal of Experimental Psychology*, **43A**, 761–91.

Hiatt, J. L. and Gartner, L. P. (1982). *Textbook of head and neck anatomy*. New York: Appleton-Century Crofts.

Hill, H. and Bruce, V. (1993). Independent effects of lighting, orientation and inversion on the hollow face illusion. *Perception*, **22**, 887–97.

Hill, H. and Bruce, V. (1994). A comparison between the hollow face and hollow potato illusions. *Perception*, **23**, 1335–7.

Hill, H. and Bruce, V. (1996). Effects of lighting on matching facial surfaces. *Journal of Experimental Psychology: Human Perception and Performance*, **22**, 986–1004.

Hill, H., Bruce, V., and Akamatsu, S. (1995). Perceiving the sex and race of faces: the role of shape and colour. *Proceedings of the Royal Society of London*, **B261**, 367–73.

Hjortsjö, C.-H. (1969). *Man's face and mimic language*. Studentlitteratur, Malmö, Sweden.

Hume, D. (1757). *Four dissertations*. IV: Of the standard of taste. Millar, London.

Izard, C. E. (1977). *Human emotions*. New York: Plenum.

Jahoda, G. (1954). Political attitudes and judgments of other people. *Journal of Abnormal and Social Psychology*, **49**, 330–4.

Johansson, G. (1975). Visual motion perception. *Scientific American*, **232**, June, 76–89.

Johnson, D. R. and Moore, W. J. (1989). *Anatomy for dental students*. 2nd edition. Oxford University Press.

Johnson, M. (1996). Babies build brains. *MRC News*, **71**, 30–3.

Johnson, M. H., Dziurawiec, S., Ellis, H., and Morton J. (1991). Newborns' preferential tracking of face-like stimuli and its subsequent decline. *Cognition*, **40**, 1–19.

Johnston, R. A. and Bruce, V. (1990). Lost properties? Retrieval differences between name codes and semantic codes for famous people. *Psychological Research*, **52**, 62–7.

Jones, D. (1995). Sexual selection, physical attractiveness, and facial neoteny. *Current Anthropology*, **36**, 723–48.

Kaufman, M. H. (1988). *Death marks and life masks of the famous and infamous*. Scotland's Cultural Heritage Unit, Edinburgh.

Kemp, R., Pike, G., White, P., and Musselman, A. (1996). Perception and recognition of normal and negative faces—the role of shape from shading and pigmentation cues. *Perception*, **25**, 37–52.

Kleinke, C. L. (1986). Gaze and eye contact: a research review. *Psychological Bulletin*, **100**, 78–100.

Knight, B. and Johnston, A. (1997). The role of movement in face recognition. *Visual Cognition*, **4**, 265–74.

Kobayashi, H. and Kohshima, S. (1997). Unique morphology of the human eye. *Nature*, **387**, 767–8.

Kuhl, P. K. and Meltzoff, A. N. (1982). The bimodal perception of speech in infancy. *Science*, **218**, 1138–41.

Lades, M., Vorbruggen, J. C., Buhmann, J., Lage, J., von der Malsburg, C., Wurtz, R. P., and Konen, W. (1993). Distortion invariant object recognition in the dynamic link architecture. *IEEE Transactions on Computers*, **42**, 300–11.

Lander, K., Christie, F., and Bruce, V. (1997). The role of dynamic information in the recognition of famous faces. Manuscript submitted for publication.

Landau, T. (1989). *About faces*. Doubleday Anchor, New York.

Langlois, J. H. and Roggman, L. A. (1990). Attractive faces are only average. *Psychological Science*, **1**, 115–21.

Langlois, J. H., Roggman, L. A., Casey, R. J., Ritter, J. M., Rieser-Danner, L. A., and Jenkins, V. Y. (1987). Infant preferences for attractive faces: rudiments of a stereotype? *Developmental Psychology*, **23**, 363–9.

Latto, R. (1995). The brain of the beholder. In *The artful eye* (ed. R. Gregory, J. Harris, P. Heard and D. Rose) pp. 66–94. Oxford University Press.

Lavater, J. C. (1793). *Essays on physiognomy: for the promotion of the knowledge and the love of mankind*. Robinson, London.

Lebert, F., Pasquier, F., Steinling, M., Cabaret, M., Caparros-Lefebvre, D., and Petit, H. (1994). SPECT data in a case of secondary Capgras delusion. *Psychopathology*, **27**, 211–14.

LeDoux, J. E. (1995). Emotion: clues from the brain. *Annual Review of Psychology*, **46**, 209–35.

Leonard, C. M., Voeller, K. K. S., and Kuldau, J. M. (1991). When's a smile a smile? Or how to detect a message by digitizing the signal. *Psychological Science*, **2**, 166–72.

Lenneberg, E. H. (1967). *Biological foundations of language*. John Wiley and Son, New York.

Lewicki, P. (1986). Processing information about covariations that cannot be articulated. *Journal of Experimental Psychology: Learning, Memory, and Cognition*, **12**, 135–46.

Lewin, R. (1993). *Human evolution: an illustrated introduction*, (3rd edn). Blackwell, Oxford.

Lewis, S. W. (1987). Brain imaging in a case of Capgras' syndrome. *British Journal of Psychiatry*, **150**, 117–21.

Liggett, J. (1974). *The human face*. Constable, London.

Lindsay, P. H. and Norman, D. A. (1977). *Human Information Processing*, 2nd Edition. New York: Academic Press.

Linney, A. D., Grindrod, S. R., Arridge, S. R., and Moss, J. P. (1989). 3-Dimensional visualization of computerized-tomography and laser scan data for the simulation of maxillo-facial surgery. *Medical Informatics*, **14**, 109–21.

Lockley, R. M. (1964). *The private life of the rabbit*. Andre Deutsch Ltd, London.

Lombroso, C. (1911). *Crime. Its causes and remedies* (trans. H. P. Horton). Heinemann, London.

McArthur, L. Z. and Apatow, K. (1983/4). Impressions of baby-faced adults. *Social Cognition*, **2**, 315–42.

McCabe, V. (1984). Abstract perceptual information for age level: A risk factor for maltreatment? *Child Development*, **55**, 267–76.

McCabe, V. (1988). Facial proportions, perceived age, and caregiving. In *Social and applied aspects of perceiving faces* pp. 89–99 (ed. T. R. Alley), Lawrence Erlbaum Associates Ltd, Hillsdale, New Jersey.

McGurk, H. and MacDonald, J. (1976). Hearing lips and seeing voices. *Nature*, **264**, 746–8.

McLynn, F. (1991). *Charles Edward Stuart: A tragedy in many acts*. Oxford University Press.

McNeill, D. (1985). So you think gestures are nonverbal? *Psychological Review*, **92**, 350–71.

McWeeny, K. H., Young, A. W., Hay, D. C., and Ellis, A. W. (1987). Putting names to faces. *British Journal of Psychology*, **78**, 143–51.

Mark, L. S. and Todd, J. T. (1983). The perception of growth in three dimensions. *Perception and Psychophysics*, **33**, 193–6.

Meltzoff, A. N. and Moore, M. K. (1977). Imitation of facial and manual gestures by human neonates. *Science*, **198**, 75–8.

Miller, G. A. and Niceley, P. (1955). An analysis of perceptual confusions among some English consonants. *Journal of the Acoustical Society of America*, **27**, 338–52.

Milner, A. D. and Goodale, M. A. (1996). *The visual brain in action*. Oxford Psychology Series, 27. Oxford University Press, Oxford.

Montepare, J. M. and McArthur, L. Z. (1986). The impact of age-related variations in facial characteristics on children's age perceptions. *Journal of Experimental Child Psychology*, **42**, 303–14.

Mori, H. (1977). *Japanese Portrait Sculpture*. Translated and adapted by W. Chie Ishibashi.-Tokyo, New York; Kodansha International, 1977.

Morris, J. S., Frith, C. D., Perrett, D. I., Rowland, D., Young, A. W., Calder, A. J., and Dolan, R. J. (1996). A differential neural response in the human amygdala to fearful and happy facial expressions. *Nature*, **383**, 812–15.

Newcombe, F. and Russell, W. R. (1969). Dissociated visual perceptual and spatial deficits in focal lesions of the right hemisphere. *Journal of Neurology, Neurosurgery, and Psychiatry*, **32**, 73–81.

Newcombe, F., Ratcliff, G., and Damasio, H. (1987). Dissociable visual and spatial impairments following right posterior cerebral lesions: clinical, neuropsychological and anatomical evidence. *Neuropsychologia*, **25**, 149–61.

Newcombe, F., Mehta, Z., and de Haan, E. H. F. (1994). Category specificity in visual recognition. *The neuropsychology of high-level vision: collected tutorial essays* (ed. M. J. Farah and G. Ratcliff), pp. 103–32. Lawrence Erlbaum, Hillsdale, New Jersey.

Nothdurft, H.-C. (1993). Faces and facial expressions do not pop out. *Perception*, **22**, 1287–98.

O'Toole, A. J., Abdi, H., Deffenbacher, K. A., and Valentin, D. (1993). Low dimensional representation of faces in higher dimensions of the face space. *Journal of the Optical Society of America A*, **10**, 405–11.

O'Toole, A. J., Deffenbacher, K. A., Valentin, D., and Abdi, H. (1994). Structural aspects of face recognition and the other race effect. *Memory and Cognition*, **22**, 208–24.

O'Toole, A. J., Peterson, J., and Deffenbacher, K. A. (1996). An other-race effect for categorising faces by sex. *Perception*, **25**, 669–76.

Parke, F. (1982). Paramaterised models for facial animation. *IEEE: Computer Graphics and Applications*, **2**, 61–8.

Pearson, D. E. and Robinson, J. A. (1985). Visual communication at very low data rates. *Proceedings of the IEEE*, **73**, 795–812.

Pearson, D. E., Hanna, E., and Martinez, K. (1990). Computer-generated cartoons. In *Images and Understanding*. Cambridge University Press. (ed. H. Barlow, C. Blakemore and M. Weston-Smith), pp. 46–60.

Perkins, D. (1975). A definition of caricature, and caricature and recognition. *Studies in the Anthropology of Visual Communication*, **2**, 1–24.

Pernkopf, E. (1963). *Atlas of topographical and applied human anatomy*. Philadelphia: Saunders.

Perrett, D. I., May, K. A., and Yoshikawa, S. (1994). Facial shape and judgements of female attractiveness. *Nature*, **368**, 239–42.

Pevzner, S., Bornstein, B., and Loewenthal, M. (1962). Prosopagnosia. *Journal of Neurology, Neurosurgery, and Psychiatry*, **25**, 336–8.

Pittenger, J. B. and Shaw, R. E. (1975). Ageing faces as viscal-elastic events: implications for a theory of nonrigid shape perception. *Journal of Experimental Psychology: Human Perception and Performance*, **1**, 374–82.

Pittenger, J. B., Shaw, R. E. and Mark, L. S. (1979). Perceptual information for the age level of faces as a higher order invariant of growth. *Journal of Experimental Psychology: Human Perception and Performance*, **5**, 478–493.

Ramachandran, V. S. (1995). 2-D or not 2-D: that is the question. In *The artful eye*, (ed. R. Gregory, J. Harris, P. Heard and D. Rose), pp. 249–67. Oxford University Press.

Rhodes, G. (1996). *Superportraits: caricature and recognition*. Psychology Press, Hove.

Rhodes, G. and Tremewan, T. (1996). Averageness, exaggeration, and facial attractiveness. *Psychological Science*, **7**, 105–10.

Rhodes, G., Brennan, S., and Carey, S. (1987). Identification and ratings of caricatures: implications for mental representations of faces. *Cognitive Psychology*, **19**, 473–97.

Rose, D. (1995). A portrait of the brain. In R. Gregory, J. Harris, P. Heard and D. Rose (Eds.), *The artful eye* (pp. 28–51). Oxford: Oxford University Press.

Rowland, D. A. and Perrett, D. I. (1995). Manipulating facial appearance through shape and colour. *IEEE Computer Graphics and Applications*, **15**, (5), 70–6.

Rumsey, N. and Bull, R. (1986). The effects of facial disfigurement on social interaction. *Human Learning*, **5**, 203–8.

Rumsey, N., Bull, R., and Gahagan, D. (1982). The effect of facial disfigurement on the proxemic behaviour of the general public. *Journal of Applied Social Psychology*, **12**, 137–50.

Salter, F. (1996). Carrier females and sender males: an evolutionary hypothesis linking female attractiveness, family resemblance, and paternity confidence. *Ethology and Sociobiology*, **17**, 211–20.

Scaife, M. and Bruner, J. S. (1975). The capacity for joint visual attention in the infant. *Nature*, **253**, 265–6.

Scott, S. K., Young, A. W., Calder, A. J., Hellawell, D. J., Aggleton, J. P., and Johnson, M. (1997). Impaired auditory recognition of fear and anger following bilateral amygdala lesions. *Nature*, **385**, 254–7.

Searcy, J. H. and Bartlett, J. C. (1996). Inversion and processing of component and spatial-relational information in faces. *Journal of Experimental Psychology: Human Perception and Performance*, **22**, 904–15.

Shaw, R. E., McIntyre, M., and Mace, W. (1974). The role of symmetry in event perception. In R. B. MacCleod and H. L. Pick (Eds.), *Perception: Essays in honor of James J Gibson*. Ithaca New York: Cornell University Press.

Shepherd, J. (1986). An interactive computer system for retrieving faces. In *Aspects of face processing*, (ed. H. D. Ellis, M. A. Jeeves, F. Newcombe and A. Young), pp. 398–409. Martinus Nijhoff, Dordrecht.

Shepherd, J. (1989). The face and social attribution. In *Handbook of research on face processing* (ed. A. W. Young and H. D. Ellis), pp. 289–320. North Holland, Amsterdam.

Sinha, P. and Poggio, T. (1996). I think I know that face... *Nature*, **384**, 404.

Stevenage, S. (in press). Which twin are you? A demonstration of induced categorical perception of identical twin faces. *British Journal of Psychology*.

Stevenson, S. (1976). *A face for any occasion: some aspects of portrait engraving*. Scottish National Portrait Gallery, Edinburgh.

Sumby, W. H. and Pollack, I. (1954). Visual contribution to speech intelligibility in noise. *Journal of the Acoustical Society of America*, **26**, 212–15.

Summerfield, Q. and McGrath, M. (1984). Detection and resolution of audio-visual incompatibility in the perception of vowels. *Quarterly Journal of Experimental Psychology*, **36A**, 51–74.

Summerfield, Q., MacLeod, A., McGrath, M., and Brooke, M. (1989). Lips, teeth, and the benefits of lipreading. In *Handbook of research on face processing* (ed. A. W. Young and H. D. Ellis), pp. 223–33. North Holland, Amsterdam.

Tanaka, J. W. and Farah, M. J. (1993). Parts and wholes in face recognition. *Quarterly Journal of Experimental Psychology*, **46A**, 225–45.

Thompson, D. W. (1917). *On growth and form*. Cambridge University Press.

Thompson, P. (1980). Margaret Thatcher: A new illusion? *Perception*, **9**, 483–4.

Thomson, D. M. (1986). Face recognition: More than a feeling of familiarity? In *Aspects of face processing*, (ed. H. D. Ellis, M. A. Jeeves, F. Newcombe, and A. Young), pp. 118–22. Martinus Nijhoff, Dordrecht.

Thornhill, R. and Gangestad, S. W. (1993). Human facial beauty: averageness, symmetry, and parasite resistance. *Human Nature*, **4**, 237–69.

Tytler, G. (1982). *Physiognomy in the European novel: faces and fortunes*. Princeton University Press, Princeton.

Tyler, C. W. (1997). An eye placement principle in 500 years of portraits. *Investigative Ophthalmology and Visual Science*, **38**, S488.

Valentine, T. (1991). A unified account of the effects of distinctiveness, inversion, and race in face recognition. *Quarterly Journal of Experimental Psychology*, **43A**, 161–204.

Valentine, T. and Bruce, V. (1986). The effects of distinctiveness in recognising and classifying faces. *Perception*, **15**, 525–35.

Vanezis, P., Blowes, R. W., Linney, A. D., Tan, A. C., Richards, R., and Neave, R. (1989). Application of 3-D computer-graphics for facial reconstruction and comparison with sculpting techniques. *Forensic Science International*, **42**, 69–84.

Wade, N. J. (ed.) (1983). *Brewster and Wheatstone on vision*. Academic Press, London.

Wade, N. (1990). *Visual allusions*. Lawrence Erlbaum Associates Ltd, Hove.

Wagenaar, W. A. (1988). *Identifying Ivan: A case study in legal psychology*. Harvester Wheatsheaf, Hemel-Hempstead.

Walker, R., Findlay, J. M., Young, A. W., and Lincoln, N. B. (1996). Saccadic eye movements in object-based neglect. *Cognitive Neuropsychology*, **13**, 569–615.

Walker, S., Bruce, V., and O'Malley, C. (1995). Facial identity and facial speech processing: familiar faces and voices in the McGurk effect. *Perception and Psychophysics*, **57**, 1124–33.

Waters, K. and Terzopoulos, D. (1992). The computer synthesis of expressive faces. *Philosophical Transactions of the Royal Society of London*, **B335**, 87–93.

Watt, R. J. (1994). A computational examination of image segmentation and the initial stages of human vision. *Perception*, **23**, 383–98.

Wells, P. A. (1987). Kin recognition in humans. In *Kin recognition in animals* (ed. D. J. C. Fletcher and C. D. Michener), pp. 395–415. Wiley, Chichester.

Weston, S. (1992). *Going back: return to the Falklands*. Penguin Books, London.

Wollaston, W. H. (1824). On the apparent direction of eyes in a portrait. *Philosophical Transactions of the Royal Society, London*, 247–56.

Woodworth, R. S. and Schlosberg, H. (1954). *Experimental psychology:* revised edn. Henry Holt, New York.

Young, A. W., Hay, D. C., and Ellis, A. W. (1985). The faces that launched a thousand slips: everyday difficulties and errors in recognising people. *British Journal of Psychology*, **76**, 495–523.

Young, A. W., Hellawell, D. J., and Hay, D. C. (1987). Configural information in face perception. *Perception*, **16**, 747–59.

Young, A. W., de Haan, E. H. F., Newcombe, F., and Hay, D. C. (1990). Facial neglect. *Neuropsychologia*, **28**, 391–415.

Young, A. W., Hellawell, D. J., and Welch, J. (1992). Neglect and visual recognition. *Brain*, **115**, 51–71.

Young, A. W., Newcombe, F., de Haan, E. H. F., Small, M., and Hay, D. C. (1993*a*). Face perception after brain injury: selective impairments affecting identity and expression. *Brain*, **116**, 941–59.

Young, A. W., Reid, I., Wright, S., and Hellawell, D. J. (1993*b*). Face-processing impairments and the Capgras delusion. *British Journal of Psychiatry*, **162**, 695–8.

Zajonc, R. B. (1980). Feeling and thinking: preferences need no inferences. *American Psychologist*, **35**, 151–75.

Name index

(Names in *italics* represent portraits or other illustrations)

Subject index

(Where several page numbers are given, those in **bold** represent the more major section of text)